Nondestructive Testing (NDT)

Nondestructive Testing (NDT)

Editor

Giuseppe Lacidogna

MDPI • Basel • Beijing • Wuhan • Barcelona • Belgrade • Manchester • Tokyo • Cluj • Tianjin

Editor
Giuseppe Lacidogna
Structural, Geotechnical and
Building Engineering
Politecnico di Torino
Torino
Italy

Editorial Office
MDPI
St. Alban-Anlage 66
4052 Basel, Switzerland

This is a reprint of articles from the Special Issue published online in the open access journal *Applied Sciences* (ISSN 2076-3417) (available at: www.mdpi.com/journal/applsci/special_issues/nondestructive_testing).

For citation purposes, cite each article independently as indicated on the article page online and as indicated below:

LastName, A.A.; LastName, B.B.; LastName, C.C. Article Title. *Journal Name* **Year**, *Volume Number*, Page Range.

ISBN 978-3-0365-2292-0 (Hbk)
ISBN 978-3-0365-2291-3 (PDF)

Contents

About the Editor vii

Preface to "Nondestructive Testing (NDT)" ix

Marco Domaneschi, Gianni Niccolini, Giuseppe Lacidogna and Gian Paolo Cimellaro
Nondestructive Monitoring Techniques for Crack Detection and Localization in RC Elements
Reprinted from: *Appl. Sci.* **2020**, *10*, 3248, doi:10.3390/app10093248 1

Katsufumi Hashimoto, Tomoki Shiotani and Masayasu Ohtsu
Application of Impact-Echo Method to 3D SIBIE Procedure for Damage Detection in Concrete
Reprinted from: *Appl. Sci.* **2020**, *10*, 2729, doi:10.3390/app10082729 17

Alessandro Grazzini, Sara Fasana, Marco Zerbinatti and Giuseppe Lacidogna
Non-Destructive Tests for Damage Evaluation of Stone Columns: The Case Study of Sacro Monte in Ghiffa (Italy)
Reprinted from: *Appl. Sci.* **2020**, *10*, 2673, doi:10.3390/app10082673 33

Elsa Garavaglia, Anna Anzani, Fabio Maroldi and Fabio Vanerio
Non-Invasive Identification of Vulnerability Elements in Existing Buildings and Their Visualization in the BIM Model for Better Project Management: The Case Study of Cuccagna Farmhouse
Reprinted from: *Appl. Sci.* **2020**, *10*, 2119, doi:10.3390/app10062119 49

Zaifu Zhan, Shen Wang, Fuping Wang, Songling Huang, Wei Zhao and Zhe Wang
Simulation of Three Constitutive Behaviors Based on Nonlinear Ultrasound
Reprinted from: *Appl. Sci.* **2020**, *10*, 1982, doi:10.3390/app10061982 73

Claudia Barile, Caterina Casavola, Vincenzo Moramarco, Carmine Pappalettere and Paramsamy Kannan Vimalathithan
Bonding Characteristics of Single- and Joggled-Lap CFRP Specimens: Mechanical and Acoustic Investigations
Reprinted from: *Appl. Sci.* **2020**, *10*, 1782, doi:10.3390/app10051782 91

Huiting Huan, Lixian Liu, Andreas Mandelis, Cuiling Peng, Xiaolong Chen and Jinsong Zhan
Mechanical Strength Evaluation of Elastic Materials by Multiphysical Nondestructive Methods: A Review
Reprinted from: *Appl. Sci.* **2020**, *10*, 1588, doi:10.3390/app10051588 105

Dimitrios G. Dimogianopoulos, Panagiotis J. Charitidis and Dionysios E. Mouzakis
Inducing Damage Diagnosis Capabilities in Carbon Fiber Reinforced Polymer Composites by Magnetoelastic Sensor Integration via 3D Printing
Reprinted from: *Appl. Sci.* **2020**, *10*, 1029, doi:10.3390/app10031029 121

Jihyun Jun, Young-Dae Shim and Kyung-Young Jhang
Stress Estimation Using the Acoustoelastic Effect of Surface Waves in Weak Anisotropic Materials
Reprinted from: *Appl. Sci.* **2019**, *10*, 169, doi:10.3390/app10010169 137

Michael Stamm, Helge Pfeiffer, Johan Reynaert and Martine Wevers
Using Acoustic Emission Measurements for Ice-Melting Detection
Reprinted from: *Appl. Sci.* **2019**, *9*, 5387, doi:10.3390/app9245387 153

Marco Civera, Matteo Ferraris, Rosario Ceravolo, Cecilia Surace and Raimondo Betti
The Teager-Kaiser Energy Cepstral Coefficients as an Effective Structural Health Monitoring Tool
Reprinted from: *Appl. Sci.* **2019**, *9*, 5064, doi:10.3390/app9235064 . . . . **165**

Shunmin Yang, Mingquan Wang and Lu Yang
The Rapid Detection Technology of Lamb Wave for Microcracks in Thin-Walled Tubes
Reprinted from: *Appl. Sci.* **2019**, *9*, 3576, doi:10.3390/app9173576 . . . . **193**

Shuai Han, Heng Li, Mingchao Li and Timothy Rose
A Deep Learning Based Method for the Non-Destructive Measuring of Rock Strength through Hammering Sound
Reprinted from: *Appl. Sci.* **2019**, *9*, 3484, doi:10.3390/app9173484 . . . . **209**

Rongning Cao, Meng Ma, Ruihua Liang and Chao Niu
Detecting the Void behind the Tunnel Lining by Impact-Echo Methods with Different Signal Analysis Approaches
Reprinted from: *Appl. Sci.* **2019**, *9*, 3280, doi:10.3390/app9163280 . . . . **223**

Dongdong Wen, Mengbao Fan, Binghua Cao, Zhian Xue and Ping Wang
A PEC Thrice Subtraction Method for Obtaining Permeability Invariance Feature in Conductivity Measurement of Ferromagnetic Samples
Reprinted from: *Appl. Sci.* **2019**, *9*, 2745, doi:10.3390/app9132745 . . . . **241**

About the Editor

Giuseppe Lacidogna

Giuseppe Lacidogna achieved the National Academic Qualification in Italy as Full Professor of Structural Mechanics in 2018. He is Fellow of the European Academy of Sciences, Brussels, Belgium and Officer of the Engineering Division of the same Academy. He has been Associate Professor in Structural Mechanics at Politecnico di Torino, Italy, since January 2011.

Professor Lacidogna is Member of the Academic Board of the Doctorate Course in Civil and Environmental Engineering of Politecnico di Torino, and has been Director of the Doctorate Course in Structural Engineering (2016-2018).

He is author of more than 300 publications, including: eight monographs, more than 120 papers in refereed international journals, and 24 book chapters. He received a Certificate Merit Award from the European Society for Experimental Mechanics (EuraSEM) and has been inserted in the Italian Scientists List, Engineering-Mechanics Area, Macro Area Engineering.

Preface to "Nondestructive Testing (NDT)"

Because of the unavoidable aging and deterioration of materials, the efficient maintenance of civil structures and infrastructures is becoming a major issue for public and private institutions. As a matter of fact, the structural conditions of built heritage affect both safety and economic aspects. In this context, structural health monitoring (SHM) is emerging as a crucial research field, able to provide vital information regarding the damage levels of structures and materials. In particular, by exploiting the most advanced technologies and techniques, nondestructive testing (NDT) is the ideal candidate for detecting defects and structural issues, both at the laboratory and full-scale levels, in a non-invasive way. Among others, acoustic emission, vibration-based identification methods, digital image correlation, tomography techniques, sonic-ultrasonic tests, Raman and terahertz spectroscopy, electromagnetic analysis, etc., allow us to evaluate the state of damage and its evolution over time. The aim of this book is to collect the newest contributions by eminent authors in the field of NDT-SHM, both at the material and structure scale. It therefore provides novel insight at experimental and numerical levels on the application of NDT to a wide variety of materials (concrete, steel, masonry, composites, etc.) in the field of Civil Engineering and Architecture.

Giuseppe Lacidogna
Editor

applied sciences

MDPI

Article

Nondestructive Monitoring Techniques for Crack Detection and Localization in RC Elements

Marco Domaneschi, Gianni Niccolini, Giuseppe Lacidogna * and Gian Paolo Cimellaro

Department of Structural, Geotechnical and Building Engineering, Politecnico di Torino, 24, Corso Duca degli Abruzzi, 10129 Torino, Italy; marco.domaneschi@polito.it (M.D.); gianni.niccolini@polito.it (G.N.); gianpaolo.cimellaro@polito.it (G.P.C.)

* Correspondence: giuseppe.lacidogna@polito.it; Tel.: +39-011-090-4871

Received: 28 March 2020; Accepted: 2 May 2020; Published: 7 May 2020

check for updates

Featured Application: Damage assessment of a reinforced concrete (RC) by means of different nondestructive testing (NDT) techniques. Joined application of a PZT sensor network for AE-based local damage detection and a FOS setup for global strain mapping. Coupling of local and global methods through DIC-based strain localization to reduce the uncertainty on the strain field description. Potential for AE- and strain-based damage detection in large structures of interest to civil engineering.

Abstract: This paper presents the structural and damage assessment of a reinforced concrete (RC) beam subjected to a four-point bending test until yielding of reinforcing steel. The deterioration progress was monitored using different nondestructive testing (NDT) techniques. The strain was measured by distributed fiber optic sensors (FOSs), embedded prior to concrete pouring. The initiation and propagation of cracks were monitored by acoustic emission (AE) sensors attached to the surface of the material. The recorded AE activity results in good agreement with FOS strain measurements. The results of the integrated monitoring system are confirmed by visual observation of the actual crack pattern. At different loading steps, digital image correlation (DIC) analysis was also conducted.

Keywords: reinforced concrete; four-point bending test; structural health monitoring; nondestructive testing techniques; fiber optic sensors; acoustic emission monitoring; digital image correlation

1. Introduction

During their service life structural elements can experience variable and increasing demand conditions (e.g., traffic increase in transportation infrastructures). They are also exposed to environmental effects, that can also degrade their capacity and affect their performance. In particular, reinforced concrete (RC) components can develop cracking due to tensile conditions that are normally absorbed by steel reinforcements. However, cracking can lead to the exposure of the steel bars to the aggression of external agents, such as chlorides, triggering corrosion and strength losses. In this incremental risk scenario for existing infrastructures, the role of the structural health monitoring (SHM) becomes crucial in order to detect unusual behaviors and damage [1–3].

The use of fiber optic sensors (FOSs) for SHM is proposed in several research works, including the seminal one that dates to 1990 by Glossop et al. [4]. FOSs can be used in aggressive environments, they show geometric adaptability, independence from electrical and magnetic field interference, and high resolution [5]. These characteristics make them excellent for implementation in the civil engineering field to detect anomalies and cracks in static conditions, while, in dynamic conditions, they can also be used to assess modal parameters [6–8].

Acoustic emission (AE) sensors have been widely used to detect early stage damage before it results in failure. One of the earliest applications of AE was related to monitoring rotating machinery in the late 1960s. A comprehensive and critical review on the field of application of AE to condition monitoring and diagnostics of different mechanical components can be found in Mba and Rao (2006) [9]. Behnia et al. [10] present a comprehensive review of the acoustic emission (AE) technique for its applications in concrete structure health monitoring. Methods of AE are also developed for large structures in field application, e.g., for walls, bridge decks and reinforced concrete multi-story buildings (Carpinteri et al., 2007, 2011; McLaskey et al., 2010; Shiotani et al., 2009) [11–14].

Although AE sensors are normally developed using piezoelectric technology, there are many examples where optical fibers have been introduced as an alternative to piezoelectric sensors (e.g., Liang et al., 2009 [15]). However, there are a few applications where AE sensors are implemented in addition to distributed FOS sensors for strain detection, in order to integrate the monitoring and damage detection results obtained separately from the two systems. A representative example is one by Li et al. [16], where the results of an experimental investigation on corrosion monitoring of a steel reinforced mortar block through combined acoustic emission and fiber Bragg grating strain measurement are presented. Ansari [17] provided a short review of long gage interferometric and acoustic sensors with representative examples on the implementation of serially multiplexed long gage interferometric sensors and multiplexing of optical fiber acoustic emission sensors. Moving to composite fiber/epoxy materials, Park et al. [18] present an application for micro-failure evaluation using embedded fiber-optic sensors and acoustic emission piezoelectric sensors.

Detection of crack-induced AEs by a piezoelectric (PZT) sensor network focuses on individual cracks and displacements providing a very detailed damage description, but it suffers from limited detection range due to high attenuation of ultrasonic waves. Vice versa, global sensors act as a fully distributed sensor network to be interrogated at any point along the fiber length or integrating the response along the FOS length so as to cover a larger area. The disadvantage of global damage detection systems is their poor damage location capability. Therefore, the concurrent application of local and global methods combines damage detection possibilities and advantages of both systems.

Verstrynge et al. [19] applied a global FOS setup for AE-based damage detection in concrete elements. This line-integrating technique did not allow AE source location—carried out in that experiment through concurrent application of PZT transducers—but it has the potential for AE detection in large structures of interest to civil engineering.

Here, AE-based local damage detection through a PZT sensor network is coupled with a FOS setup for global strain mapping. FOSs act as a distributed sensor network to be interrogated at different points along the fiber length. Local and global methods are herein coupled through a digital image correlation (DIC)-based strain localization covering an intermediate-sized area. Hence, DIC-based strain data is herein used to reduce uncertainty on the FOS-based strain field description.

The integrated SHM system has been implemented in an RC beam specimen subjected to a four-points loading test to monitor the state of cracking in terms of initiation and the thickness propagation. The outcomes of the integrated AE and FOS monitoring systems are confirmed by comparison with the final crack pattern. At different loading steps, digital image correlation (DIC) analysis was also conducted in specific areas of the beam specimen to reconstruct the strain field. A numerical model has been implemented to investigate the mechanical behavior of the beam up to collapse and to confirm the monitoring outcomes.

The following sections describe the methodology and the SHM systems; then, the laboratory specimen and setup are discussed. Finally, the results and the conclusive remarks are presented.

2. Experimental Setup and Monitoring Systems

The laboratory test was conducted on a reinforced concrete beam under gradually increasing loading in the four-point bending configuration, as shown in Figure 1. The reinforced concrete beam is 4 m long and was designed to have a ductile behavior to be able to follow a progressive and gradual propagation of cracks. The cross section of the beam is 150 mm wide and 300 mm high. The longitudinal reinforcements are composed of two bars of 14 mm in diameter at the bottom and two bars of 18 mm in diameter at the top of the section, with a concrete cover of 30 mm. Stirrups of 8 mm in diameter as shear reinforcement have been installed with 70 mm spacing at the bearings, while in the remaining beam shear reinforcements are spaced every 140 mm. According to the Italian ministerial decree, Updating of the Technical standards for construction (Gazzetta Ufficiale, D.M. 17 Gennaio 2018), C20/25 and B450C have been selected for the concrete and steel materials. The maximum aggregate size has been fixed to 10 mm with a water–cement weight ratio of about 0.5 and 300 kg/m^3 of cement.

The load frame and the front side of the beam specimen are shown in Figure 2 (top), with a servo hydraulic actuator fixed at the center of the frame. A stepwise increasing load was applied downwards in displacement control by means of steel roller bearings in a four-point bending configuration. The load was measured with a 1000 kN load cell at the bottom of the actuator, while the mid span vertical displacement was measured through a linear variable differential transformer (LVDT) device, compensated with two LVDTs measurement points at the beam supports. Figure 2 (bottom) shows the beam specimen at the back side with the speckle pattern applied to a central surface of about 0.8 m length.

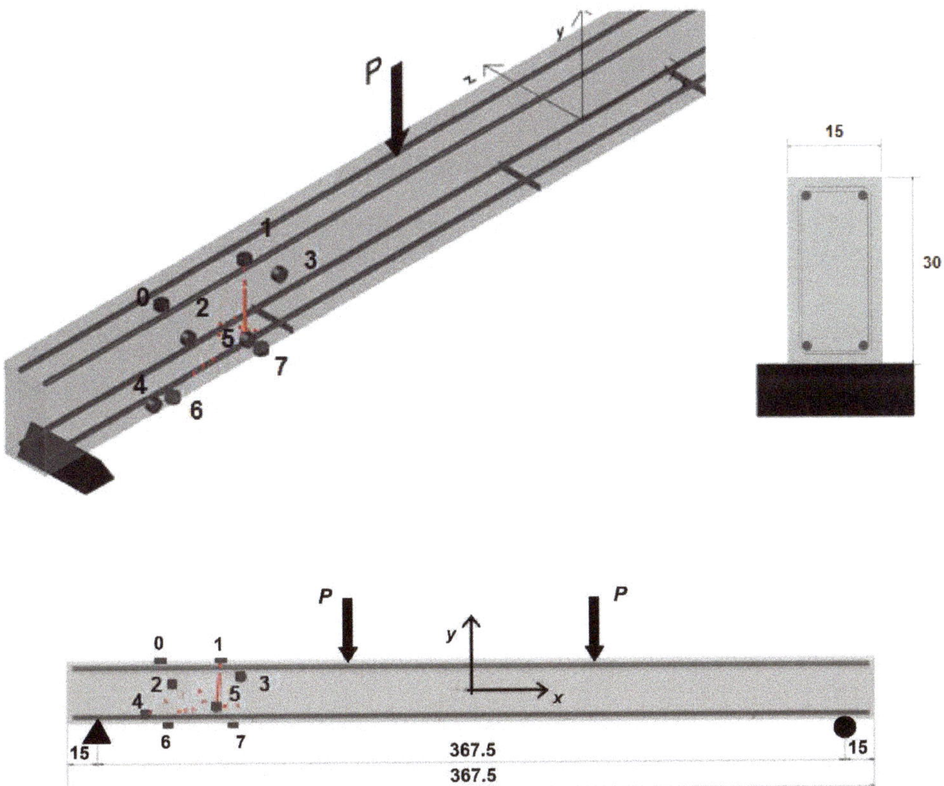

Figure 1. Specimen geometry and acoustic emission (AE) sensor locations. The eight AE sensors are numbered from 0 to 7. Dimensions are in cm.

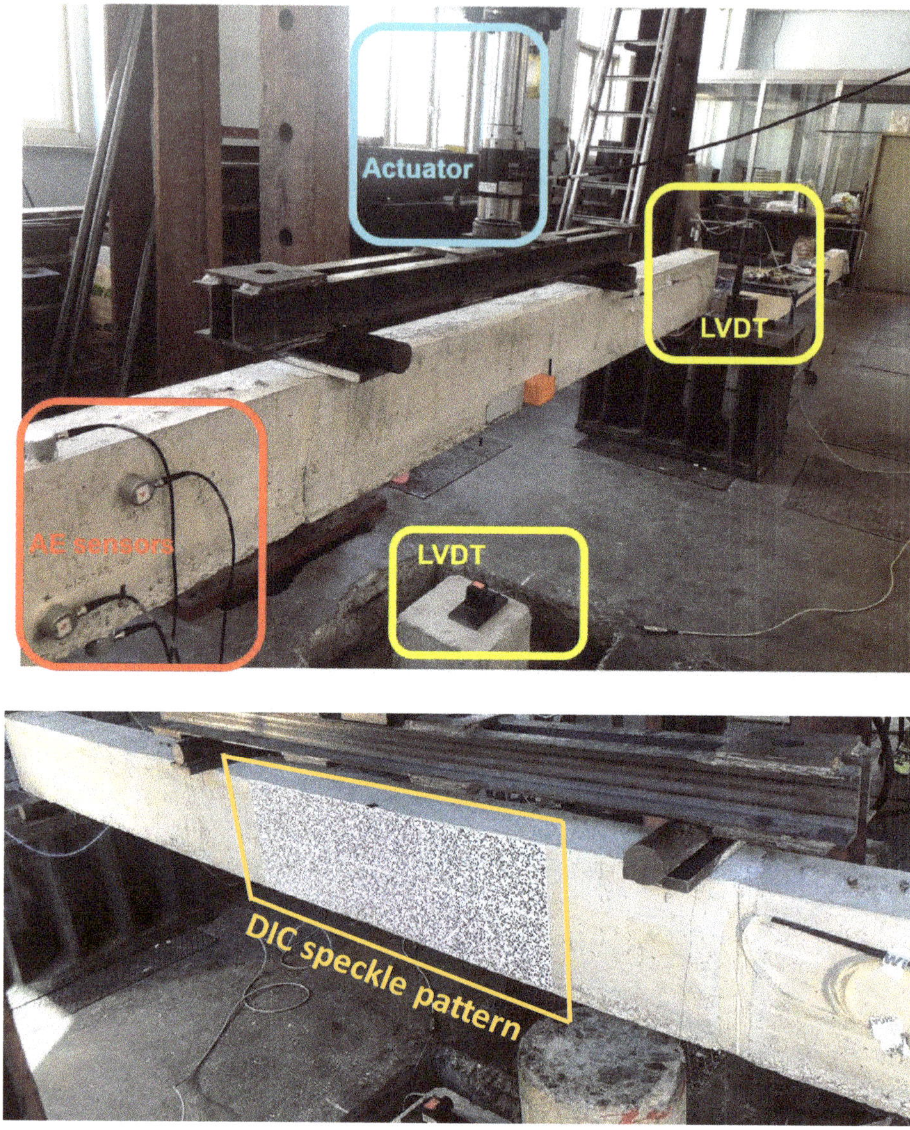

Figure 2. Front (**top**) and back (**bottom**) sides of the test setup showing the servo hydraulic actuator, the linear variable differential transformer (LVDT) sensors, the location of some AE sensors, and the speckle pattern on the central beam surface for digital image correlation (DIC) applications.

3. AE Sensors

Different damage mechanisms and levels can be verified during the four-point bending test of the RC beam, such as tensile cracking followed by shear cracking, yielding of reinforced steel, and crushing. The fracture propagation mainly includes two stages: accumulation of uncorrelated tensile cracks between the central loads, and appearance of diagonal shear cracks, starting on the bottom surface and developing to the top, interconnecting the previously formed tensile cracks. Statistical analysis of the AE signal features can reveal trends that can be ascribed to different damage stages. For example, shear events originate signals with longer rise time (the time delay between onset and peak amplitude) and higher amplitude.

Eight 60 kHz resonant AE sensors were attached at one-fourth of the span's length (refer to Figures 1 and 2 respectively for beam design and application). When the sensors are hit by a wave (due to crack growth or another reason), the pressure on their surface is converted to electric voltage, exploiting the piezoelectric effect. If the voltage was beyond a given threshold (60 dB to filter out the signal noise in the specific case), the signal was digitized and stored in an 8 channel Lunitek AEmission system with 5 MHz sampling rate per channel. The software associated to the acquisition board permits automatic extraction and storage of some basic parameters for each individual AE signal waveform (feature-based AE analysis), such as amplitude, duration and ring-down counts (RDC), i.e., the total number of times the signal exceeds the threshold level. RDC divided by duration gives the average signal frequency.

4. FOS Sensors

Different FOS distributed sensors have been embedded in the concrete beam to collect strain and temperature measurements. In particular, strain sensors have been applied to the steel reinforcements at the top (FOS #1) and the bottom (FOS #2) of the cross section along the whole beam length (Figure 3).

When the optical fiber is strained in the longitudinal direction, the backscattered light of Brillouin is subjected to a frequency shift proportional to the strain. Equation (1) expresses the strain ε as function of the Brillouin frequency shift:

$$\varepsilon = \frac{\upsilon_B(\varepsilon) - \upsilon_B(0)}{c}, \tag{1}$$

where $\upsilon_B(\varepsilon)$ is the Brillouin frequency with strain, while $\upsilon_B(0)$ is the Brillouin frequency without strain. The constant c is the proportional coefficient of strain that depends on the sensor characteristics. Basic principles of Brillouin systems are detailed in Bao et al. [20].

Although sensitive along the entire length, the fiber optic sensors measure at discrete points, that are spaced by constant value called the sampling interval. The measured parameter is an average over a certain length called spatial resolution and corresponds to the pulse width used to interrogate the fiber. For the present research, a SHM FOS system usually employed in full scale real structures has been used, thus the sampling interval and the spatial resolution have been assumed to be 0.2 and 0.5 m, respectively.

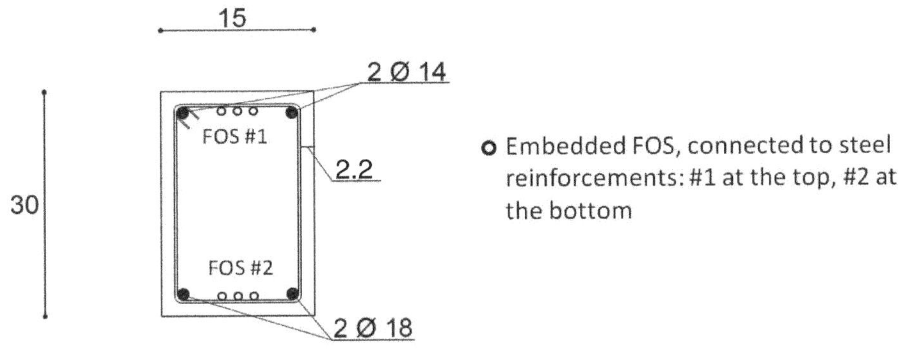

Figure 3. Fiber optic sensor (FOS) installation (dimensions in cm).

5. DIC Technique

DIC photogrammetry is a non-contact, optical measuring method adapted for extracting surface displacement and geometry profiles at different stages or times from images acquired through a camera (or multi-camera). It has been proposed for structural inspection and monitoring, and has been effectively applied previously to analyze different structures at different scales [21].

To perform a DIC measurement, a stochastic pattern (e.g., black dots on a white background or white dots on a black background), must be applied to the surface of interest and the relative position of each of them is identified as the surface deforms over time. Each image can be considered as a matrix of natural integers where white pixels have a 0 grayscale level, while level 100 is associated to black pixels. Since a single value is not a unique point, a neighborhood of pixels is used (i.e., facets or subsets). These facets include several dots of the pattern used and are typically squares with sides of 10–50 pixels. The main principle of DIC is to match the same physical point between a reference image and several deformed images based on gray-scale variations of continuous patterns [22].

The DIC computation is herein performed through NCORR software [23], which is an open-source subset-based 2D DIC package that uses modern DIC algorithms with additional improvements.

6. Test Results

The global behavior in terms of load and mid-span vertical displacement is illustrated in Figure 4. Vertical load has been applied in multiple steps up to the yielding limit of reinforcing bars, and then removed. The first cracking point can be identified between steps #1 and #2. Then, the ductile beam behavior allowed crack propagation up to step #7 when the longitudinal reinforcing steel bars at the bottom of the cross section start yielding. The loading phase of the four-points bending test was stopped at step #8 and loads removed from the beam (unloading phase).

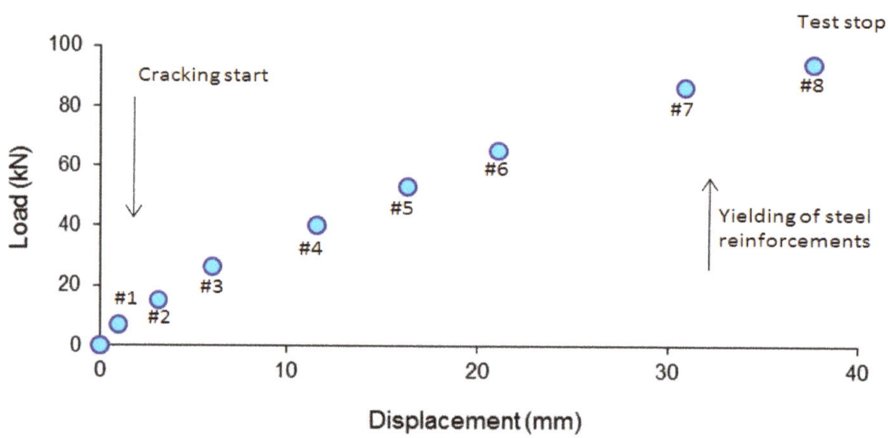

Figure 4. Load versus mid-span deflection at the different load steps, from #0 to #8.

7. AE Measurements

7.1. Total AE Activity

The applied load history along with the accumulated AE events are shown in Figure 5. The number of AE events started to increase rapidly during 4th and 7th load steps at approximately 2100 and 4500 s, meaning intensive cracking phenomena occurred at the time of increased central deflection. The major cluster of events comes at the unloading stage, from approximately 6900 s, where a possible AE source mechanism is the friction in aggregates and reinforcing bars due to cracks reclosing.

The AE activity is first investigated considering the individual sensor recordings and the related statistical distributions of the signal parameters, i.e., peak amplitude, duration and average frequency. Figure 6a shows the distributions of the received signals by sensors AE0 and AE1, and Figure 6b shows those received by sensors AE4 and AE5. Considering the mean values of the distributions, recordings of AE0 and AE1 (located on the top surface of the beam) are characterized by lower frequency, lower amplitude and longer signal duration than AE4 and AE5 (located next to the bottom surface), namely $\langle f \rangle = 17$ kHz vs. 31 kHz, $\langle A \rangle = 3.3$ mV vs. 4.5 mV, and $\langle \Delta t \rangle = 0.8$ ms vs. 0.4 ms. These findings can be explained in terms of amplitude loss (signal attenuation)—stronger for high frequencies—and, possibly, of spreading of AE wave packets traveling in a dispersive medium [24], such as the concrete to some extent. Hence, the measured effects appear to be descriptive of AE activity predominantly originating near AE4 and AE5 sensors, consistent with a crack pattern developing from the bottom surface of the beam.

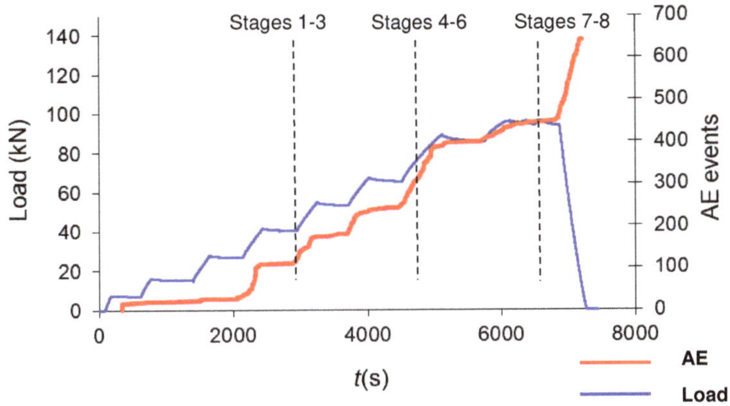

Figure 5. Step load (blue line) and cumulated AE signals time history (red line). Dashed lines correspond to the 3rd, 6th and 8th load stages.

7.2. The b-Value Analysis

It is reasonable that as the load increases, larger-scale fracture events occur giving rise to AE signals of larger amplitude. The signal amplitudes are converted in a magnitude scale by using the equation $M = \log_{10}(V/V_0)$, where V is the peak of voltage signal expressed in microvolts, and $V_0 = 1\ \mu V$ is the reference voltage (conversion to decibels is possible through $M = A_{dB}/20$). The signal magnitudes are studied by their cumulative distribution, using the b-value analysis. For a population of N events, the b-value is calculated as:

$$\log_{10} N(M) = a - bM, \tag{2}$$

where $N(M)$ is the number of signals with magnitude higher than M, and a and b are fitting constants.

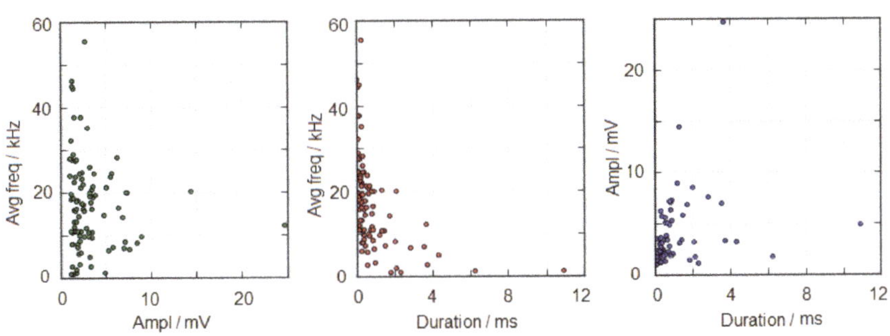

Figure 6. *Cont.*

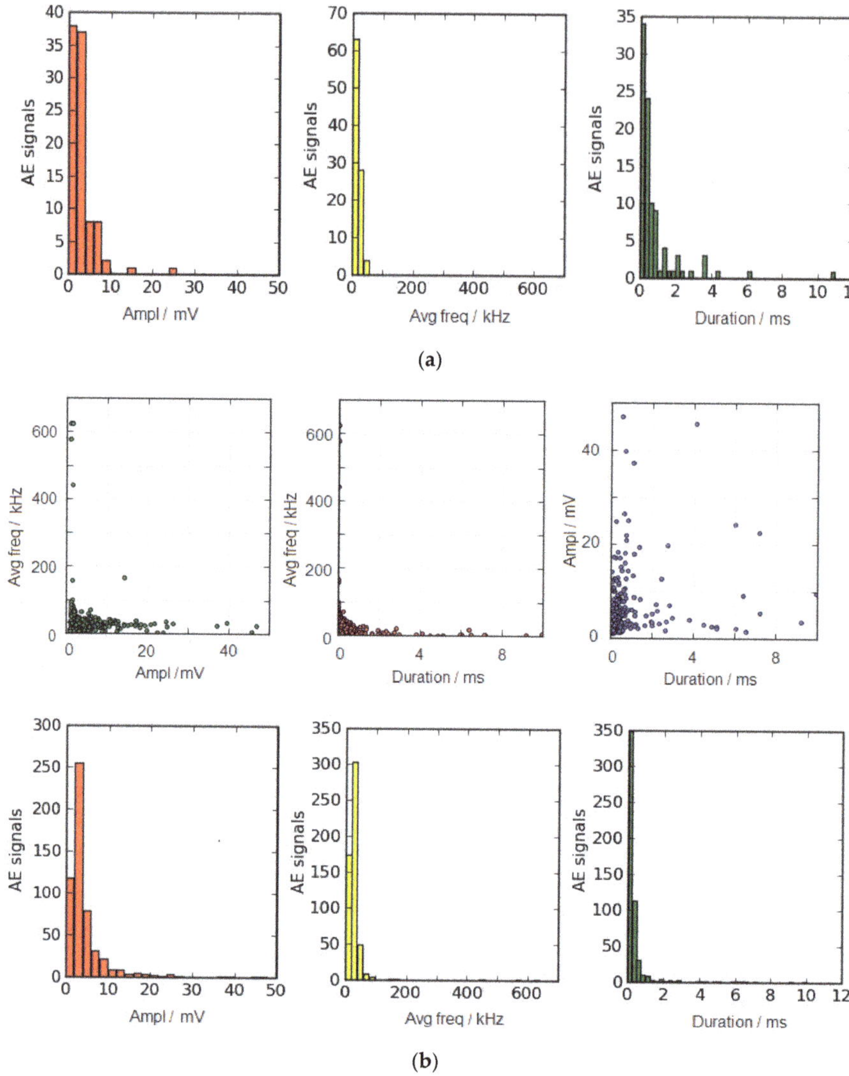

Figure 6. (**a**) AE features relationships and distributions of received signals by sensors AE0 and AE1. The mean values of distributions represented by histograms from left to right are: $<A> = 3.3$ mV, $<f> = 17$ kHz, and $<\Delta t> = 0.8$ ms. (**b**) AE features relationships and distributions of received signals by sensors AE4 and AE5. The mean values of distributions represented by histograms from left to right are: $<A> = 4.5$ mV, $<f> = 31$ kHz, and $<\Delta t> = 0.4$ ms.

The *b*-value expresses the absolute gradient of the cumulative magnitude distribution as illustrated in Figure 7. As the fracture process becomes more intense, the percentage of high-amplitude events increases relative to the low-amplitude ones in the total population, resulting in drops of the *b*-value [25].

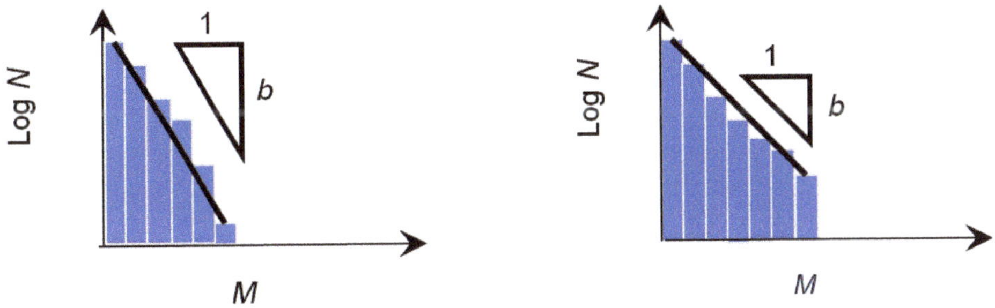

Figure 7. AE magnitude distribution for different damage conditions: low damage (high *b*-value) and high damage (low *b*-value).

Taking into account the number N of recent events to calculate every single *b*-value, possible decreasing trends can be identified and associated with damage severity and change in structural performance state, especially during laboratory loading tests where a single minimum is observed just prior to specimen failure. The recommended value number of events for computation (see e.g., [24]) is typically $N = 100$. Smaller values would yield too strong oscillations, masking the actual trend. Instead, using too large N, any small occurrence producing a small number of AE signals would be undermined in the average of the large population. Here, four partitioning criteria are followed to calculate the evolving *b*-values during the test, namely using disjoint subsets of $N = 100$ and 150 events (Figure 8 (left)), or overlapping subsets of $N = 100$ events with step 25 and a loading stage-based partition (Figure 8 (right)) [25,26]. The most important parameters to be considered in a *b*-value graph are the general trend and the minima of the curve. Comparing all the time series, nothing more than an oscillating trend can be identified, with *b*-values mainly ranging between 1.2 and 1.6 (with an isolated minimum at 1.1 reported in two cases) as descriptive of minor to moderate damage levels reached during the test. As herein illustrated—and widely recognized—the *b*-value, although being sensitive to damage, does not necessarily decrease monotonically. That is due to the presence of both healthy material encountered during crack propagation and steel reinforcements that limit crack growth in concrete matrix. Namely, fracture propagates during steel deformation or concrete sliding with respect to steel. Fracture generates signals related to early damage, which are mixed with those of the already heavily damaged part, as stated in [27–30].

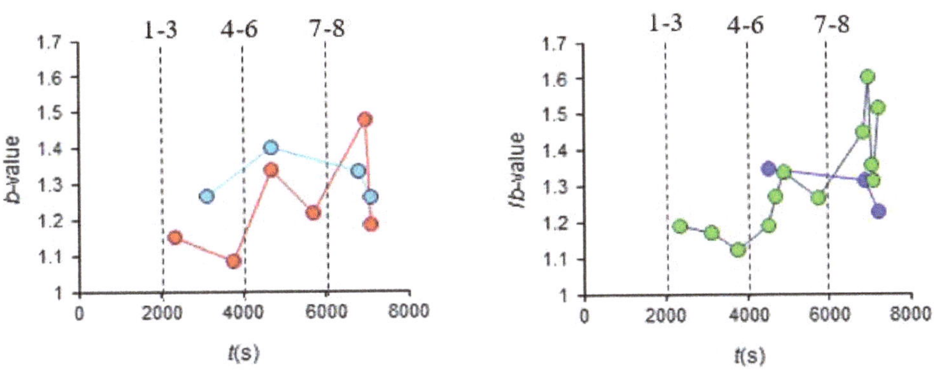

Figure 8. *b*-value time histories with dashed lines referring to the load stages of Figure 5. Using disjoint subsets of $N = 100$ (red dots) and $N = 150$ signals (cyan dots) (**left**). *Ib*-value with overlapping subsets of $N = 100$ events with step 25 (green dots) and *b*-value by a stage-based partition (blue dots) (**right**).

7.3. AE Source Location

Due to the configuration of the experiment, the tensile surface is identified at the bottom of the beam cross sections. Therefore, crack initiation reasonably occurs close to the bottom surface. This is revealed by Figure 9, where the height of localized AE sources is depicted for different loading steps. The fitting line identifies the average height of localized events, exhibiting a slightly ascending trend. At the lowest load level, the average location is close to the bottom surface (−0.06 m). As the cracks propagate towards the top, the AE sources naturally move to higher y-locations, reaching 0.1 m.

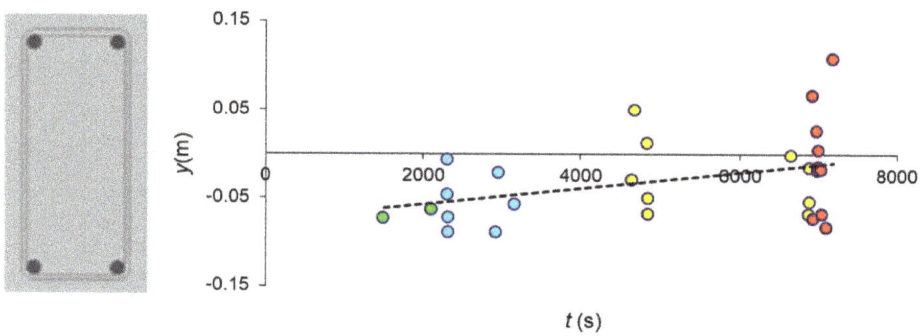

Figure 9. Height of localized AE events for different loading stages. Adopted reference system (x, y, z) as in Figure 1. Green color refers to load stages #1–#4, blue to #5 and #6, yellow to #7 and #8, and red to #9.

The determined locations of AE events concerning different loading stages are depicted together with sensor positions in Figure 10. Location data are only available in 2D, along the beam's lateral axis due to the almost 2D sensor setup, and therefore, information regarding depth position of AE sources cannot be provided. Typically, AE source location is carried out along lines between detecting sensors, i.e., AE0, AE1, AE4 and AE5. Hence, location data do not exactly match the crack initiation from the bottom surface of the beam. One can see localized AE sources starting from the left and accumulating vertically towards the top surface, as the signature of coalescence of uncorrelated tensile cracks and the appearance of diagonal shear cracks. Red circles, representing localized events at the unloading stage, are probably due to friction between aggregates and reinforcing bars during reclosing of the cracks.

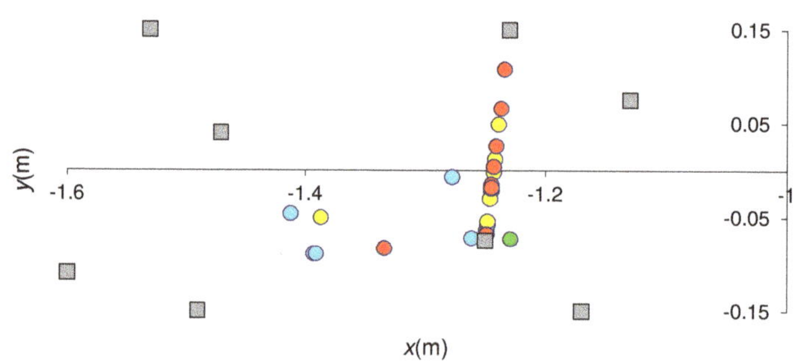

Figure 10. Location of AE sources and AE sensors (numbered gray squares). Adopted reference system (x, y, z) as in Figure 1 and color map as in Figure 9.

8. FOS Strain Measurements

In line with the configuration of the experiment, the tensile state at the bottom of the beam cross sections is revealed by Figure 11, which depicts the history of the internal positive strain measured by the embedded FOS #2 running about 30 mm from the bottom of the external surface. Vice versa, the expected negative strain values in the compressed zone are measured by the FOS #1, running 30 mm from the top external surface (Figure 3).

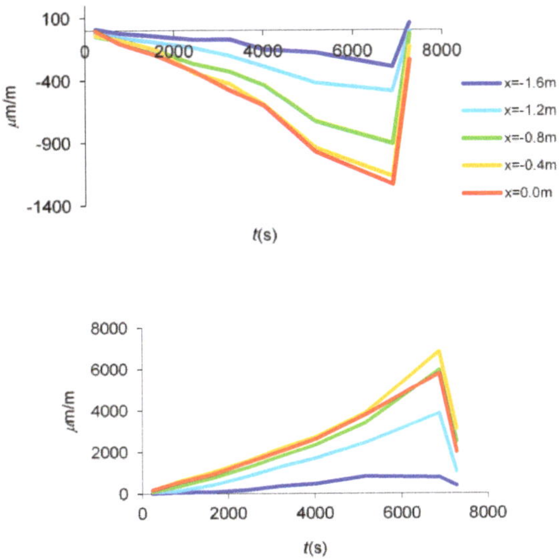

Figure 11. FOS strain measurements at different loading stages along the beam length: FOS #1 (**up**), FOS #2 (**down**).

8.1. AE vs. FOS Measurements

Figure 12a,b compares the strain history (depicted by a blue line) measured close to the bottom surface by FOS #2, respectively at the horizontal locations of −1600 and −1200 mm, with the accumulated AE activity (red line) recorded by the nearest AE sensor, respectively AE4 and AE5.

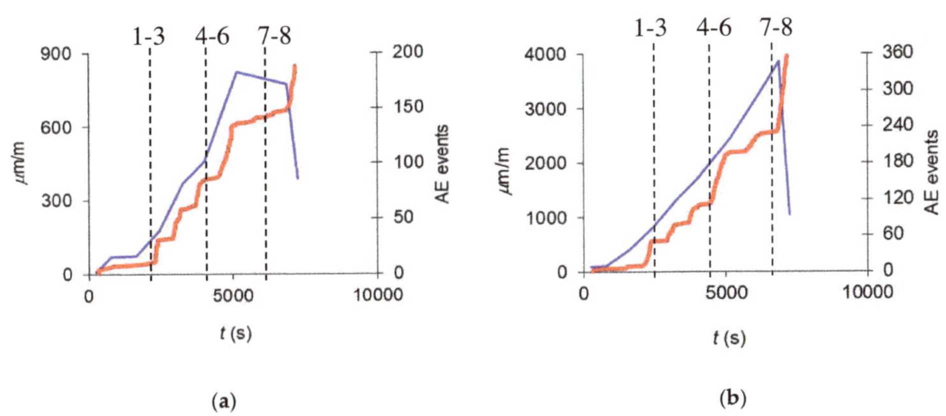

Figure 12. Cumulative AE activity (red) and FOS strain (blue) time histories for: (**a**) fiber optic n.9 at x = −1.6 m and AE sensor n.4; (**b**) fiber optic n.9 at x = −1.2 m and AE sensor n.5. The dashed lines identify the different loading stages.

Figure 13a,b compares the strain history measured close to the top surface by FOS #1, at x = −1600 and −1200 mm, with the accumulated AE activity recorded by the nearest AE sensors, namely AE0 and AE1. Both Figures 12 and 13 illustrate noticeable similarity between AE and FOS graphs.

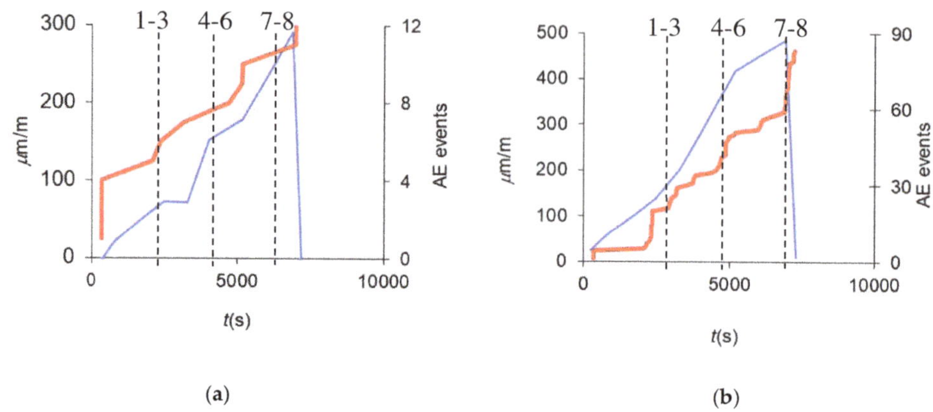

Figure 13. Cumulative AE activity and FOS strain time histories for: (**a**) fiber optic n.7 at x = −1.6 m and AE sensor n.0; (**b**) fiber optic n.7 at x = −1.2 m and AE sensor n.1. The dashed lines identify the different loading stages.

8.2. Cracks Pattern

The results of FOS and AE monitoring are compared with the actual pattern of cracks directly observable in the laboratory. Figure 14 shows the cracks at loading step #7 where the crack development at the bottom side of the beam can be noticed. Furthermore, the transition of crack development from the vertical shape (due to pure bending moment at the center of the beam) to the inclined shape at the extremity (due to shear effects at the support) is highlighted.

Figure 14. Crack pattern at loading step #7 and numbered AE sensors.

9. DIC Strain Assessment

A comparison between DIC results in terms of the strain field on the central span of the beam specimen and the FOS #2 measurements has also been presented in this section. Figure 15 (top) presents the strain field computed through DIC [22] at loading steps #4 (left) and #8 (right), while the FOS strains at the bottom side of the specimen are reported for the same loading steps at Figure 15 (bottom). Focusing on the 0.8 m central length of interest, the FOS #2 sensor highlights a reasonable constant tensile strain of about 1700 μm/m at step #4 and 7000 μm/m at step #8. These results have been computed considering a sampling interval and spatial resolution of 0.2 and 0.5 m, respectively, typically used for large real structures.

The DIC strain field at the bottom side results in the range 1000–3500 μm/m at step #4, and 4000–12,000 μm/m at step #8. Therefore, a satisfactory comparison between the two adopted techniques can be verified.

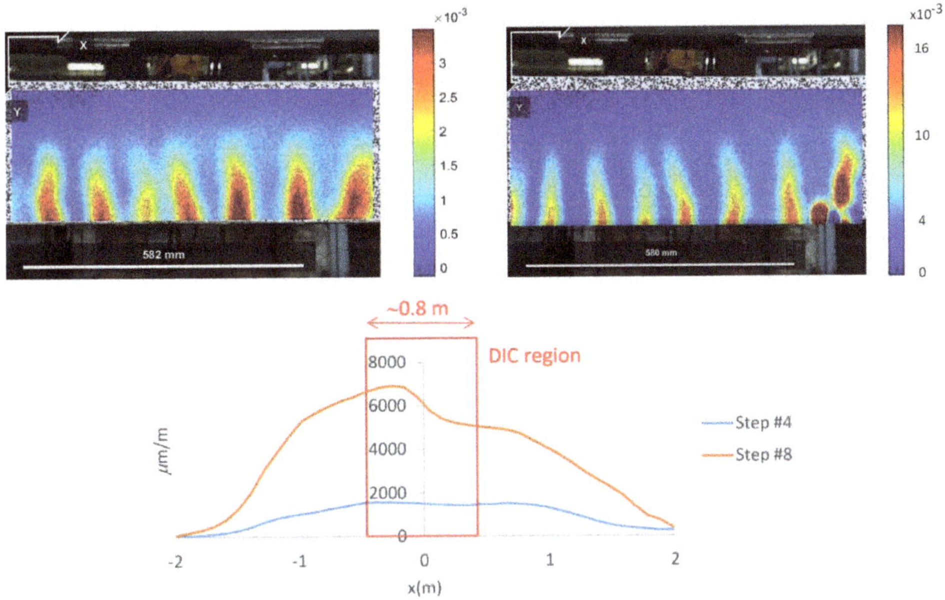

Figure 15. DIC (**top**) and FOS #2 (**bottom**) outcomes at loading steps #4 (top-left) and #8 (top-right) in the central area of the beam specimen (between the vertical loads—see Figure 2).

10. Numerical Analysis

A numerical model of the beam has been prepared through the applied element method (AEM) to perform nonlinear static analysis and reproduce the laboratory observations up to collapse. The AEM is a rather new method, akin to the discrete element method, which is capable of predicting with a reasonable degree of accuracy the continuum and with a high degree of accuracy the discrete behavior of structures during the collapse. The AEM has been proved to track the structural collapse behavior passing through all the application load stages. Within the AEM, the structure is modeled as an assembly of 8-points hexahedral elements. Therefore, each element is assumed rigid (6 degrees-of-freedom) and has a 3-D physical solid shape. Two adjacent elements are assumed to be connected by one normal and two shear springs distributed around the elements' edges on the interface. Each group of springs represents the entire stresses and deformations of a certain volume. More details on the theoretical aspects related to AEM and its comparison with finite element method can be found in Grunwald et al. [31].

The four-points bending test on the reinforced concrete beam has been reproduced through nonlinear static analysis in AEM. The results have been used to anticipate the structural behavior and to compare with the measurements [32]. Different AEM models have been developed by considering a sensitivity analysis to identify a reasonable discretization for the beam. Accordingly, materials have been assumed to be C20/25 and B450C for concrete and steel respectively, with nonlinear constitutive laws following the models by Menegotto and Pinto for steel and Maekawa and Okamura for concrete [33]. A 5 cm side-length element mesh has been selected for the comparison with SHM systems as the reasonable compromise between computational efforts and accuracy.

Considering the global behavior (Figure 16a), the AEM model performs satisfactorily by reasonably anticipating the force–displacement response. Moving to the local behavior, the selected mesh results also are able to correctly trace the bending and shear cracks. Figure 16b, in particular, reports the development of a crack at step #8. The two regions where the AE and DIC techniques have been applied can be observed. The strain concentration due to crack opening, as detected by the DIC technique (Figure 15, top-right) in the central region and by the AE sensors at the extremity, is correctly reproduced by the AEM simulation. It is worth underlining how the screen effect of voids, generated by micro-fractures and distributed damage in the concrete matrix, does not allow the AE sensors to detect the cracks generated at the mid-span of the beam.

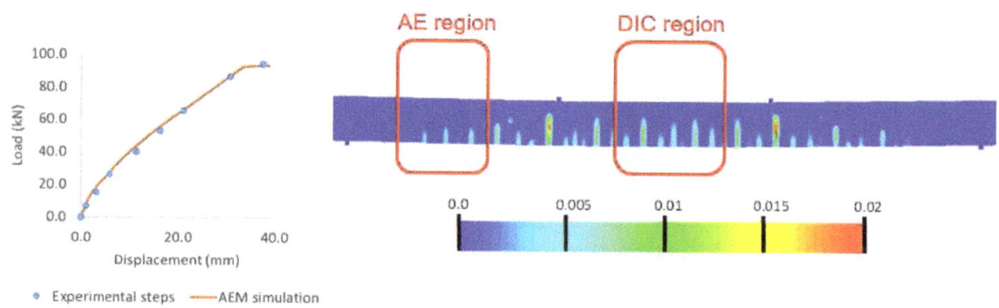

(a) (b)

Figure 16. (**a**) Force–displacement response up to collapse, (**b**) development of cracks at step #8.

11. Conclusions

The implementation of integrated SHM schemes of nondestructive testing techniques is presented in this paper for a reinforced concrete beam tested in laboratory through a four-points bending arrangement. The beam is subjected to loading stages to induce characteristic concrete cracking and steel reinforcements yielding. All of these phases have been identified in a global behavior curve in terms of force–displacement measurements.

The strain measures of the embedded fiber optic system have been compared to the acoustic emission sensors' outcomes. Furthermore, visual observation of the actual cracks pattern has been also considered in parallel with DIC analysis in the central region of the beam specimen.

The recorded AE activity resulted in good agreement with FOS strain measurements. Furthermore, the results of the AE–FOS integrated SHM system are confirmed by the actual cracks pattern. Focusing on the pure bending moment region of the specimen, the strain field computed by DIC also resulted in good agreement with the FOS tensile strains at the bottom side of the specimen.

A numerical model has also been prepared to investigate the mechanical behavior of the beam up to collapse. The analysis has been focused on both the global behavior in terms of force–displacement response, and also the local characterization of cracks. A satisfactory match between the numerical outcomes and the monitoring data has been observed.

Therefore, the integrated use of different nondestructive testing techniques as proposed in this paper is a reasonable and reliable SHM strategy for damage detection and localization in RC elements. This study confirms how the application of different nondestructive testing techniques can be useful for specific civil engineering applications, e.g., for large-sized structural elements where visual inspection is not always possible and accurate. Thus, the simultaneous use of nondestructive testing techniques such as acoustic emission, embedded fiber optic sensors, and digital image correlation, possibly aided by visual inspection, seems to be a step toward the realization of a reliable real-time structural alert system.

Author Contributions: Conceptualization, M.D., G.N., G.L., G.P.C.; investigation and data curation, M.D., G.N.; writing—review and editing, M.D., G.N., G.L.; supervision and project administration, G.P.C. All authors have read and agreed to the published version of the manuscript.

Funding: The research is partially supported by the European Research Council under the Grant Agreement n° ERC_IDEalreSCUE_637842 of the project IDEAL RESCUE-Integrated Design and control of Sustainable Communities during Emergencies.

Acknowledgments: Maurizio Morgese assisted the laboratory analyses in partial fulfillment for the requirements of the Bachelor's Degree in Civil Engineering at Politecnico di Torino, under the guidance of the Authors. His contribution is gratefully acknowledged. The technical support of SMARTEC (CH) and MastrLab at Politecnico di Torino-DISEG is gratefully acknowledged.

Conflicts of Interest: The authors declare no conflict of interest.

References

1. Farrar, C.R.; Czarnecki, J.J.; Sohn, H.; Hemez, F.M. *A Review of Structural Health Monitoring Literature 1996–2001*; Report Number: LA-13976-MS; Los Alamos National Laboratory: Los Alamos, NM, USA, 2002; pp. 1–301.
2. Farrar, C.R.; Worden, K. An introduction to structural health monitoring. *Philos. Trans. R. Soc. A* **2006**, *365*, 303–317. [CrossRef]
3. Mechbal, N.; Uribe, J.S.; Rébillat, M. A probabilistic multi-class classifier for structural health monitoring. *Mech. Syst. Signal* **2015**, *60–61*, 106–123. [CrossRef]
4. Glossop, N.D.W.; Dubois, S.; Tsaw, W.; Leblanc, M.; Lymer, J.; Measures, R.M.; Tennyson, R.C. Optical fibre damage detection for an aircraft composite leading edge. *Composites* **1990**, *21*, 71–80. [CrossRef]
5. Glišić, B.; Inaudi, D. *Fibre Optic Methods for Structural Health Monitoring*; Wiley Online Library: Hoboken, NJ, USA, 2007; pp. 1–262.
6. Ansari, F. Practical implementation of optical fiber sensors in civil structural health monitoring. *J. Intell. Mater. Syst. Struct.* **2007**, *18*, 879–889. [CrossRef]
7. Casciati, S.; Domaneschi, M.; Inaudi, D. Damage assessment from SOFO dynamic measurements. In Proceedings of the SPIE 17th International Conference on Optical Fibre Sensors, Bruges, Belgium, 23–27 May 2005; Volume 5855, pp. 1048–1051.
8. Domaneschi, M.; Sigurdardottir, D.; Glišić, B. Damage detection based on output-only monitoring of dynamic curvature in concrete-steel composite bridge decks. *Struct. Monit. Maint. Int. J.* **2017**, *4*, 1–15.
9. Mba, D.; Rao, R.B.K.N. Development of acoustic emission technology for condition monitoring and diagnosis of rotating machines: Bearings, pumps, gearboxes, engines, and rotating structures. *Shock Vib. Dig.* **2006**, *38*, 3–16. [CrossRef]
10. Behnia, A.; Chai, H.K.; Shiotani, T. Advanced structural health monitoring of concrete structures with the aid of acoustic emission. *Constr. Build. Mater.* **2014**, *65*, 282–302. [CrossRef]
11. Carpinteri, A.; Lacidogna, G.; Pugno, N. Structural damage diagnosis and life-time assessment by acoustic emission monitoring. *Eng. Fract. Mech.* **2007**, *74*, 273–289. [CrossRef]
12. McLaskey, G.C.; Glaser, S.D.; Grosse, C.U. Beamforming array techniques for acoustic emission monitoring of large concrete structures. *J. Sound Vib.* **2010**, *329*, 2384–2394. [CrossRef]
13. Shiotani, T.; Aggelis, D.G.; Makishima, O. Global monitoring of large concrete structures using acoustic emission and ultrasonic techniques: Case study. *J. Bridge Eng.* **2009**, *14*, 188–192. [CrossRef]
14. Carpinteri, A.; Lacidogna, G.; Niccolini, G. Damage analysis of reinforced concrete buildings by the acoustic emission technique. *Struct. Control Health* **2011**, *18*, 660–673. [CrossRef]
15. Liang, S.; Zhang, C.; Lin, W.; Li, L.; Li, C.; Feng, X.; Lin, B. Fiber-optic intrinsic distributed acoustic emission sensor for large structure health monitoring. *Opt. Lett.* **2009**, *34*, 1858–1860. [CrossRef] [PubMed]

16. Li, W.; Xu, C.; Ho, S.C.; Wang, B.; Song, G. Monitoring concrete deterioration due to reinforcement corrosion by integrating acoustic emission and FBG strain measurements. *Sensors* **2017**, *17*, 657. [CrossRef]
17. Ansari, A. Fiber optic health monitoring of civil structures using long gage and acoustic sensors. *Smart Mater. Struct.* **2005**, *14*, S1–S7. [CrossRef]
18. Park, J.M.; Lee, S.I.; Kwon, O.Y.; Choi, H.S.; Lee, J.H. Comparison of nondestructive micro-failure evaluation of fiber-optic Bragg grating and acoustic emission piezoelectric sensors using fragmentation test. *Compos. Part A Appl. Sci. Manuf.* **2003**, *34*, 203–216. [CrossRef]
19. Verstrynge, E.; Pfeiffer, H.; Wevers, M. A novel technique for acoustic emission monitoring in civil structures with global fiber optic sensors. *Smart Mater. Struct.* **2014**, *23*, 065022. [CrossRef]
20. Bao, X.; Chen, L. Recent progress in Brillouin scattering based fiber sensors. *Sensors* **2011**, *11*, 4152–4187. [CrossRef]
21. Nonis, C.; Niezrecki, C.; Yu, T.Y.; Ahmed, S.; Su, C.F.; Schmidt, T. Structural health monitoring of bridges using digital image correlation. In Proceedings of the SPIE Smart Structures and Materials + Nondestructive Evaluation and Health Monitoring, San Diego, CA, USA, 17 April 2013; Volume 869507, pp. 1–13.
22. Chu, T.; Ranson, W.; Sutton, M.A. Applications of digital-image-correlation techniques to experimental mechanics. *Exp. Mech.* **1985**, *25*, 232–244. [CrossRef]
23. Blaber, J.; Adair, B.; Antoniou, A. Ncorr: Open-Source 2D Digital Image Correlation Matlab Software. *Exp. Mech.* **2015**, *55*, 1105–1122. [CrossRef]
24. Lympertos, E.M.; Dermatas, E.S. Acoustic emission source location in dispersive media. *Signal Process.* **2007**, *87*, 3218–3225. [CrossRef]
25. Shiotani, T.; Yuyama, S.; Li, Z.W.; Ohtsu, M. Application of AE improved *b*-value to quantitative evaluation of fracture process in concrete materials. *J. Acoust. Emiss.* **2001**, *19*, 118–132.
26. Rao, M.V.M.S.; Prasanna-Lakshmi, K.J. Analysis of *b*-value and improved *b*-value of acoustic emissions accompanying rock fracture. *Curr. Sci.* **2005**, *89*, 1577–1582.
27. Colombo, S.; Main, I.; Forde, M. Assessing damage of reinforced concrete beam using "*b*–value" analysis of acoustic emission signals. *J. Mater. Civ. Eng.* **2003**, *15*, 280–286. [CrossRef]
28. Carpinteri, A.; Lacidogna, G.; Manuello, A.; Niccolini, G. A study on the structural stability of the Asinelli Tower in Bologna. *Struct. Control Health* **2016**, *23*, 659–667. [CrossRef]
29. Carpinteri, A.; Lacidogna, G.; Corrado, M.; Di Battista, E. Cracking and crackling in concrete-like materials: A dynamic energy balance. *Eng. Fract. Mech.* **2016**, *155*, 130–144. [CrossRef]
30. Niccolini, G.; Borla, O.; Accornero, F.; Lacidogna, G.; Carpinteri, A. Scaling in damage by electrical resistance measurements: An application to the terracotta statues of the Sacred Mountain of Varallo Renaissance Complex (Italy). *Rend. Lincei Sci. Fis. Nat.* **2015**, *26*, 203–209. [CrossRef]
31. Grunwald, C.; Khalil, A.A.; Schaufelberger, B.; Ricciardi, E.M.; Pellecchia, C.; De Iuliis, E.; Riedel, W. Reliability of collapse simulation—Comparing Finite and Applied Element Method at different levels. *Eng. Struct.* **2018**, *176*, 265–278. [CrossRef]
32. Domaneschi, M.; Cimellaro, G.P.; Marano, G.C.; Morgese, M.; Pellecchia, C.; Khalil, A.A. Numerical simulations of collapse tests on RC beams. In Proceedings of the 10th International Conference on Bridge Maintenance, Safety and Management—IABMAS 2020, Sapporo, Japan, 11–15 April 2020.
33. Applied Science International LLC. *Extreme Loading for Structures 2018*; Applied Science International LLC.: Durham, NC, USA, 2018; Available online: https://www.appliedscienceint.com/extreme-loading-for-structures/ (accessed on 30 March 2020).

applied sciences

MDPI

Article

Application of Impact-Echo Method to 3D SIBIE Procedure for Damage Detection in Concrete

Katsufumi Hashimoto [1,*], Tomoki Shiotani [1] and Masayasu Ohtsu [2]

1 Department of Civil and Earth Resources Engineering, Graduate School of Engineering, Kyoto University, Kyoto 615-8540, Japan; shiotani.tomoki.2v@kyoto-u.ac.jp
2 Center for the Promotion of Interdisciplinary Education and Research, Graduate School of Engineering, Kyoto University, Kyoto 615-8540, Japan; ohtsu.masayasu.4v@kyoto-u.ac.jp
* Correspondence: hashimoto.katsufumi.8a@kyoto-u.ac.jp; Tel.: +81-75-383-3496

Received: 25 February 2020; Accepted: 12 April 2020; Published: 15 April 2020

Abstract: In this study, to visualize damage and defects, such as cracks and voids in concrete, the SIBIE (stack imaging of spectral amplitudes based on impact echo) procedure is applied and numerically improved to construct a three-dimensional (3D) model of elastic wave propagation behavior. A unit of arrayed accelerometers is installed to detect multi-channel signal waveforms in the frequency domain. The resonant frequencies due to reflections at each node in 3D lattice nodes are computed by using the distances from elastic wave input to multiple output locations. The amplitudes corresponding to the resonant frequencies in the spectrum are summed up as the reflection intensity of elastic wave at each node. The reflection intensity distribution is visualized finally in the targeted area three-dimensionally. Case studies are carried out on the proposal of the improved 3D-SIBIE procedure, applied to a concrete specimen with simulated-damage as well as in-situ highway RC (Reinforced Concrete) slabs in service. As for the signal detection, a non-contact elastic wave detecting system using a laser doppler vibrometer is also introduced to consider and validate the promising remote sensing and inspection technique for damage evaluation in concrete with the 3D SIBIE procedure.

Keywords: non-destructive testing; concrete; impact-echo; elastic wave; SIBIE; 3D visualization; accelerometer; laser doppler vibrometer; remote sensing

1. Introduction

1.1. Non-Destructive Testing for Concrete

It has been generally recognized that preventive and proactive maintenance ideas are necessary for such social infrastructure as bridges and tunnels to develop rational life-cycle scenarios. For reinforced concrete (RC) members, essential issues involve the establishment of a maintenance system with appropriate measurements before evolving the serious damage and failure. For non-destructive testing (NDT), to qualify the materials' property or structural integrity of concrete, several NDT methods have been applied.

Since sufficient and effective maintenance of infrastructure is desired under the restriction of budget, appropriate inspections by means of NDT must be applied adequately. As for the damage assessment and estimation of repair and retrofit recovery in concrete structures, in addition to developing effective NDTs, innovative methods corresponding to the damage characteristics must be established in parallel. Also, it is strongly demanded to propose appropriate management systems for ageing concrete infrastructures, adapting the aforementioned accurate and ready-to-use inspection techniques for the damage in concrete.

In concrete structures of bridges such as decks, the most problematic type of damage is fatigue damage due to numerous repetitions of excessive traffic overloads [1,2]. The final form of damage due to cyclic loads is well known as horizontally induced large cracks in steel–concrete composite slabs [3]. As it is difficult to detect interior damage, which potentially affects structural performance, by visual inspection [4], this becomes a very serious problem for highway structures composed of reinforced concrete (RC) [5]. For the RC bridge slabs in service, inspection work is generally allowed to implement the bottom side, not including the top side, to avoid controlling the traffic. Therefore, a one-side access NDT method is strongly demanded to quantify or assess the damage in RC members. When the special permission to access both sides is given for in-depth inspections or experimental purposes, tomographic approaches with elastic waves have been applied to visualize the internal damages. Specifically, AE (acoustic emission) sensors or accelerometer arrays are installed on the bottom side and artificial excitations are made on the top side so the internal damage can be visualized with such elastic wave parameters as P-wave velocity and energy attenuation [6,7]. One-side measurements with surface wave techniques have thus been employed to investigate the interior damage of concrete [8]. However, specific depths could not so far be identified with this technique. Meanwhile, the impact-echo method is the most well-used non-destructive testing for concrete structures to characterize the surface area of interest [9,10]. The SIBIE (Stack Imaging of spectral amplitudes Based on Impact-Echo) procedure is an advanced visualization technique using the data of impact echo testing [11,12]. SIBIE demonstrates a two-dimensional (2D) internal image of concrete simply with one receiver while leaving such issues to study such as 3D visualization and more accurate damage identification.

1.2. Impact-Echo Method

Impact-echo is an acoustic method for the nondestructive evaluation of concrete, which is based on the use of impact-generated elastic waves that propagate through concrete, and they are reflected by internal flaws or external surfaces [9]. Impact-echo can be used to not only determine the location and extent of flaws, such as cracks, delamination, and voids in the targeted concrete, but also to provide thickness measurements of concrete slabs as well [10].

As shown in Figure 1, in the impact-echo system, an induced elastic wave at a surface of concrete propagates and reflected by defects or the surfaces in concrete. The received signal is monitored by a transducer coupled to the surface near the elastic wave source. The time-domain waveform is transformed into the frequency-domain one using the fast Fourier transform (FFT). The frequency spectrum contains those resonance frequencies due to reflections at such interfaces as the internal defects and the surfaces of the material of interest. When the P-wave velocity in the concrete is known, the detected peak frequency can be expressed as in Equations (1) and (2), respectively.

$$f_T = \frac{C_p}{2T} \tag{1}$$

$$f_d = \frac{C_p}{2d} \tag{2}$$

where f_T is resonance frequency due to material thickness, f_d is resonance frequency due to internal damage (defect), T is the thickness of targeted member, d is defect depth, and C_p is P-wave velocity in concrete.

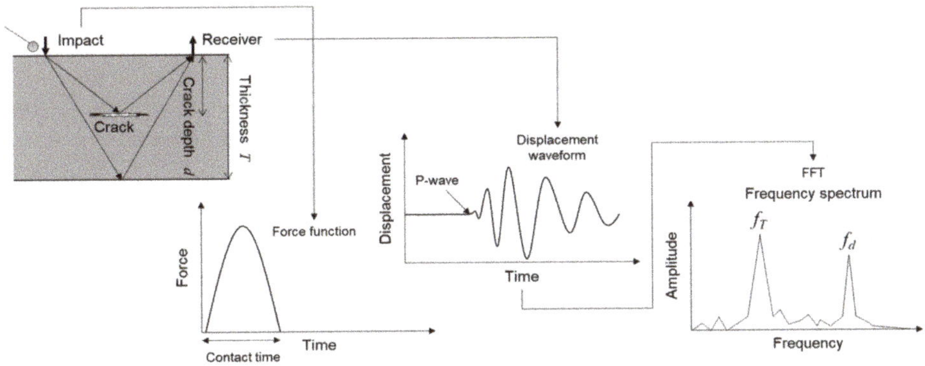

Figure 1. Schematic of impact-echo method applied to the testing interior concrete.

1.3. SIBIE Procedure

In order to visualize such defect as crack and void in concrete, SIBIE (stack imaging of spectrum amplitudes based on impact-echo) procedure has been developed [11,12]. In the procedure, a targeted cross-section is modeled and divided into square elements as shown in Figure 2. Then, resonant frequencies obtained from FFT analysis due to elastic wave reflections at each element are computed. The travel distance from the input, which is excited by hammering (tapping) concrete surface with steel sphere ball of 10 mm diameter, to the output through the elements is calculated by Equation (3).

$$R = r_1 + r_2 \tag{3}$$

where r_1 is the distance from the input (impact) point to the nodal point of an element and r_2 is the distance from the nodal point of an element to the output (signal detection) point. The resonant frequencies due to reflections at each element f_R are calculated by Equation (4).

$$f_R = \frac{C_p}{R} \tag{4}$$

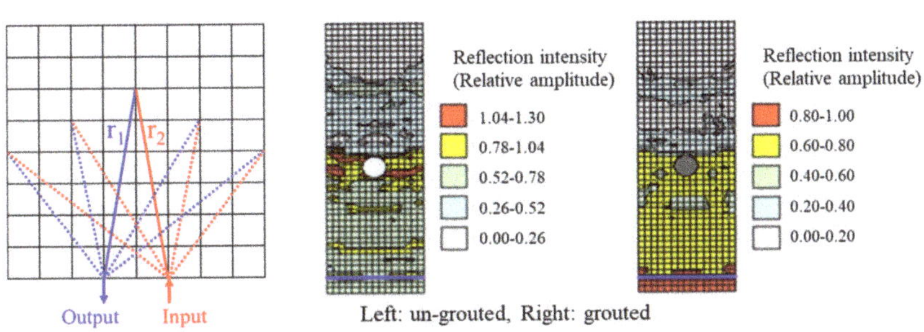

Figure 2. Mesh division and results of SIBIE (Stack Imaging of spectral amplitudes Based on Impact-Echo) procedure [3].

In the frequency spectrum, amplitudes corresponding to these resonances are summed up at each nodal point. Thus, the reflection intensity as the data stacking result at all the analytical elements is prepared for the visualized contour image. After the SIBIE analysis, a 2D view of the contour image is displayed, and the high reflected zone is visually identified as the estimated location of anomaly namely damage.

The SIBIE procedure has been applied for the case studies in the author's previous research. The results for detecting grouted and un-grouted PC (pre-stressed concrete) sheaths is illustrated as in Figure 2 [11]. The size of each mesh is of dimensions 10 × 10 mm. In the figures, it can be observed that both grouted and un-grouted sheaths were clearly identified with the values of reflection intensities. The contour values are given as the relative amplitude, which is obtained by normalizing data with a maximum amplitude provided from the FFT result, of the P-wave intensity computed from frequency spectrum of received signal.

In the above-mentioned results, after the SIBIE procedure, it is possible to identify the grouting condition only when the sheath location is preliminarily known. When the sum of distance R in Equation (3) from input point to a nodal point and the nodal point to output point is the same (Figure 2), it is understandable that the same reflection intensities are given. As the location is based on Equation (3), there are many candidates of nodes with the same travel distance, and therefore the conventional the SIBIE procedure with only one receiver cannot occasionally provide correct geometrical information for asymmetrically distributed damage to the vertical axis e.g., in Figure 2. To upgrade the procedure from 2D to 3D with multiple receivers, in this study, an array with multiple sensors over a targeted area is employed. Specifically, four accelerometers and laser doppler vibrometer system, which is innovatively introduced for remote sensing technique, are used to improve the identification of damage in concrete. A concrete slab of existing structures and an experimentally casted slabs embedded by artificial damage with styrofoam are subjected to the expanded SIBIE procedure.

1.4. Remote Sensing for Elastic Wave Detection with Laser Doppler Vibrometer

As for remotely detecting elastic waves propagated through concrete, non-contact acoustic systems with a laser doppler vibrometer (LDV) for detecting signals have been reported [13–15]. These research approaches provide remote and rapid operations for inspection works based on the elastic wave method. In this study, a laser doppler vibrometer is also introduced, in addition to conventional accelerometers, to detect the elastic wave in concrete. The signal output from LDV is generally a continuous analog voltage that is directly proportional to the velocity component at targeted surface along the laser beam direction. The system specification of LDV used in this study is given as digital velocity decoder with 0.5 Hz–22 kHz frequency band and velocity resolution is <0.02 (μm/sec)/√Hz. The signal detectable distance of LDV is from 0.1 to 30 m. This enables the non-contact vibration measurements of concrete surface and obtain the signal information required for the SIBIE analysis in addition to the conventional contact signal acquisition system with accelerometers.

2. 3D SIBIE Procedure

2.1. Sensor Array with Accelerometers and Laser Doppler Vibrometer

To develop an improvement for the SIBIE procedure in this study, the measurement and the computation steps in the procedure are extended to three-dimensional (3D) model, using a multi-sensor (accelerometer) array unit for on-site measurement as shown in Figure 3. The unit is developed to access only underneath an RC deck in service with 30 × 30 cm grid composing of 4 sensors, arranging each apex of 300 mm square. The thickness of the deck is approximately 400 mm, composed of 180 mm RC (reinforced concrete), 70 mm FRC (fiber reinforced concrete) overlay, and 50 mm asphalt pavement.

The multi-sensor inputs and outputs provide the spectral amplitudes of resonance frequencies due to 3D propagation of elastic wave reflections at each element (voxels) in depth (thickness of the targeted concrete) direction of 2D plane (300 × 300 mm), which is the volume of interest for 3D SIBIE analysis. Additionally, to achieve remote-controlled non-destructive testing, the center of the array unit is targeted by laser doppler vibrometer to obtain the signal as well as the accelerometers in the laboratory test. Figure 4 shows the application overview and the received signal when introducing the laser doppler vibrometer to detect the signals. As the waveform signal detected by LDV demonstrates sufficient amplitude, this can be used as the received signal through the concrete by the SIBIE procedure.

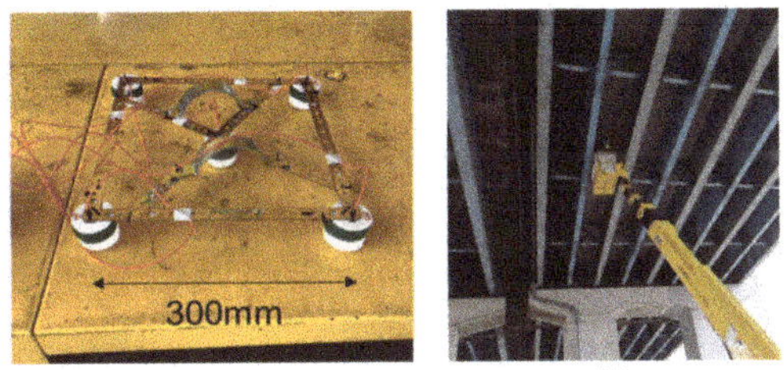

(a) (b)

Figure 3. (**a**) Accelerometer array unit; (**b**) Overview of on-site implementation.

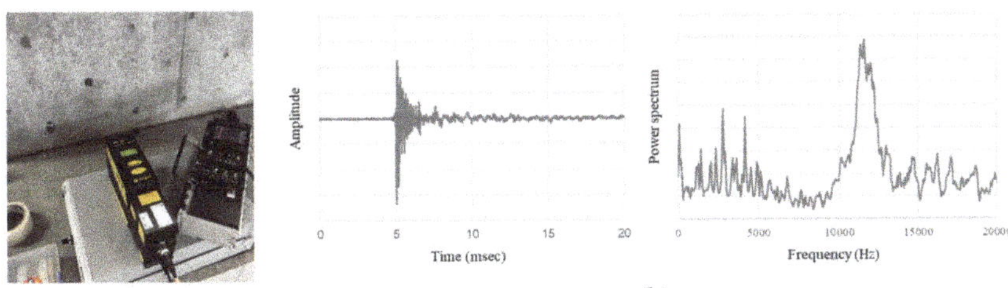

(a) (b)

Figure 4. (**a**) Laser doppler vibrometer; (**b**) Detected waveform with application of laser doppler vibrometer for laboratory test.

2.2. Computing Procedure for 3D SIBIE

The following steps show one-side access measurement and the computation procedure, developed in this study for the 3D SIBIE procedure. Figure 5 displays the targeted volume of interest and cross sections employed for the computation.

1. Place the sensor array on one surface of targeted concrete and hammer with steel sphere ball of 10 mm diameter near each sensor (n = 4) for 10 times respectively.
2. Divide 300 (x-direction) × 300 mm (z-direction) in depth (y-direction) for volume of interest into 10 mm meshes.
3. Employ the nodal points on the cross sections of x = 0, x = 300, z = 0, z = 300, x = z, x + z = 300 to mitigate computational load.
4. Calculate R (travel length of elastic wave) from each nodal point to input and output points (R is the shortest linear path as explained).
5. Search for f_R frequencies due to reflection of propagating P-wave with velocity of C_p.
6. Normalize the amplitude from 0 to 1 based on the highest peak value in the frequency spectrum to cancel the variability of each manual hammering impact.
7. Stack amplitudes according to f_R and quantify reflection intensity values at each nodal point.
8. Visualize 3D geometry with the stacking data based on local polynomial regression method.

Figure 6 shows the example of frequency spectrum successfully obtained from the received signal followed by FFT analysis. Sampling rate is set at 100 kHz in the following measurement. Those sampling rates are appropriately determined to discuss the FFT (Fast Fourier Transform) results in this study since the waveform is sampled at 10000 points for FFT to achieve the sufficient resolution less than 50 Hz.

Meanwhile, some recorded waveform due to an impact of low energy may cause the low S/N (signal-to-noise ratio) signal and FFT indeterminate peaks. The procedure of FFT signal data stacking with/without including low S/N waveforms are shown in Figure 7. Tapping 10 times according to the above-mentioned steps for 3D SIBIE method, FFT peak stacking results in giving an incorrect data with unclear frequency spectrum when some signals into a sensor are not entirely high S/N, wherein n = 3, 6, 8, 9 for the example. As a consequence, between the process of 5 and 6 for the development of the 3D SIBIE procedure, low S/N signals are removed from the computation to guarantee data quality. The FFT signal data stacking result without signal normalization procedure is indicated numerically as in Figure 8, where the process of 6 is excluded. It is obvious that the spectrum of n:3 is different from others in 5 kHz implying that this is attributed to some noises. As shown in Figure 8a, the stacked amplitude of 10 raw data without the signal normalization is influenced by the noise signal of n:10 but now found in Figure 8b. The normalization for each frequency spectrum before stacking is thus very effective on the data acquisition process of the 3D SIBIE procedure.

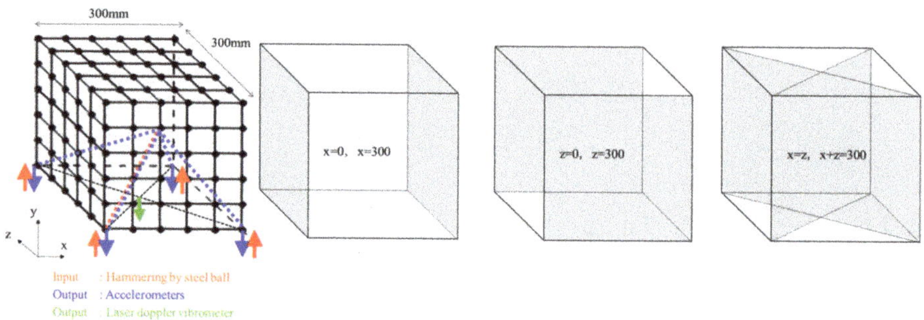

Figure 5. Volume of interest and cross section for 3D SIBIE procedure.

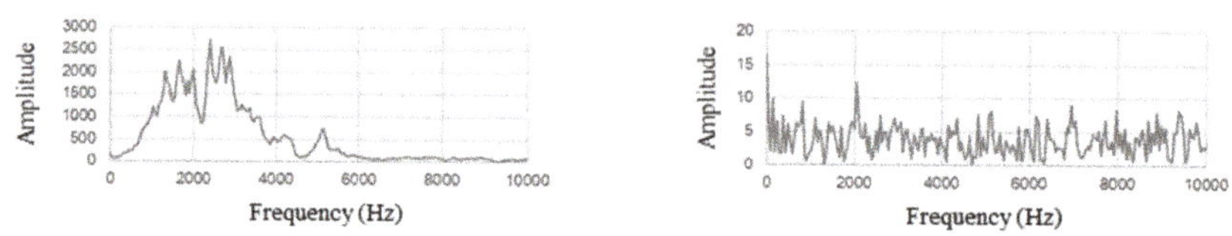

(a) FFT result with High S/N signal

(b) FFT results with Low S/N signal

Figure 6. An example of FFT (Fast Fourier Transform) result of low S/N signal.

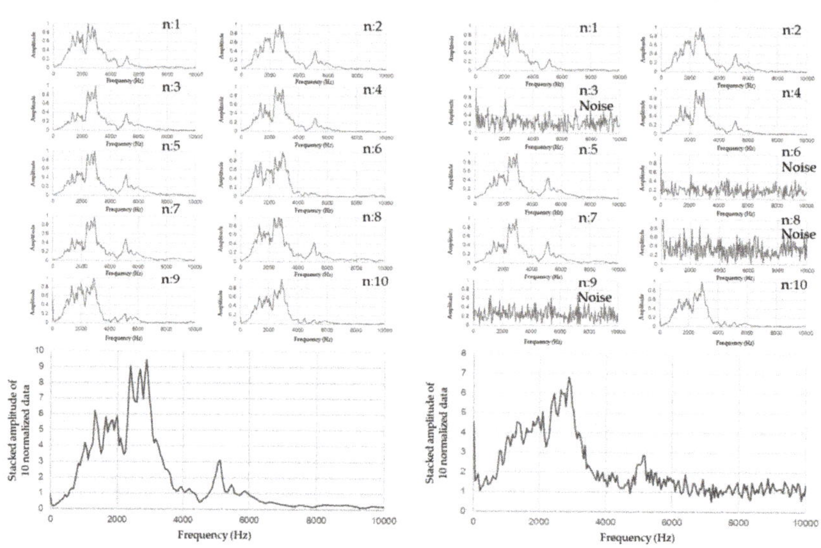

Figure 7. FFT signal data stacking procedure.

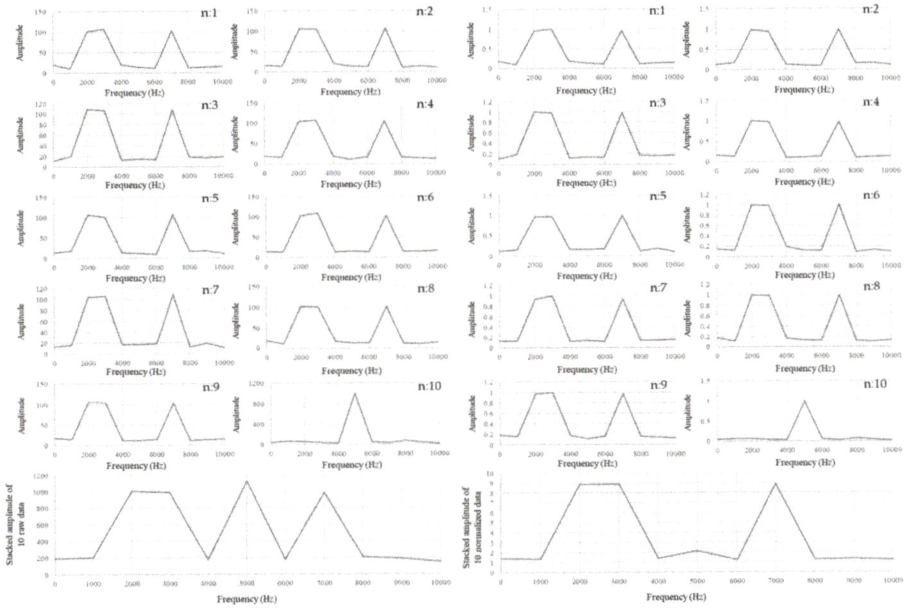

a) without each signal normalization procedure b) with each signal normalization procedure

Figure 8. FFT signal data stacking procedure with/without each signal normalization procedure.

3. Examination of 3D SIBIE Procedure

3.1. Application of Accelerometer Array for Detecting Signals

3.1.1. Reinforced Concrete Bridge Deck

The 3D SIBIE procedure has been applied for existing structures. A RC bridge deck in service was targeted. Two different areas were selected for the case study after the crack inspection work. The crack inspection procedure is schematically described in Figure 9 with 4 steps as follows.

1. Drilling a 5 mm diameter hole into the concrete deck from the surface of asphalt pavement
2. Injecting epoxy resin (in orange-colored to clearly see the crack)
3. Re-Drilling a 10 mm diameter hole at the same location after the resin is hardened
4. Inserting a flexible fiber optic borescope into concrete to take photos

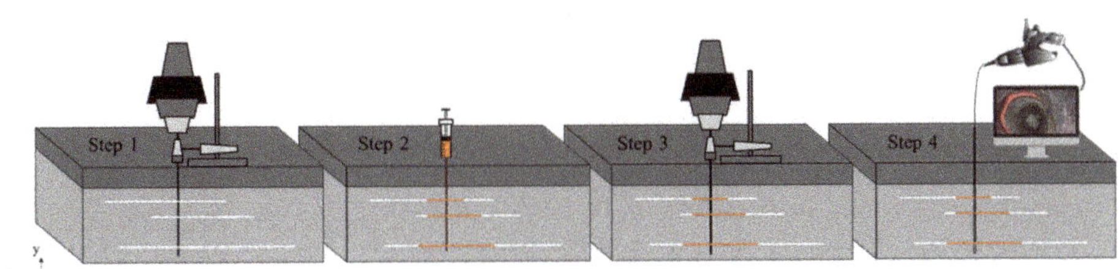

a) 1st Drilling with 5 mm diameter b) Injecting color resin c) 2nd Drilling with 5 mm diameter d) Taking photos by fiber-scope

Figure 9. Visual rack inspection procedure.

Figure 10 shows an example of the view from the fiberscope when it caught the crack filled with resin. In this way, damaged and non-damaged namely intact parts are chosen for the 3D SIBIE procedure. Figure 11 shows the 3D SIBIE result for the RC deck in the intact part and Figure 12 shows that for RC deck in the damaged part where a lateral crack in concrete and also delamination between an asphalt pavement layer and a concrete layer was found. The sensor array was placed on the bottom surface of the deck and measurement was carried out for the 3D SIBIE procedure. The top surface denotes that of asphalt pavement and the bottom shows the bottom of the concrete layer both in Figures 11 and 12. In these figures, the thickness of the concrete layer is 0–180 mm, 180–250 mm in the fiber reinforced concrete (FRC) overlay, and the asphalt pavement is 250–300 mm from the bottom. In the intact part as shown in Figure 11, no sharply defined reflection intensity is found in the volume of interest while, for the damaged part as shown in Figure 12, the color gradations show strong reflections of elastic waves due to anomaly such as cracks or voids around the bottom of the RC deck (see the bottom area in the right side sectional view). Also, three layers consisting of concrete, FRC and asphalt can be classified by the 3D visualized image. As the strong reflections are obtained from the large difference of acoustic impedance of which the individual materials have, the delamination between the asphalt and the concrete layers, which can be visually confirmed, and sub-critical cracks, which cannot be visually confirmed, are possibly attributed to this fact.

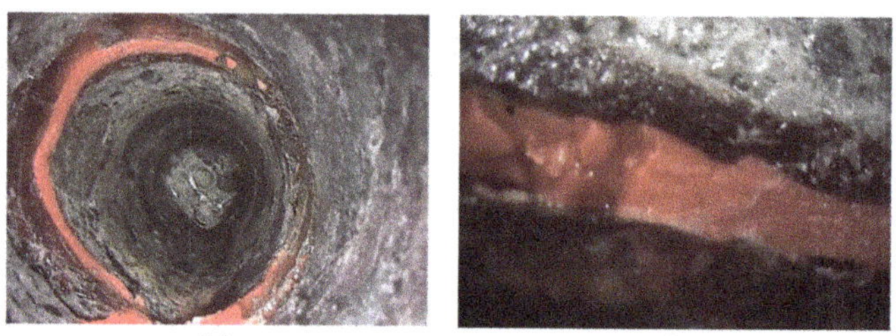

(a) (b)

Figure 10. Cracks filled with resin (**a**,**b**).

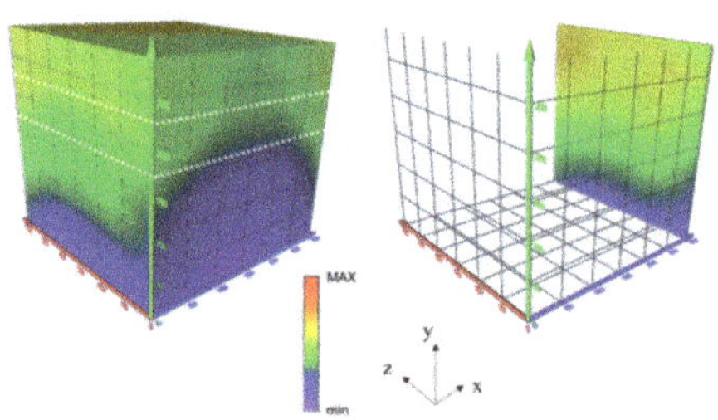

Figure 11. 3D SIBIE result at intact part.

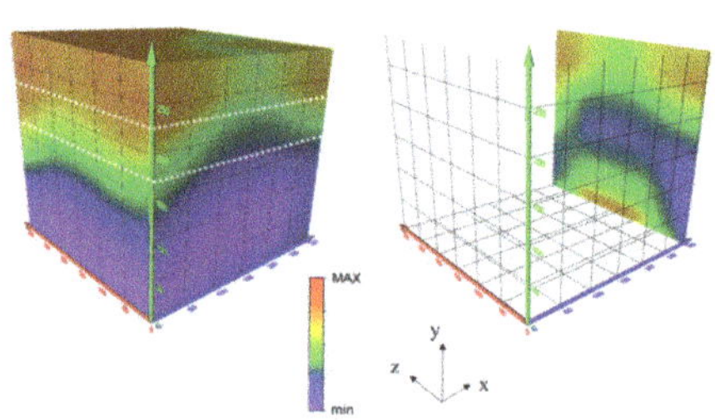

Figure 12. 3D SIBIE result at damaged part.

3.1.2. Concrete Specimen with Simulated Defect

Concrete specimen using high-early strength cement was prepared for 30 MPa design strength. The size is 300 (x-direction) × 400 (y-direction) × 300 mm (z-direction). The thickness is defined as y-direction in this study. Figure 13 shows the geometrical information of concrete specimens and sensing (accelerometers and LDV) locations. As for experimentally simulating damage in concrete, a cubic styrofoam of 300 (x-direction) × 50 (y-direction for thickness) × 150 mm (z-direction) was embedded at 175 mm depth in the y-direction in the center of specimen.

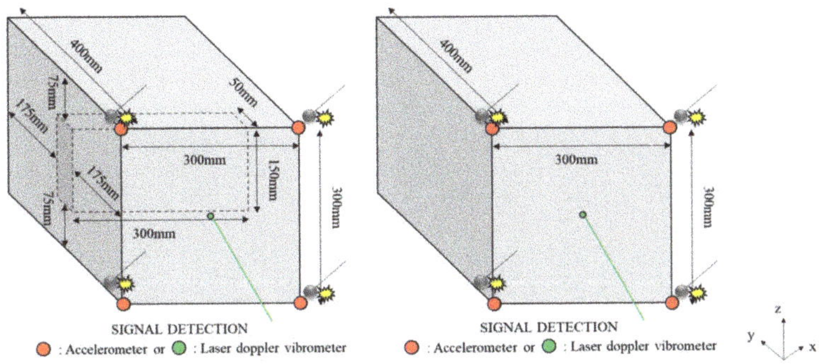

Figure 13. Concrete specimens and sensor locations (with defect and with no defect (intact)).

As shown in Figures 14 and 15, when the accelerometer array unit is employed for detecting signals, in the visualization results with the 3D SIBIE procedure for interior concrete, the location of defect (simulated damage) is clearly identified from 175 to 225 mm in the depth (y) direction in Figure 14, where higher reflection intensity is given, while no remarkable changes is not found in other depths as well as in the intact specimen as shown in Figure 15. Note, the values of reflection intensities are provided as color gradations relatively from 0 to 1 as normalized data based on the maximum value of each contour displayed. The reflection intensity at the embedded styrofoam is even lower than that of the thickness reflection at 400 mm depth in the y-direction, since the acoustic impedance of air is obviously lower than that of the styrofoam. As for the result of intact concrete in Figure 15, only the thickness reflection is demonstrably observed at 400 mm depth in the y-direction.

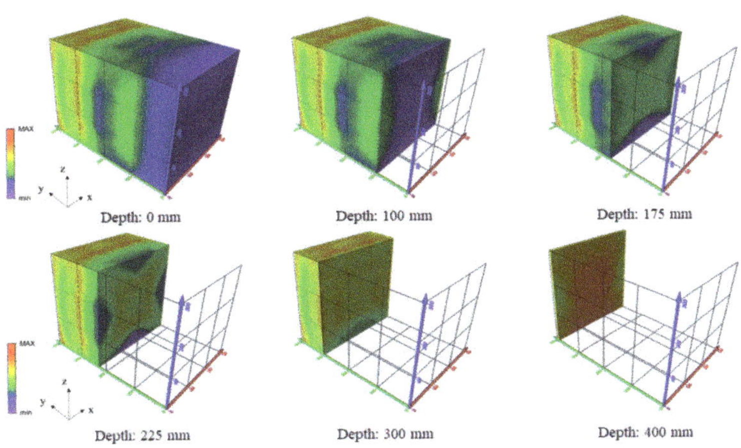

Figure 14. Result of 3D SIBIE procedure for damage-simulated concrete specimen.

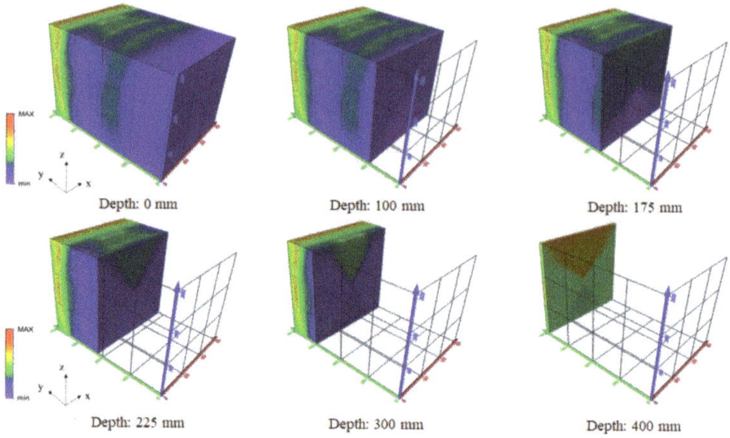

Figure 15. Result of 3D SIBIE procedure for intact concrete specimen.

3.2. *Application of Laser Doppler Vibrometer for Detecting Signals*

A laser doppler vibrometer for detecting signals has been introduced for the 3D SIBIE procedure to visualize the simulated damage in the concrete specimen. When LDV signal detection is applied and the data are introduced to the 3D SIBIE procedure, the visualized images in Figure 16 successfully indicate the defect position with higher reflection intensity at the middle depth of the specimen than that at the other area in the targeted volume of interest. On the other hand, when it is applied to the intact specimen as shown in Figure 17, an intensive reflection is only obtained from the thickness at 400 mm depth in the y-direction, as is the case of the accelerometers.

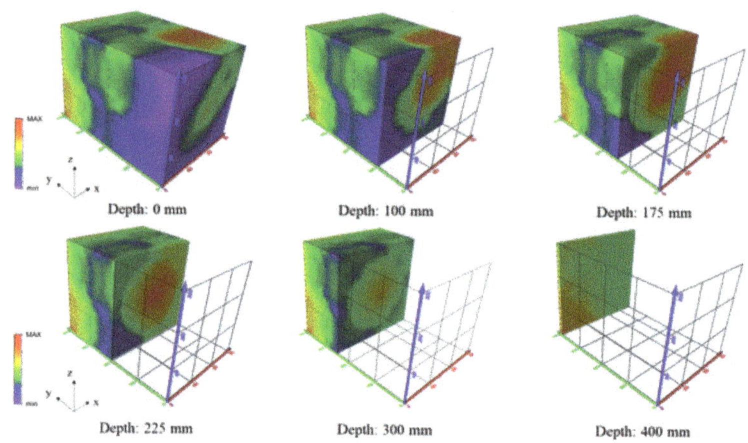

Figure 16. 3D SIBIE result with LDV (Laser Doppler Vibrometer) signal detection for damage-simulated concrete specimen.

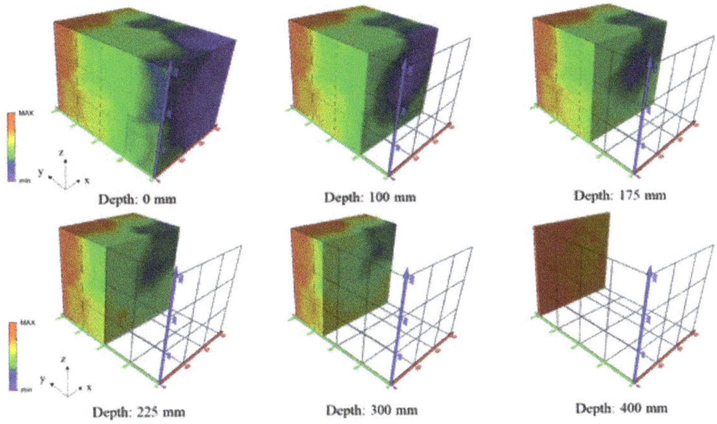

Figure 17. 3D SIBIE result with LDV signal detection for intact concrete specimen.

3.3. 3D SIBIE Procedure Improved with Signal Amplification

As explained in this study, the 3D SIBIE procedure is developed based on 3D model with the use of the multiple-sensor array. The procedure is successfully applied for the case studies of the targeted concrete specimen for the laboratory test by employing the accelerometer array unit and the laser doppler vibrometer for detecting signals as well as the case of the existing RC deck in service. As the results, the extension of the 2D computation process for the conventional SIBIE procedure to 3D coordinates enables to visualize the interior damage. The 3D SIBIE procedure has a great promise for on-site and non-contact measurement and potentially could be used to operate remote and rapid damage inspection and material assessment for concrete structures.

The 3D SIBIE procedure do not consider the attenuation of propagation in elastic waves so that the results described above in Figures 14–17, relatively high reflection intensity is found even in the intact area of interest. Since the amplitude in frequency domain of the signal detected at the receiving point with the accelerometers or LDV by reflection of elastic wave becomes smaller as its propagation distance becomes longer, it is considered that the deeper the detection depth in the targeted volume, the smaller the reflection intensity mathematically in the result of signal computing process of the SIBIE procedure.

Pile Integrity Test (PIT) utilizes a simple principle of elastic wave characters to assure the quality of the concrete pile. When elastic waves are excited at the pile head, it is assumed that elastic waves travel at the speed of C inside the concrete pile shaft, and the pile length can be determined by measuring the time lapse of T. Based on the detected signal waveform between the time of hammering and the time of receiving reflections on pile head, as shown in Figure 18, the amplitude of the reflected wave is attenuated due to the effect of geometrical or internal damping of the materials during the propagation.

Thus, in the applied technique for in-situ measurement, an exponential amplification function is applied across the recorded waveform to amplify the low energy reflection peaks from the pile toe and other internal defects such as voids and cracks [16–18].

In the same manner as in the signal amplification method of PIT principle, the method is also introduced for the 3D SIBIE procedure to correct the reflection intensity in the visualization results. Figure 19 shows some examples of actually recorded signals with the accelerometer sensor array, explained in Figure 13, followed by hammering at concrete surface. In order to define an exponential function for signal amplification in consideration with the signal attenuation through the targeted concrete, an exponential function is derived for each data set from the obtained waveforms as in Figure 20.

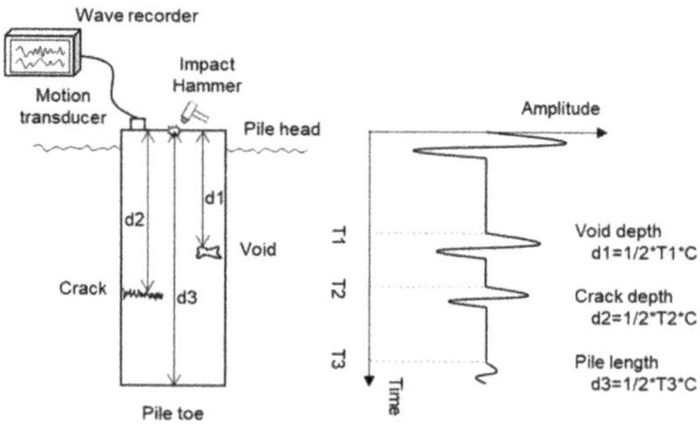

Figure 18. Principle of PIT (Pile Integrity Test).

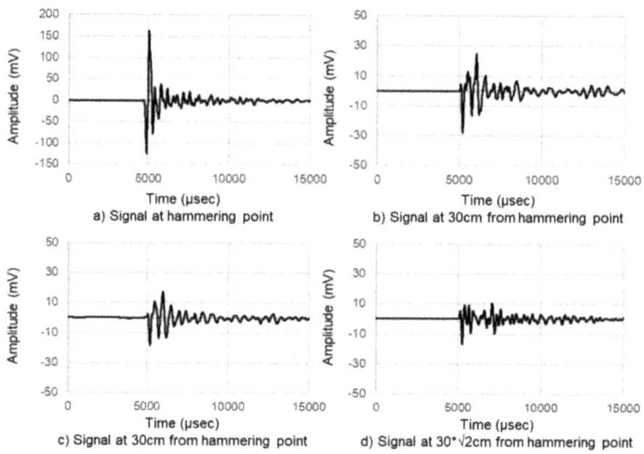

Figure 19. Detected signals at accelerometers with a hammer strike.

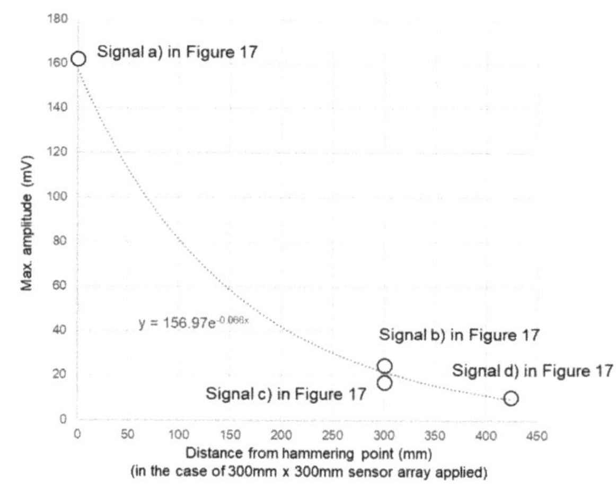

Figure 20. Signal attenuation along wave propagation in sensor distance.

The respective functions for each recorded data can be used to calibrate the attenuation of the detected signals. By using the calibrated data according to the signal amplification function, as shown in Figures 21–24, the visualization images of reflection intensity are modified by the 3D SIBIE procedure. The results indicate that the visualized images are changed due to the improved signal processing.

This means that reflection intensities at the damage-simulated part with styrofoam is sufficiently high compared to those at an intact part without the consideration of the attenuation effect during elastic waves propagation. On the other hand, as for intact cases, the relatively high reflection intensities expressed in the intact part are perfectly removed and only the high reflection at 400-mm depth corresponding to the boundary of the specimen is obtained.

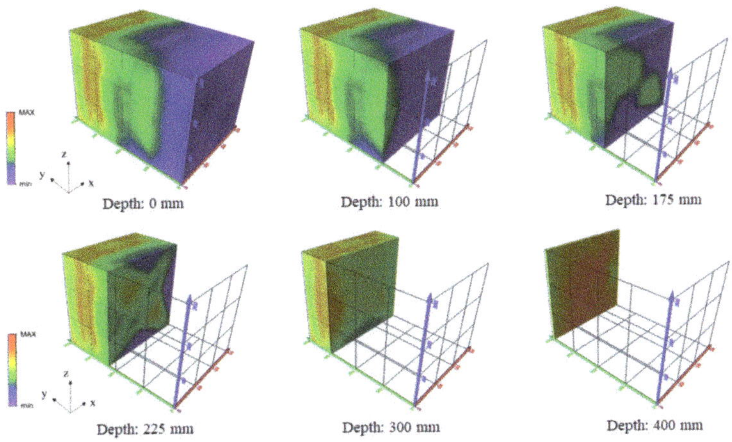

Figure 21. Improved 3D SIBIE results for damage-simulated concrete specimen.

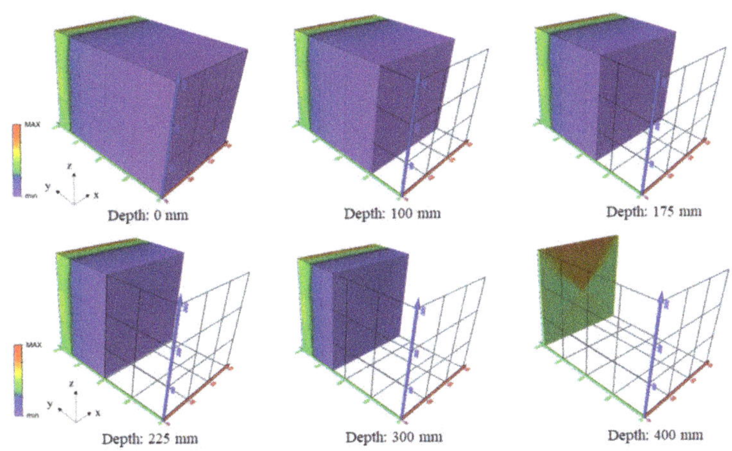

Figure 22. Improved 3D SIBIE results for intact concrete specimen.

In order to get even clearer visualized image, more elastic wave ray paths by increasing the number of hammering locations and sensors would work effectively. In addition, it is necessary to take into account the fact that the elastic wave velocity is not constant in real situation. It is assumed in the above-mentioned computations of this study that the elastic velocity is the same throughout the targeted concrete matrix. The further investigation needs to consider that the velocity could be varied according to the material property and treated as the variable number in the 3D-SIBIE procedure.

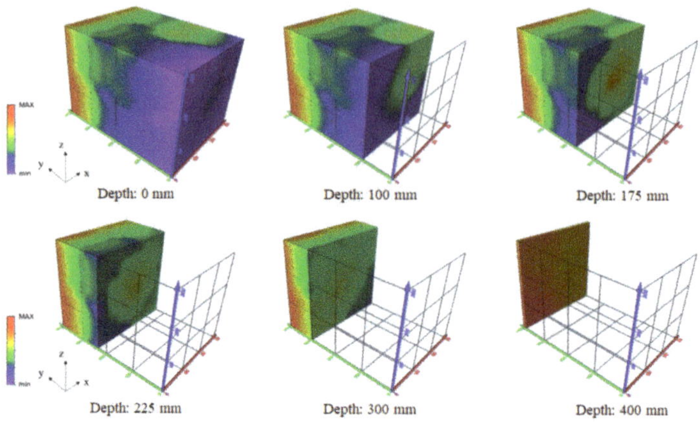

Figure 23. 3D SIBIE result with LDV signal detection for damage-simulated concrete specimen.

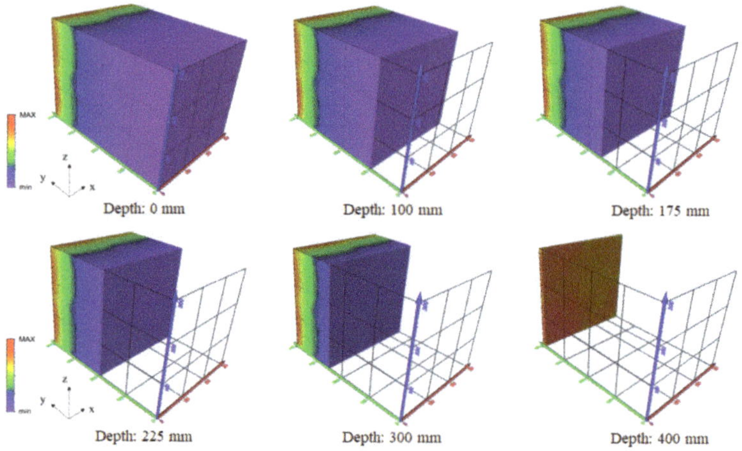

Figure 24. 3D SIBIE result with LDV signal detection for intact concrete specimen.

4. Conclusions

There is great demand for simplified and innovative techniques for the maintenance and assessment of concrete structures. Inspection works need to be ready-to-use as well as save time and costs related to damage evaluation. Developments in this area will support damage visualization to diagnose social infrastructures practically, i.e. highways, tunnels, and dams built using concrete.

In order to visualize damage, such as cracks and voids in concrete, the SIBIE (stack imaging of spectral amplitudes based on impact echo) procedure was applied and improved numerically. SIBIE was proposed in this study to be upgraded from a 2D to 3D model of elastic wave propagation behavior. Using the multiple sensor array composed of a handy mechanical device with accelerometers, the 3D SIBIE procedure was developed as a simple and ready to use measuring technique. It was also found that the laser doppler vibration sensing method could work for the remote data acquisition in the detection of elastic wave reflection for the procedure.

The procedure was successfully applied for the case studies of targeting RC deck in service as well as a concrete specimen for the laboratory test. In terms of results, the extension of the 2D computation process for the conventional SIBIE procedure to 3D coordinates enabled to estimate and display interior damages and geometrical information of materials constituting the layers of RC deck, such as asphalt pavement, fiber-reinforced overlay, and concrete. Additionally, the data computation process was improved considering the signal attenuation during the propagation, resulting in a clearer finding in terms of the damage assessment in the visualized images than that of the conventional calculation method.

It could be concluded that the 3D SIBIE procedure has a great promise for on-site measurement and can be employed to operate deterioration diagnosis for invisible damage in concrete and material assessment for concrete infrastructures accordingly.

Author Contributions: Conceptualization, K.H., T.S. and M.O.; Formal analysis, K.H.; Funding acquisition, T.S.; Investigation, K.H.; Methodology, K.H.; Visualization, K.H.; Validation, K.H. and T.S.; Supervision, T.S. and M.O.; Writing—Original Draft, K.H.; Writing—Review and Editing, T.S. All authors have read and agreed to the published version of the manuscript.

Funding: This research received no external funding.

Conflicts of Interest: The authors declare no conflict of interest.

References

1. ACI Committee 215. Consideration for design of concrete subjected to fatigue loading. *ACI J.* **1974**, *71*, 97–121.
2. Zhang, J.; Peng, H.; Cai, C.S. Field study of overload behavior of an existing reinforced concrete bridge under simulated vehicle loads. *J. Bridge Eng.* **2010**, *16*, 226–237. [CrossRef]
3. Fujiyama, C.; Maekawa, K. A computational simulation for the damage mechanism of steel-concrete composite slabs under high cycle fatigue loads. *J. Adv. Concr. Technol.* **2011**, *9*, 193–204. [CrossRef]
4. Fathalla, E.; Tanaka, Y.; Maekawa, K. Effect of crack orientation on fatigue life of reinforced concrete bridge decks. *Appl. Sci.* **2019**, *9*, 1644. [CrossRef]
5. Toutlemonde, F.; Ranc, G. Fatigue tests of cracked reinforced concrete slabs for estimating the service life of composite bridge decks. *Revue Française de Génie Civil* **2001**, *5*, 483–494. [CrossRef]
6. Hashimoto, K.; Shiotani, T.; Nishida, T.; Miyagawa, T. Application of elastic-wave tomography to repair inspection in deteriorated concrete structures. *J. Disaster Res.* **2017**, *12*, 2017. [CrossRef]
7. Chai, H.K.; Momoki, S.; Kobayashi, Y.; Aggelis, D.G.; Shiotani, T. Tomographic reconstruction for concrete using attenuation of ultrasound. *NDT E Int.* **2011**, *44*, 206–211. [CrossRef]
8. Chai, H.K.; Aggelis, D.G.; Momoki, S.; Kobayashi, Y.; Shiotani, T. Single-side access tomography for evaluating interior defect of concrete. *Constr. Build. Mater.* **2010**, *24*, 2411–2418. [CrossRef]
9. Carino, N.J.; Sansalone, M. Detecting voids in metal tendon ducts using the impact-echo method. *Mater. J. Am. Concr. Inst.* **1992**, *89*, 296–303.
10. Sansalone, M.; Lin, J.M.; Streett, W.B. A New Procedure for determining the thickness of concrete highway pavements using surface wave speed measurements and the impact-echo method. In *Innovations of Nondestructive Testing*; Pessiki, S., Olson, L., Eds.; American Concrete Institute: Farmington Hills, MI, USA, 1997; p. 167.
11. Ohtsu, M.; Watanabe, T. Stack Imaging of Spectral Amplitudes based on Impact-Echo for Flaw Detection. *NDT E Int.* **2002**, *35*, 189–196. [CrossRef]
12. Ohtsu, M.; Yamada, M. and Sonoda, T. *Quantitative evaluation of SIBIE procedure and case studies. Constr. Build. Mater.* **2013**, *48*, 1248–1254. [CrossRef]
13. Akamatsu, R.; Sugimoto, T.; Utagawa, N.; Katakura, K. Proposal of non-contact inspection method for concrete structures using high-power directional sound source and scanning laser doppler vibrometer. *Jpn. J. Appl. Phys.* **2013**, *52*, 07HC12. [CrossRef]
14. Kurahashi, S.; Mikami, K.; Kitamura, T.; Hasegawa, N.; Okada, H.; Kondo, S.; Nishikino, M.; Kawauchi, T.; Shimada, Y. Demonstration of 25-Hz-inspection speed laser remote sensing for internal concrete defects. *J. Appl. Remote Sens.* **2018**, *12*, 015009. [CrossRef]
15. Furusawa, A.; Takenaka, Y.; Nishimura, A. Proposal of laser-induced ultrasonic guided wave for corrosion detection of reinforced concrete structures in Fukushima daiichi nuclear power plant decommissioning site. *Appl. Sci.* **2019**, *9*, 3544. [CrossRef]
16. Rausche, F.; Kung, S.R.; Linkins, G. Comparison of pulse echo and transient response pile integrity test methods. *Transp. Res. Rec.* **1991**, *1331*, 21–27.

17. Rausche, F.; Linkins, G.; Kung, S.R. Pile integrity testing and analysis. In Proceedings of the International Conference on the Application of Stress-Wave Theory to Piles, The Hague, The Netherlands, 21–24 September 1992; pp. 613–617.
18. Ni, S.H.; Lehmann, L.; Charng, J.J.; Lo, K.F. Low-strain integrity testing of drilled piles with high slenderness ratio. *Comput. Geotech.* **2006**, *33*, 283–293. [CrossRef]

applied sciences

MDPI

Article

Non-Destructive Tests for Damage Evaluation of Stone Columns: The Case Study of Sacro Monte in Ghiffa (Italy)

Alessandro Grazzini *, Sara Fasana, Marco Zerbinatti and Giuseppe Lacidogna

Department of Structural Geotechnical and Building Engineering, Politecnico di Torino, 10138 Torino, Italy; sara.fasana@polito.it (S.F.); marco.zerbinatti@polito.it (M.Z.); giuseppe.lacidogna@polito.it (G.L.)

* Correspondence: alessandro.grazzini@polito.it

Received: 29 February 2020; Accepted: 10 April 2020; Published: 13 April 2020

Abstract: The Italian Sacri Monti are heritage sites with some unique characteristics; they are a successful symbiosis between nature and art and are unconfined structures, therefore always being accessible but exposed to atmospheric agents, with many relevant consequences with regard to conservation problems. The paper discusses some aspects related to the application of non-destructive techniques (NDT) for the interpretation of degradation phenomena occurring in stone structural elements. Ultrasonic and impact tests were used to evaluate the structural properties of the stone columns in the Via Crucis portico, within the monumental complex of the Sacro Monte in Ghiffa (Piedmont, Italy), in order to determine their conditions of maintenance and to evaluate the portico static stability. Ultrasonic tests made it possible to obtain the value of the dynamic elastic modulus, which was variable at different points of the columns due to the diversified level of material damage. The impact test, performed with an instrumented hammer in the same points of the ultrasonic test, enables, by comparison, a deeper knowledge of the surface resistance of damaged columns. These results are the first step in a research path that will require further laboratory tests to better calibrate the diagnostic techniques applied to different levels of damage to surface materials.

Keywords: non-destructive techniques (NDT); historical buildings; maintenance; stone columns; ultrasonic test; impact test

1. Introduction

Historical architectural heritage requires specific maintenance and restoration interventions, especially in the presence of prolonged deterioration over time. Restoration works may control deteriorations, which can affect monuments, without secondary effects on treated surfaces of the historical elements [1–5]. Similarly, diagnostics, which is a fundamental phase of knowledge in order to deepen the critical aspects of structure assessments and degradation causes, must operate with technologies that can be used in a non-destructive and non-invasive way for the historical buildings [6–13]. The main purposes related to the general framework of the material maintenance conditions are those to refine methodological paths for such contexts, to have well-founded analytical support and to reduce the error range in the qualitative and quantitative evaluation of the damage process. Ultrasonic tests are one of the non-destructive techniques most often used in the field of diagnostics on historical buildings [14–16]. They measure the wave speed through the material thickness, which is in relation to the dynamic elastic modulus. The results obtained by tests on architectural heritage are representative, if compared with known physical characteristics of similar materials, of the degradation inside the structural section of historical elements. In addition, the impact tests, performed with an instrumented hammer in the same points of the ultrasonic test, enable the evaluation of the surface degradation level [17].

This research concerns the mechanical characterization, through non-destructive techniques (NDT), of the 13 stone columns (granitoid rock) that support the Via Crucis portico inside the monumental complex of the Sacro Monte in Ghiffa. The Sacri Monti of northern Italy are a UNESCO heritage site that includes nine complexes that constitute groups of chapels and other buildings with many artistic masterpieces. All buildings are part of a high-quality natural environment. The first complex was founded in the late 15th century in Varallo Sesia, and the others were constructed up to the end of 17th century and dedicated to different aspects of the Christian faith [18]. In addition to their symbolic spiritual meaning, they are of great beauty by virtue of the skill with which they have been integrated into the surrounding natural landscape of hills, forests and lakes. They also house much important artistic material in the form of wall paintings and statuary.

Many buildings within the Sacri Monti monumental complexes suffer from different conditions of degradation, linked in particular to the climatic conditions of moisture, rain, damp action from capillary rising, freeze–thaw cycles that cause the detachment of plaster and the degradation of the stone elements [19]. A research group of the Politecnico di Torino has been interested for years in the study of non-destructive diagnostic techniques for the evaluation of degradation levels of the architectural and structural elements of historical buildings in the Sacri Monti complexes [1,20–23].

In the specific case of the Sacro Monte at Ghiffa, the Via Crucis portico (Figure 1) is supported by 13 stone columns with evident intense superficial material degradation. The portico was built between 1752 and 1761. It is marked by 14 spans and has columns and capitals made by stones derived both from local "granitic stones". It is possible that the origin of the stone of columns is not from historical quarries but from erratic boulders found near the construction site (aplite is the current interpretation, but other investigations are in progress), referred to in slang as "trovanti", while for the capitals there is a strong similarity with a serizzo quarried near Locarno (Valle Verzasca, Switzerland) (Figure 2). Considering a macroscopic mineralogical analysis with a 10x magnifying glass, the rock is mainly made up of more than 90% of crystalline clear silica minerals (quartz and feldspar), with little biotitic dark mica and rare light mica. The main accessory mineral is a small red-vined garnet that characterizes the stone. The veins present confirm the low quality of the stone, almost certainly extracted from makeshift material, greatly influencing the variability of mechanical characteristics. Perhaps, also for the lower quality, it is very prone to degradation phenomena, worsened by atmospheric moisture, the presence of salts, freeze–thaw cycles and thermal expansion, which is also due to strong summer radiation, which can produce erosion and exfoliation on the stone surface [24,25] (Figure 3). The material degradation is increasing from the inside to the outside of the structural section of the columns, and the result is a decay of their compressive strength, the residual value of which needs to be investigated.

The aim of the ongoing research is to deepen the knowledge of the mechanical characteristics of the columns of the Via Crucis portico resulting from degradation, to estimate their residual strength. In order to improve the maintenance and conservation planning of the historical buildings belonging to the Sacri Monti monumental complex, some non-destructive tests were carried out on the columns of the portico to investigate the elastic modulus of the stone. Specifically, the ultrasonic technique was used, associated with that of impacts for a more precise comparison of the stone surface degradation. From the value of the dynamic elastic modulus, it was possible to estimate the compressive strength of each single column using experimental reports. As a matter of fact, the ultrasonic and impact tests showed a great variability of the elastic modulus already within the same column and between columns. This variability was the demonstration of a material degradation distributed unevenly within the same stone element and between the columns.

Figure 1. The Via Crucis portico inside the Sacro Monte of Ghiffa (Piedmont, Italy).

Figure 2. The stone columns of the Via Crucis portico in the Sacro Monte at Ghiffa.

(a) (b)

Figure 3. Examples of erosion (**a**), exfoliation and detachment (**b**) problems in stone columns of the Via Crucis portico. The reference lexicon is that of the "Ente Nazionale Italiano di Unificazione" (UNI) Standard rules 11182:2006.

These results could be used in forthcoming finite element numerical modeling of the portico, in order to evaluate its stability in the hypothesis of replacing the current roof covering, made with lighter modern stones slabs, by traditional roof covering made by more heavy ones. The assessments that emerged from these diagnostic tests will also be useful for future maintenance work planned for the conservation of the Via Crucis portico; the diagnostic procedure proposed could then be extended to investigate for the various degradation problems afflicting the historical buildings of Sacri Monti monumental complexes.

The experimental results described represent a first phase of the research, which will be subsequently extended to similar types of stones obtained from quarries on which to perform compression tests in the laboratory and related non-destructive tests (ultrasonic and impact). In this way it will be possible to calibrate the non-destructive techniques, which will then be more frequently used in situ on valuable architectural elements.

2. Materials and Methods

Each column was investigated at the bottom, center and top by an ultrasonic test and subsequently at the same points by an impact test. The choice of the points was based above all on a visual examination, comparing points where the material degradation was evident with those in which the stone seemed in better condition (the columns were also struck with light plastic hammers to hear their "sound"). Both were non-destructive investigation techniques that adapt very well to analyze the architectural element under investigation. The goal was to obtain the value of the elastic modulus and to evaluate the quality of the resistant stone section subject to degradation. The test was also an opportunity to compare the results of the ultrasonic test, useful for investigating the entire length of the stone section, with those obtained from the impact test, which, on the contrary, estimates the material characteristic deformation by analyzing only the external surface on which the hit is performed.

2.1. Ultrasonic Test

The ultrasonic test device consisted of a Proceq Pundit PL-200® pulse generator (0.1–7930 μs range; resolution of 0.1 μs for pulse speed <793 μs, 1 μs for pulse speed >793 μs; impulse voltage 100–450 Vpp); two transducers of 54 kHz, one emitting and the other receiving; a pulse amplifier and an electronic device for measuring the time interval. On the basis of the time taken by the wave to travel the distance between the two transducers, the propagation speed of the elastic waves through the stone material of the column is calculated. The speed propagation of the waves in a homogeneous material depends on its density and elastic characteristics (elastic modulus, Poisson's ratio), closely related to the quality and compressive strength of the material itself. The speed of the wave is in fact low if the material has voids, cracks or detachments that slow down the path; vice versa, it is high if the material is compact and homogeneous, demonstrating good quality. The measurement of the propagation time of the ultrasonic pulses was carried out with the use of two transducers, applied on the surface of the elements involved in the control.

Direct tests were carried out by positioning the transducers on the two opposite faces of the element under investigation, as indicated in Figure 4. The coupling of the transducer on the stone surface was facilitated through the application of a thin layer of modeling clay in order to avoid air interpositions and facilitate the transmission of the wave in the material. The measurements were conducted in accordance with the methods proposed in American Society for Testing and Materials (ASTM) rules D2845 1995 and in Ente Nazionale Italiano di Unificazione, Comité Européen de Normalisation, (UNI EN) rules 12504-4: 2005 [26,27]. The tests are based on the determination of the propagation speed (v) of the pulses of the ultrasonic longitudinal waves in the material, the values of which, in the first approximation, are related to the dynamic modulus (E_d) of the material, which consequently is related to its mechanical strength [16]:

$$E_d = v^2\rho\frac{(1+\nu)\,(1-2\nu)}{(1-\nu)} \tag{1}$$

Figure 4. Ultrasonic test carried out on columns of Via Crucis, Ghiffa.

In Equation (1), ρ is the density of material, and ν is Poisson's ratio. In this case, from scientific literature sources, a ρ value equal to 2630 daN/m^3 and a ν value equal to 0.18 were considered [28].

2.2. Impact Test

The impact test was performed using an instrumented hammer capable of generating an impact, with a known mass and predetermined energy, against the surface of the stone. The software that analyzes the data is able to return the force–time diagram of each impact. This type of test, unlike other impulsive methods, allows the evaluation of the evolution over time of the forces acting in the impact. For example, the sclerometer test detects only the elastic energy returned by the concrete after the impact. On the other hand, the impact test also evaluates the given energy, the dissipated energy, the impact time and the maximum force. It therefore represents a non-destructive technique for evaluating the elastic and inelastic properties of materials. [17,29]. The analysis of the diagram (F, t) can obtain the value of the homogeneous material elastic modulus through Hertz's impact theory [29–31] (see Appendix A).

The electric impact hammer (Figure 5) is a device produced by Piezotronic PCB Inc Company (Depew, NY, USA). It is characterized by a variable force from 44.48 N to 4448.26 N, and a force sensor sensitivity of 2.47 mV/N. It performs the impact of a 10 mm diameter steel ball with a total mass of 207 grams, connected to a piezoelectric pulse transducer. The impact is generated by the hammer by means of a predetermined energy, characterized by a level of amplifier and impedance adapter. The acquisition is performed by a Pimento model multichannel LMS signal analyzer (Figure 5).

Figure 5. Setup of the impact test.

The hammer was positioned perpendicular to the surface to be investigated. Perpendicularity alignment was achieved thanks to four fixed metal brackets. The impact of the small impact mass was generated by pressing a trigger on the electric hammer. The intensity of the impact is very low and therefore not invasive even for surfaces of artistic value. The elastic modulus of the stone was determined according to the classical theory of the elastic collision between two bodies elaborated by Hertz [17,29–31], using the following equations:

$$E_s = \frac{E^* E_0 \left(1 - \nu^2\right)}{E_0 + E^* \left(\nu_0^2 - 1\right)} \quad (2)$$

where E^* was obtained by:

$$E^* = \sqrt{\left(\frac{2,87}{T}\right)^5 \frac{m^3}{RA_1}} \quad (3)$$

The parameters in Equations (2) and (3) are the following: A1 is the area beneath the diagram (*F*, *t*) (see Figure 6), *T* represents the time in which the contact occurs between the two surfaces; *R* is the radius of curvature (10 mm) of the impact mass *m* (0.207 Kg), E_0 is the mass elastic modulus (222,000 MPa) and ν_0 is its Poisson's ratio (0.2), while, with regard to the stone material of the columns, a Poisson's ratio ν of 0.18 was considered [28].

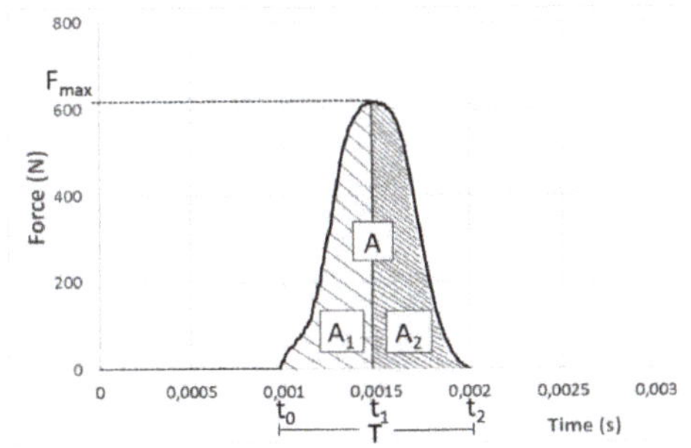

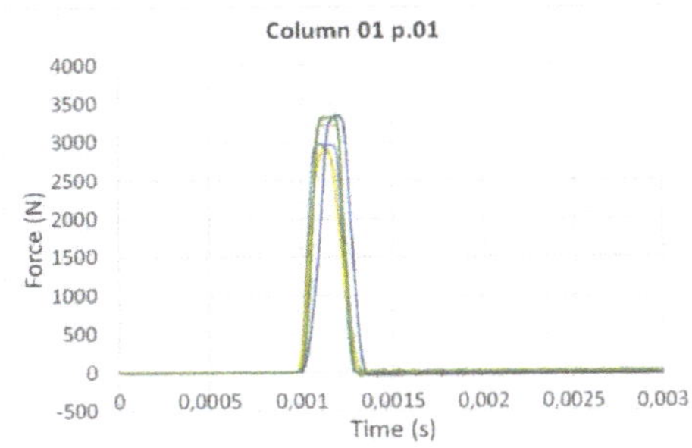

Figure 6. (**a**) Generic force–time diagram of an impact test; (**b**) test results by impact test on column 1 of the portico.

3. Results

The diagnostic study of the structural characteristics of the columns of the Via Crucis portico was mainly based on the execution of ultrasonic tests, which were followed by impact tests at the same points. While the ultrasonic tests enable the evaluation of the dynamic elastic modulus of the columns by measuring the speed pulses, the impact tests investigate the material degradation on the surface of the columns (the detachment points of material, for example).

3.1. Ultrasonic Tests

The ultrasonic tests were carried out on all the columns of the Via Crucis portico on 1–3 points each (depending on the visual level of degradation), in order to have an average of the speed pulses and the related dynamic modulus over the entire height of the column. As can be seen from the results of Table 1, the columns showed highly variable values of the dynamic modulus, sometimes even on the same column. This variability confirmed the diversified state of material degradation to which the columns were subjected in relation with the alteration of the original natural stone used (Figure 7). Through Correlation Curve (4), experimentally obtained by Vasconcelos et al. [16], it was possible to estimate the compressive strength, f_c, of the granitoid rock from the value of the dynamic modulus. The f_c value obtained by the experimental correlation laws represents an estimation (the values of which fall within the ranges present in the scientific literature), pending a better confirmation from subsequent tests that will be carried out in the laboratory. These tests will provide the compression stress results on samples taken from erratic stones present in the area, made of the same material as that of the columns.

$$f_c = 0.0407\ v - 36.31 \tag{4}$$

Table 1. Values of dynamic modulus by ultrasonic tests on Via Crucis portico columns.

Column	Test Point	Height (cm)	v (m/s)	E_d (MPa)	f_c (MPa) [2]
1	1	57	3269	25,878	97
	2	98	2503	15,170	66
	3	230	2805	19064	78
2	1	65	3411	28,183	103
	2	123	2466	14,728	64
	3	223	2494	15,067	65
3	1	70	2910	20,506	82
	2	230	2939	25,406	96
4	1	60	4024	39,219	127
	2	232	3908	36,985	123
5	1	68	2724	17,972	75
	2	228	2738	18,156	75
6	1	55	2766	18,525	76
7	1	86	3402	28,033	102
	2	206	2948	21,046	84
8	1	86	2774	18,644	77
	2	229	3223	25,158	95
9	1	146	2495	15,080	65
10	1	100	2522	15,406	66
	2	228	2756	18,402	76
11	1	101	2698	17,627	73
	2	204	2773	18,629	77
12	1	90	2965	21,300	84
	2	190	3178	24,468	93
13	1	35	4638	52,111 [1]	152 [1]
	2	130	2840	19,532	79
	3	224	3149	24,024	92

[1] Basic element of the column made of different stone materials of better quality. [2] Estimated according to the correlation curve of Vasconcelos et al. [13].

3.2. Impact Tests

The purpose of the impact test was to verify the surface resistance of the stone elements, considering their surface degradation. The impact test, when performed on homogeneous material such as concrete or stone, can provide through Equation (2) values of the elastic modulus completely comparable to those obtained by the ultrasonic technique [29]. It should be remembered that the value obtained with the impact test represents a static elastic modulus (E_s), the conversion of which into a dynamic value deserves separate research considering the many different formulas present in scientific literature for the different types of stones and rocks [32]. To obtain a correct and reliable correlation formula for each type of stone, a large number of tests must be performed in the laboratory and correlated with non-destructive tests; this correlation could then be calibrated as a suitable diagnostic technology to be applied on specific architectural elements. However, since the test is based on an energetic evaluation relative to the impact on the surface of the element (Figure 8), in presence of a surface degradation, like the one found in situ, it does not provide elastic modulus values that can be correlated with those obtained by ultrasonic tests.

(a)

(b)

Figure 7. Different types of surface material degradation on the columns of the Via Crucis portico: (**a**) erosion; (**b**) exfoliation.

Figure 8. Impact test carried out on a column of Via Crucis portico.

Figure 9 shows, for example, the time–force curves between two different points of the same test on column 3. In presence of surface degradation, the curves are characterized by lower peak force and wider response times. The ratio between the returned A_2 and supplied energy A_1 (Figure 6a) provided indications regarding the detachment of the degraded matter crust [17,33].

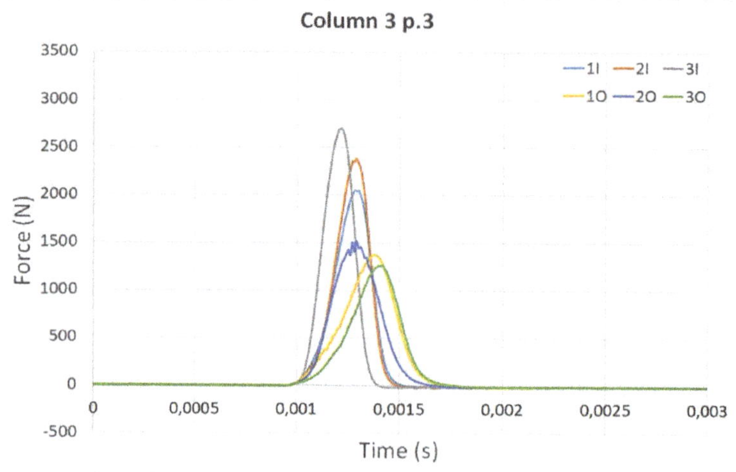

Figure 9. Time–force diagrams of third point of column 3 (I = point inside of the portico; O = outside of the portico).

With reference to the values of the dynamic elastic modulus, Figure 10 reports, for example, the dispersion of the data collected in the impact tests on the columns at Ghiffa and the same comparison obtained from the similar type of homogeneous and non-degraded stone tested in the laboratory.

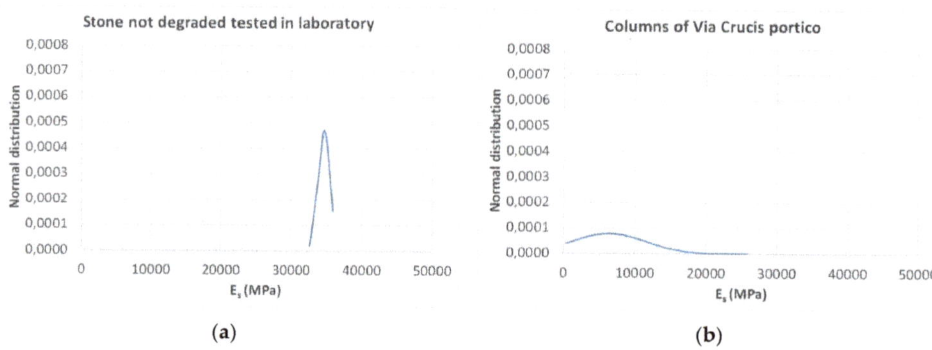

Figure 10. Gaussian distribution E_S of non-degraded stone (**a**) and of the same type of stone degraded in the columns of the Via Crucis portico at Ghiffa (**b**).

From Figure 10, a significant difference can be seen between the two Gaussian curves. The one relating to laboratory tests on a sample of new stone similar to that used in the Ghiffa columns revealed a significantly lower standard deviation σ than that detected with the impact tests performed in situ on the columns of the Via Crucis portico. The differences are due to the considerable diversification of the material degradation affecting the surface of the columns. Similarly, Table 2 shows the results provided by impact tests on the new stone in the laboratory and on the old in situ columns of the Via Crucis portico.

Table 2. Impact tests on new and old granitic stones.

Stone	$E_{s\ average}$ (MPa)	σ (MPa)
New stone in the laboratory	34,711	843
Old stone of the columns	6394	5100

4. Discussion

The in situ diagnostic tests and their comparison in the laboratory show how dispersive the results of the ultrasonic and impact tests can be when applied to architectural elements in a degraded state. The surface roughness of ancient in situ elements affects the impact test results, not allowing a good correlation with the values of the dynamic elastic modulus obtained from the ultrasonic tests. On the contrary, the same tests performed in the laboratory on a sample of new and unaltered stone show a very good correlation between the dynamic elastic modulus obtained from the ultrasonic tests and the static modulus obtained from the impact test, both with a Gaussian curve characterized by very small standard deviations.

The ultrasonic tests recorded speed waves average values of about 3000 m/s (Table 3), which the scientific and experimental literature [16,28] classifies as medium quality material. However, there is a high variability of the values derived from the material degradation level (not excluding the original alteration of the quarried material, from upper part of the extraction site, or by single big rocks like the "trovanti"), from possible detachments or exfoliation of surface material. The speeds measured on the same type of not degraded stone obtained from the quarry detected higher values (albeit with little difference) and statistically more stable speed values (Table 3). The standard deviation value was representative of the material degradation of the stone surface on which the ultrasonic and impact tests were performed. In this case, both were characterized by a high value of the standard deviation due to the diversified level of material degradation that characterizes the columns of the Via Crucis portico (Figures 10b and 11b). Vice versa, as is visible in Tables 2 and 3 and in Figures 10a and 11a, corresponding to the same stone, though intact and without any material alteration, there was a very low value of the standard deviation and a good correlation between the values of the elastic modulus obtained from both types of non-destructive tests.

Table 3. Ultrasonic tests on new and old stones.

Stone	$V_{average}$ (m/s)	$E_{d\ average}$ (MPa)	σ (MPa)
New stone in the laboratory	3498	29,640	611
Old stone of the columns	2961	21,623	6183

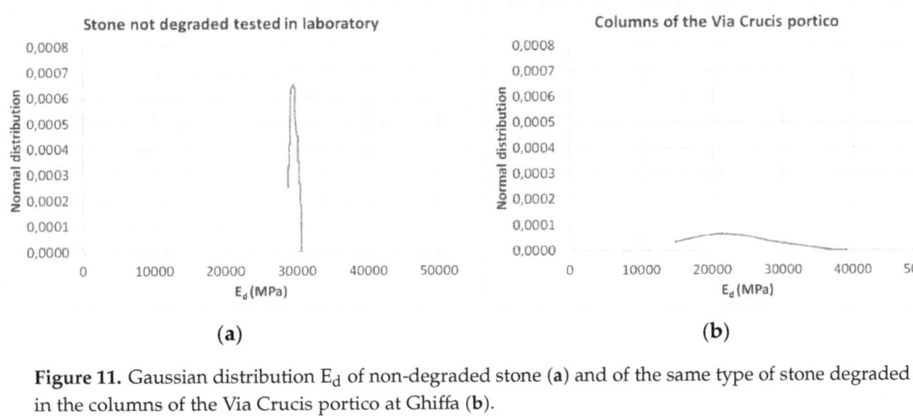

Figure 11. Gaussian distribution E_d of non-degraded stone (**a**) and of the same type of stone degraded in the columns of the Via Crucis portico at Ghiffa (**b**).

5. Conclusions

The experimental research carried out in situ to determine the residual strength of the degraded stone columns of the Via Crucis portico at the Sacro Monte of Ghiffa is a first step in a research path that aims to calibrating non-destructive techniques for the diagnosis of heritage monument degradation. Ultrasonic and impact techniques were used, critically analyzing their correlation in the presence of surface material degradation. The ultrasonic test highlighted the variability of the dynamic elastic modulus within the same columns, while the impact test highlighted the local points of surface material degradation. Many other tests in the laboratory and in situ will be needed, on different types of stones, to correlate the values of dynamic elastic modulus obtained by the ultrasonic technique with the static one that resulted from the impact tests. Subsequent laboratory tests will be also planned to calibrate the impact test applied to different levels of surface material damage, in order to find correlations with the values of dynamic elastic modulus obtained from ultrasonic tests.

Author Contributions: Conceptualization, A.G., M.Z., S.F. and G.L.; *In situ* tests, S.F., M.Z. and A.G.; Writing A.G., S.F.; Supervision, M.Z. and G.L.; Project administration, M.Z. All authors have read and agreed to the published version of the manuscript.

Funding: This research was funded by MAIN10ANCE Interreg Project.

Acknowledgments: The authors would like to thank Giovanni Bricca and Franco Grindatto for their valuable technical support during the execution of the in situ tests. The authors also wish to thank Josè Delgado Rodrigues for his precious visit and petrographic consultancy at the Sacro Monte of Ghiffa.

Conflicts of Interest: The authors declare no conflict of interest.

Appendix A

Hertz's theory of the elastic collision between two bodies in the absence of friction is modeled according to the scheme in Figure A1, where a sphere of radius R impacts with velocity v_0 penetrating with a depth δ in an elastic semi-infinite space [27]:

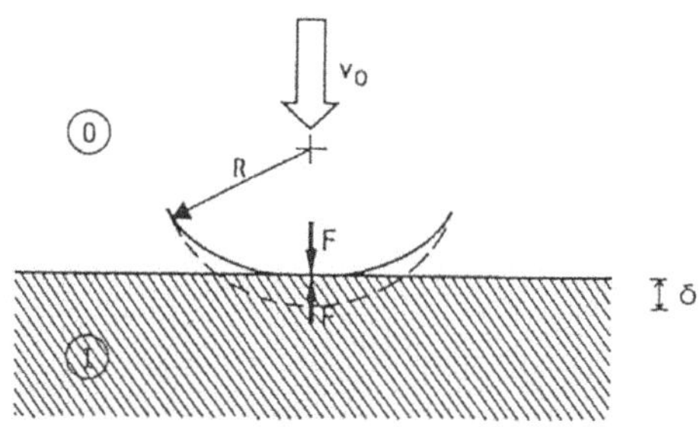

Figure A1. Impact between the moving mass and the elastic semi-infinite surface.

The depth of indentation is:

$$\delta = \left(\frac{9F^2}{16RE^{*2}}\right)^{\frac{1}{3}} \tag{A1}$$

The applied force *F* is related to the displacement δ by:

$$F = \frac{4}{3}R^{\frac{1}{2}}E^{*}\delta^{\frac{3}{2}} = K\delta^{\frac{3}{2}} \tag{A2}$$

where

$$\frac{1}{E^{*}} = \frac{1-\nu_0^2}{E_0} + \frac{1-\nu_1^2}{E_1} \tag{A3}$$

and ν_0, ν_1, E_0 and E_1 are, respectively, Poisson's ratios and the elastic moduli associated with the sphere (0) and the semi-infinite space (1).

Considering that the sphere impacts on the surface in the initial time with a speed v_0, and that v_1 is the speed in the next instant, the following equation can be written:

$$m\frac{dv}{dt} = m\frac{d^2\delta}{dt^2} = -F, \tag{A4}$$

and from (A1) and from (A4):

$$m\frac{d^2\delta}{dt^2} = -\frac{4}{3}R^{\frac{1}{2}}E^*\delta^{\frac{3}{2}}. \tag{A5}$$

Through integration, the following can be achieved:

$$\frac{1}{2}\left[v_0^2 - \left(\frac{d\delta}{dt}\right)^2\right] = \frac{2K\delta^{\frac{5}{2}}}{5m} \tag{A6}$$

where K is the value indicated in (A2).

The maximum depth of indentation δ^* occurs in the case of $\frac{d\delta}{dt} = 0$, for which:

$$\delta^* = \left(\frac{5mv_0^2}{4K}\right)^{\frac{2}{5}} = \left(\frac{15mv_0^2}{16E^*R^{\frac{1}{2}}}\right)^{\frac{2}{5}}. \tag{A7}$$

The diagram (F, t) in the compression phase can be obtained by means of a second integration:

$$t = \frac{\delta^*}{v_0}\int \frac{d(\delta/\delta^*)}{\left[1-(\delta/\delta^*)^{\frac{5}{2}}\right]^{\frac{1}{2}}}. \tag{A8}$$

The instant t^*, corresponding to the maximum depth of the deformation δ^* during the impact, referred to as time t_1 in Figure 6a, is given by the following equation:

$$t^* = \frac{\delta^*}{v_0}\int \frac{d(\delta/\delta^*)}{\left[1-(\delta/\delta^*)^{\frac{5}{2}}\right]^{\frac{1}{2}}} = 1,435\left(m^2/RE^{*2}v_0\right)^{\frac{1}{5}}. \tag{A9}$$

The overall time T of the whole impact is given by:

$$T = 2t^* = 2.87\left(m^2/RE^{*2}v_0\right)^{\frac{1}{5}}. \tag{A10}$$

From the curve (F, t) the experimental value of T is obtained, which, when inserted into (A10), provides the value of E^* (see Equation (3)). Finally, using Equation (2), the value of the elastic modulus E_s can be calculated.

References

1. Formia, A.; Serra, C.L.; Zerbinatti, M.; Tulliani, J.M. The plasters of the Sacro Monte of Varallo Sesia. From the characterization to the proposition of a restorative mix. *Case Stud. Constr. Mater.* **2014**, *1*, 46–52. [CrossRef]
2. Cizer, Ö.; Schueremans, L.; Serre, G.; Janssens, E.; Van Balen, K. Assessment of the compatibility of repair mortars in restoration projects. *Adv. Mater. Res.* **2010**, *133–134*, 1071–1076. [CrossRef]
3. Wiggins, D.E.; Klemm, A.J. Compatible repairs to stone built structures. *Proc. Inst. Civ. Eng. Eng. Hist. Herit.* **2014**, *167*, 196–207. [CrossRef]

4. Grazzini, A.; Zerbinatti, M.; Fasana, S. Mechanical characterization of mortars used in the restoration of historical buildings: An operative atlas for maintenance and conservation. *IOP Conf. Ser. Mater. Sci. Eng.* **2019**, *629*, 1–7. [CrossRef]
5. Isebaert, A.; Van Parys, L.; Cnudde, V. Composition and compatibility requirements of mineral repair mortars for stone—A review. *Const. Build. Mater.* **2014**, *59*, 39–50. [CrossRef]
6. Binda, L.; Gambarotta, L.; Lagomarsino, S.; Modena, C. A multilevel approach to the damage assessment and the seismic improvement of masonry buildings in Italy. In *Seismic Damage to Masonry Buildings*; Bernardini, A., Ed.; Balkema: Rotterdam, Holland, 1999; pp. 179–194.
7. Binda, L.; Cantini, L.; Tedeschi, C. Diagnosis of Historic Masonry Structures Using Non-Destructive Techniques. In *Nondestructive Testing of Materials and Structures*; Buyukozturk, O., Tasdemir, M.A., Gunes, O., Akkaya, Y., Eds.; Springer Netherlands: Heidelberg, Germany, 2013; pp. 1089–1102. [CrossRef]
8. Grazzini, A. The important role of diagnostics for the structural recovery of historic buildings. *Int. J. Arch. Eng. Technol.* **2019**, *6*, 7–16. [CrossRef]
9. Invernizzi, S.; Lacidogna, G.; Carpinteri, A. Structural monitoring and assessment of an ancient masonry tower. *Eng. Fract. Mech.* **2019**, *210*, 429–443. [CrossRef]
10. Carpinteri, A.; Lacidogna, G. Damage monitoring of an historical masonry building by the acoustic emission technique. *Mater. Struct.* **2006**, *39*, 161–167. [CrossRef]
11. Carpinteri, A.; Lacidogna, G.; Pugno, N. Structural damage diagnosis and life-time assessment by acoustic emission monitoring. *Eng. Fract. Mech.* **2007**, *74*, 273–289. [CrossRef]
12. Carpinteri, A.; Invernizzi, S.; Lacidogna, G. Historical brick-masonry subjected to double flat-jack test: Acoustic emissions and scale effects on cracking density. *Constr. Build. Mater.* **2009**, *23*, 2813–2820. [CrossRef]
13. Cantini, L.; Bonavita, A.; Parisi, M.A.; Tardini, C. Historical analysis and diagnostic investigations in the knowledge acquisition path for architectural heritage. In *Proceedings of the Congress in Structural Analysis of Historical Constructions—Anamnesis, Diagnosis, Therapy, Controls, Leuven, Belgium, 13–15 September 2016*; Van Balen, V., Ed.; Taylor & Francis Group: London, UK, 2016.
14. Grazzini, A. Sonic and Impact Test for Structural Assessment of Historical Masonry. *Appl. Sci.* **2019**, *9*, 5148. [CrossRef]
15. Chastre, C.; Ludovico-Marques, M. Nondestructive testing methodology to assess the conservation of historic stone buildings and monuments. In *Handbook of Materials Failure Analysis with Case Studies from the Construction Industries*; Hamdy Makhlouf, A.S., Aliofkhazraei, M., Eds.; Elsevier: New York, NY, USA, 2018; pp. 255–294, ISBN 978-0-08-101928-3. [CrossRef]
16. Vasconcelos, G.; Lourenço, P.B.; Alves, C.S.A.; Pamplona, G. Prediction of the mechanical properties of granites by ultrasonic pulse velocity and Schmidt hammer hardness. In *Proceedings of the 10th North American Masonry Conference, St. Louis, MO, USA, 3–6 June 2007*; Myers, B., Ed.; Masonry Society: Longmont, CO, USA, 2007.
17. Grazzini, A. In Situ Analysis of Plaster Detachment by Impact Tests. *Appl. Sci.* **2019**, *9*, 258. [CrossRef]
18. The Sacred Mounts of Piedmont and of Lombardy. Available online: www.sacrimonti.net (accessed on 10 February 2020).
19. Gonçalves, T.D.; Pel, L.; Delgado Rodrigues, J. Influence of paints on drying and salt distribution processes in porous building materials. *Constr. Build. Mater.* **2009**, *23*, 1751–1759. [CrossRef]
20. Bocca, P.; Grazzini, A.; Masera, D.; Alberto, A.; Valente, S. Mechanical interaction between historical brick and repair mortar: Experimental and numerical tests. *J. Phys. Conf. Ser.* **2011**, *305*, 1–10. [CrossRef]
21. Bocca, P.; Valente, S.; Grazzini, A.; Alberto, A. Detachment analysis of dehumidified repair mortars applied to historical masonry walls. *Int. J. Arch. Herit.* **2014**, *8*, 336–348. [CrossRef]
22. Grazzini, A.; Lacidogna, G.; Accornero, F. Delamination of plasters applied to historical masonry walls: Analysis by acoustic emission technique and numerical model. *IPO Conf. Ser. Mater. Sci. Eng.* **2018**, *372*, 1–7. [CrossRef]
23. Accornero, F.; Invernizzi, S.; Lacidogna, G.; Carpinteri, A. The Sacred Mountain of Varallo renaissance complex in Italy: Damage analysis of decorated surfaces and structural supports. In *Acoustic, Electromagnetic, Neutron Emissions from Fracture and Earthquakes*, 1st ed.; Carpinteri, A., Lacidogna, G., Manuello, A., Eds.; Springer: Heidelberg, Germany, 2015; pp. 249–264. [CrossRef]
24. Delgado Rodrigues, J. Conservation of stone monuments. From diagnostic to practice. In Proceedings of the MINBAR AL JAMIAA n°7—RIPAM 2005, Mekés, Maroc, 26–28 September 2005; pp. 287–295.

25. Delgado Rodrigues, J.; Ferreira Pinto, A.; Rodrigues da Costa, D. Tracing of decay profiles and evaluation of stone treatments by means of microdrilling techniques. *J. Cult. Herit.* **2002**, *3*, 117–125. [CrossRef]
26. ASTM D2845 1995. *Standard Test Method for Laboratory Determination of Pulse Velocities and Ultrasonic Elastic Constants of Rock*; American Society for Testing Materials: West Conshohocken, PA, USA, 1995.
27. UNI EN 12504-4: 2005. *Prove sul Calcestruzzo nelle Strutture—Parte 4: Determinazione della Velocità di Propagazione Degli Impulsi Ultrasonici (in Italian)*; Ente Nazionale Italiano di Unificazione: Milan, Italy, 2005.
28. Domede, N.; Parent, T.; Sellier, A. Mechanical behaviour of granite: A compilation, analysis and correlation of data from around the world. *Eur. J. Environ. Civ. Eng.* **2019**, *23*, 193–211. [CrossRef]
29. Bocca, P.; Scavia, C. The impulse method for the evaluation of concrete elastic characteristics. In Proceedings of the 9th International Conference on Experimental Mechanics, Copenhagen, Denmark, 20–24 August 1990.
30. Johnson, K.L. *Contact Mechanics*; Cambridge University Press: Cambridge, UK, 1985.
31. Deresiewicz, H. A note on Hertz Impact. *Acta Mech.* **1968**, *6*, 110–112. [CrossRef]
32. Dhawan, K.R.; Muralidhar, B. Relationship between static and two types of dynamic moduli for different rocks. *Ind. Getech. J.* **2015**, *45*, 341–348. [CrossRef]
33. Bocca, P.; Carpinteri, A.; Valente, S. On the applicability of fracture mechanics to masonry. In Proceedings of the 8th International Brick/Block Masonry Conference, Dublin, Ireland, 19–21 September 1988.

applied sciences

MDPI

Article

Non-Invasive Identification of Vulnerability Elements in Existing Buildings and Their Visualization in the BIM Model for Better Project Management: The Case Study of Cuccagna Farmhouse

Elsa Garavaglia [1,*], Anna Anzani [2], Fabio Maroldi [1] and Fabio Vanerio [1]

1 Department of Civil and Environmental Engineering, Politecnico di Milano, 20133 Milano, Italy; fabio.maroldi@polimi.it (F.M.); fabio.vanerio@polimi.it (F.V.)
2 Department of Design, Politecnico di Milano, 20158 Milano, Italy; anna.anzani@polimi.it
* Correspondence: elsa.garavaglia@polimi.it

Received: 21 February 2020; Accepted: 15 March 2020; Published: 20 March 2020

check for updates

Featured Application: The procedure could be applied, on a municipal or territorial scale, in order to proceed with a rapid assessment of the vulnerability of the existing historical heritage.

Abstract: Due to the conjunction between the European and African plates, complex tectonic phenomena take place in the Mediterranean basin. These phenomena cause more or less violent seismic resentments in the countries facing the basin itself. The diffused built historical heritage, characteristic of villages in the Mediterranean countries, is the most vulnerable toward seismic action, and in case of a catastrophic event can cause the loss of human lives. In Italy, the protection of historic buildings is a significant issue, and many regions promoted policies to ensure the safety of the diffused built heritage. Research groups work in synergy to develop procedures for the vulnerability assessment of existing buildings and to define appropriate action plans. This research presents a little or not at all invasive procedure for investigating vulnerability. This procedure is easily replicable and able to support techniques already in use with innovative aspects such as laser scanning of the entire complex and visual identification of vulnerable elements through the BIM (building information modeling) methodology. The procedure applicability is shown in the study of a Milanese farmhouse that has been financed by Fondazione CARIPLO, Bandi 2017 *Arte e Cultura-Beni culturali a rischio*, Project PRE.CU.R.S.OR.

Keywords: existing buildings; seismic vulnerability assessment; non-invasive experimental technics; BIM models

1. Introduction

The Mediterranean basin, a cradle of many civilizations and a repository of a remarkable built heritage, is located at the junction between the European and African tectonic plates. This conjunction constitutes the so-called Faglia Giulia that runs along the entire Italian Apennine ridge (INGV Database of individual seismogenetic sources DISS_v3, http://diss.rm.ingv.it/diss/), a set of more or less long faults, often almost parallel to each other and juxtaposed, sometimes transversal, whose movements are the cause of seismic resentments in the peninsula. In recent years, the land movements recorded on the Italian Apennine ridge have caused considerable losses in the historical heritage of the affected regions, not to mention the consequent loss of human lives due to the induced collapses. The reason for these disasters is linked to the high vulnerability that characterizes the Italian historical heritage, as well as the historical heritage of all countries in the Mediterranean basin, which is not exempt from vulnerability to seismic actions.

A campaign to secure the diffused built heritage requires a careful assessment of the buildings' vulnerability and an economic commitment on the part of their owners. Since 2014, the ReLUIS Network of Seismic Engineering University Laboratories (http://www.reluis.it/) has been working to survey the existing heritage and achieve a reliable vulnerability classification. With the creation of the Casa Italia department (http://www.casaitalia.governo.it/it/) in 2016, a campaign was launched to raise public and administrative awareness of the need to secure public and private buildings, also encouraged by the introduction of possible tax benefits.

In recent years, to support and integrate government initiatives, seismic research has developed numerous investigation methods aimed at the census of aggregate buildings and the subsequent identification of minimum units and structural units on which base the seismic behavior can be analyzed.

In complex urban situations, such as those present in historical centers, this census is not easy. In recent decades, numerous studies have adopted Earth geo-referencing (GIS) technology. Such tool allows association of an object not only to a geo-referenced location that is easy to browse, but also to some of its peculiar characteristics. In the seismic field, this methodology has made it possible to associate each geo-referenced object with the evaluated level of vulnerability. Consequently, vulnerability maps of public and private buildings of entire municipalities could be drawn up [1–12].

To achieve both more detailed investigations on buildings seismic behavior and 3D representation of the level of vulnerability characterizing some construction elements, recent research focuses on the use of Building Information Modeling (BIM).

The BIM methodology is used in the identification phase of vulnerable structural elements; as a possible self-diagnosis process in the pre- and post-earthquake phase if associated with a health monitoring system; as a tool for the creation of an emergency hub to control and shutdown particular facilities (e.g., gas pipes) in the event of an earthquake [13].

As for the use of BIM in the field of seismic vulnerability, the structural analysis, seismic response assessment, and fragility curves associated with different levels of damage are implemented in a BIM environment in order to obtain a 3D graphical display of vulnerable or damaged elements as well as their associated level of vulnerability or damage. A cost analysis also allows the development of clearer intervention strategies for both the designers and the clients [13–16].

The latter also includes the research proposal presented in this paper. The investigation procedure, replicable on different buildings, is based on a quick assessment of the vulnerability obtained with:

- The application of reliable methodologies already known in the literature,
- The use of laser scanner surveys,
- The implementation of the results in a BIM environment.

The implementation of the results in a BIM environment aims to obtain an effective graphical representation of the different aspects of building vulnerability, their severity and possible co-presence, thanks to the opportunity offered by the BIM methodology to perform queries on probable occurrence scenarios (in this case vulnerability scenarios), including the level of vulnerability found on each interrogated element. This methodology is a first approach to the graphical representation of historical buildings vulnerability. It can already be useful for planning a possible targeted diagnostic campaign and for designing suitable interventions of seismic adaptation or improvement addressed to solve the problems that have emerged. The implementation of the economic evaluation of the diagnostic and seismic improvement interventions associated with every possible investigation scenario will be the future objective of this research.

The proposed procedure has been applied to safeguard Cascina Cuccagna, a historic farmhouse in the urban fabric of Milan city. The safety works have been partially financed by Fondazione Cariplo in the framework of the calls for proposals Art and Culture-Cultural Heritage at risk.

The paper is made of three sections: the first section presents the investigation procedure for the vulnerability assessment of historic buildings. The second section introduces Building Information Modeling (BIM) as a method of graphic detection of building vulnerabilities and of their impact on the buildings' overall ability to withstand an earthquake of a certain magnitude. The third section presents an application of the proposed method to Cascina Cuccagna, a historic building located in the city of Milan, Italy.

2. The Procedure

The knowledge of a building and of its history is an essential prerequisite for any safeguarding action; therefore, all possible historical documentation summarizing the different construction phases of the building should be retrieved and, if possible, their typical construction techniques.

Another form of knowledge comes from a detailed survey of the building in its current state, to be compared with a previous survey, if available.

The next phase of knowledge will require visual and photographic investigations of the building and the outline of its possible crack pattern. The identification of the loads acting on the structural elements will also be important, as well as the evaluation of the mechanical qualities of the masonry, wooden elements, and floors.

The results of this first phase of investigation provide enough elements to face an initial and rapid assessment of possible vulnerabilities.

2.1. *Historical Survey of the Building*

In order for an intervention to be structurally effective, compatible with the existing built patrimony, and respectful of its material, constructive and documentary characters, an articulated and interdisciplinary survey has to be carried out. A fundamental contribution to the study of an existing building or a constructed centre is the historical survey, which should take account of indirect and direct sources of information.

Analysis of indirect sources include cataloguing of the data collected through documentaries, archives, and bibliographic information, with the aim to reconstruct the evolution of the complex, its building, and transformation phases and its load history as an history of the earthquakes that occurred in the past. When known, the seismic history of the building is particularly relevant with respect to the safety and intervention aspects. Of great relevance are also the iconographic elements (views, ancient documents) giving information on the historical housing, allowing the evaluation of the transformations of a territory and the comparison with its present configuration.

A compared study is fundamental, especially when the research objects are minor centers or rural architectures. If the documental (indirect) sources do not allow a complete reconstruction of the complex evolution, the historical information coming from the documents should be verified in situ and accompanied by a direct analysis of the buildings. Direct sources, which sometimes are the sole available if historic documents have been lost, come from the observation of the building itself as a data source and include various methods for approaching the knowledge of the building and its chronological phases. They can take advantage from a multidisciplinary approach which critically combines the information coming from the detection of the adopted constructive techniques and materials, traces of previous interventions, epigraphic and heraldic documentations if present, decorative apparatus, and a stratigraphic survey when possible.

2.2. *Geometry and Material Knowledge: Help Comes from the Laser Scanner Survey*

In the vulnerability assessment, the laser scanner is certainly a useful type of survey. This methodology is able to provide a 3D survey of the entire building, its precision depending on the finesse of the instrument and the density of the measured points. The output is a point cloud that represents an operational guide for computer aided design (CAD) and BIM graphic rendering programs capable of transforming it into more classic, very detailed, reproductions such as plans, elevations and sections (Figure 1).

As we will see, the laser scanner survey also allows the construction of a 3D model of the entire building that serves as a basis for the Building Information Modeling (BIM), here used in the identification of vulnerable elements and their associated degree of vulnerability.

The knowledge process of an existing building should necessarily involve the detection of its constructive materials and techniques, aimed at the reduction and control of physical, chemical, and mechanical material decay. As mentioned, graphic representation is a fundamental tool in the preliminary phase of knowledge as well as an accurate survey of the building in all its formal, material, and degradation aspects, as well as in the subsequent preservation project.

The material survey is a fundamental tool of inquiry into the building and should be coupled with the historical research and the geometric survey. It should thoroughly study the construction consistency, the material origin and the production technique, the opening features which can help dating the construction phases, and should note its features and peculiarities such as alignments, wall thickness diversities, discontinuities, and variations of the wall apparatus, which are always indicative of some significant historical-artistic or simply of technical-constructive choice.

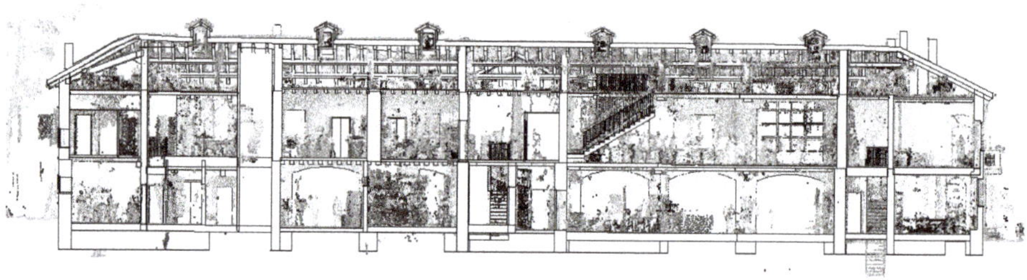

(a)

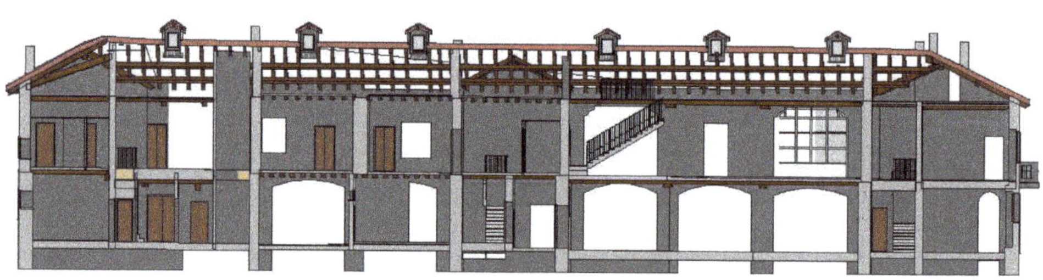

(b)

Figure 1. CAD (Computer-Aided Design) geometric survey obtained from a points cloud: (**a**) cloud resolution; (**b**) CAD re-drawing of the transversal section.

2.3. Redrawing the Crack and Deformation Pattern

The building assessment requires the architectural redrawing of the complex starting from a traditional survey or from the points cloud obtained from the laser scanner instrumental survey. This redrawing could highlight:

- Possible deformation anomalies in the horizontal elements, symptom of high stress, decrease in load-bearing capacity, constraints failure, etc.;

- Possible out of plane of vertical elements, symptom of not contrasted horizontal thrusts, lack of interlocking between walls, local instability of unconnected walls, etc.;
- Presence of not contrasted thrusts in the roof or incidence of vaults on load-bearing walls;
- Misalignments between walls and between inter-stories.

These anomalies could threaten the building safety and require a visual check and a more accurate photographic survey.

For all the different parts of the building, the survey should include drawings of the crack patterns, a necessary step to formulate hypotheses on its possible causes and to suggest possible diagnostic investigations to be carried out for a more thorough verification of the problems found.

In the case of a historical asset, a material and pathological survey should also be carried out in order to plan preservation interventions compatible with the historicity of the building.

2.4. Mechanical Characteristics of Structural Elements and Loads Analysis

Certainly, the knowledge level reached with rapid investigations and non-destructive tests can only be limited, but it already provides the professional with information on where and how to obtain more precise knowledge levels.

The loads analysis should be carried out taking into account the actions to which the structure is currently subjected, considering any change in its intended use and the replacement of structural elements, floors, roofing membranes, or other transformations undergone by the building during its service life.

If a detailed examination of the flooring system is not possible, the acting loads can be estimated according to the constructive typologies used at the building time or due to whatever change has occurred.

For better knowledge of the mechanical characteristics of the structural elements, more or less invasive experimental tests both in the laboratory and on site will be necessary. The connection between floors and walls, and between walls perpendicular to each other should be investigated on-site. However, the damage level of wooden elements can already be detected through a visual investigation of their crack and deformation patterns, of the constraints quality (if detectable), and of any rottenness, biological, and/or parasitic attack. This could be part of the qualitative vulnerability evaluation and provide useful indications for a selective and targeted diagnosis.

The masonry mechanical qualities will also have to be evaluated experimentally, but reasonable hypotheses on the type of masonry texture present in the building can help to establish its quality and behavior toward in plane and out of plane actions. The historical investigation, associated with the knowledge of the construction techniques typical of the area, will allow achievement of truthful assumptions about the type of masonry, while a possible geo-radar investigation could help understand if the masonry is made of a single or multiple leaves.

A rapid method of characterizing the masonry quality has been developed by Borri et al. [17–20]. Having at their disposal a rich catalogue of masonry typologies typical of Italian regions [21,22] (https://www.abacomurature.it/), the authors have developed a methodology that, on the basis of a sample visual survey of the masonry surface, can provide an index of masonry quality. This classifies masonry into three categories against its possible response to vertical or horizontal—in plane or out of plane—forces: good (A), medium (B), very bad (C) (Table 1 and Figure 2).

The masonry quality index (IQM) calculation is based on the rules of good building. It takes into account the units' quality (type of brick/stone and its geometry), the mortar quality, the presence of staggered joints, in prospect and section (if visible), and the presence of headers or other forms of connection of the wall leaves. Depending on their recorded quality, each of these parameters is assigned a weight. This weight is established on an empirical basis (expert judgment) and on a statistical basis where possible (Figure 2).

The final evaluation associates the masonry with a certain IQM value as shown in Table 1.

Table 1. Attribution of masonry category.

Actions on the Masonry	Class A	Class B	Class C
Vertical	$5 \leq IQ \leq 10$	$2.5 \leq IQ < 5$	$0 \leq IQ < 2.5$
Horizontal perpendicular	$7 \leq IQ \leq 10$	$4 < IQ < 7$	$0 \leq IQ \leq 4$
Horizontal coplanar	$5 < IQ \leq 10$	$3 < IQ \leq 5$	$0 \leq IQ \leq 3$

2.5. *Initial Vulnerability Assessment*

The redrawing of the building in plan and in elevation allows us to: define the surface of each floor and each inter-storey average height; identify the resistant elements against horizontal actions in two orthogonal directions (x and y) and their dimensions in plan (total length and average thickness of the resistant elements in both directions); identify the points of transmission of vertical and horizontal loads (thrusts of vaults and coverage).

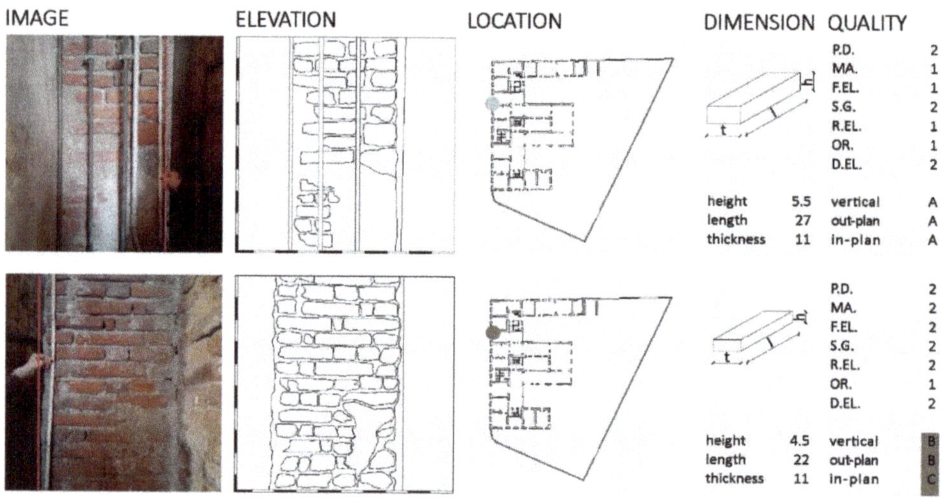

Figure 2. Identification of the masonry quality.

These data are the basis for an initial assessment of the building vulnerability.

Having identified the seismic zone and the parameters that characterize its response spectrum for the reference limit state, Borri et al. [19,20] propose a methodology that allows control of the overall building vulnerability on the basis of:

- An expert judgement on the vulnerability level of the entire building formulated on the basis of the presence, or absence, of 10 possible vulnerability aspects and the severity level found (Table 2).
- Three simplified numerical checks: simplified global check (VG), which allows the definition of the ground acceleration that could lead to collapse; simplified local check (VL), which allows the determination of the acceleration of activation of a given mechanism (overturning, wedge overturning, buckling); simplified static check (VS), which allows the determination of vertical loads only and the safety factor.

Table 2. C/Q Classification/Quantification) seismic method; structural deficiencies to be considered for vulnerability identification (synthesis) [20].

N	Description of Possible Vulnerable Evidences
1	State of conservation. Extent and severity of significant cracks or serious decay.
2	Masonry quality. Assessment of compliance with the rule of art and number of walls with poor workmanship.
3	Linkage. Connections or metal tying between outer and inner walls and with floors and roofs.
4	Floors and roofs. Span or insufficient support.
5	Structural regularity (only on stiff floors). Presence of plane irregularities.
6	Inclusion of a building in a block or a line. Corner or head position, presence of protruding or elevating elements not contrasted by other buildings.
7	Elevation regularity. Mass or stiffness variation between two subsequent levels.
8	Foundation and soil. Evidence of foundation settlements.
9	Local structural defects. Thrusting elements, unsupported walls or pillars, superfetation, hammering with adjacent buildings.
10	Vulnerable non-structural elements. Chimneys, parapets, balconies, eaves, veils, badly connected significantly heavy ceilings.

The results of the three simplified checks are compared with the safety levels required by the Italian NTC2018 standards for the protection of human life limit state (SLV) for the site under examination, the considered type of building, and the envisaged class of use. These results are expressed in terms of conventional safety factor (*SF*). In other words: for each of the three checks, *VG*, *VL*, *VS*, and for each verified element, there will be a *SF* factor (e.g., $SF_{VL,xi}$ $SF_{VL,yi}$ factors related to the *VL* check on element *i* in the direction *x* and *y*); the minimum safety factor obtained in the check under examination will be divided by the safety factor present in the standard and referred to the SLV state.

$$SF_{Vj}(\%) = \frac{SF_{Vj,min}}{SF_{Vj,\ SLV-Law}} \quad \text{with } j = \text{G, L, S,} \tag{1}$$

$SF_{Vj}(\%)$ expressed in percentage, is the conventional safety factor for the check *Vj*.

$SF_{Vj,min}$ indicates the minimum safety factor among all the safety factors calculated for the check *Vj* and for all the elements of the structural system, $SF_{Vj,\ SLV-Law}$ is the safety factor indicated by the code for the check *Vj*.

The minimum among all the safety factors SF_{Vj} will determine the vulnerability class of the system [20]

$$SF(\%) = \min\left\{SF_{Vj}(\%)\right\} \quad \text{with } j = \text{G, L, S,} \tag{2}$$

The combination of the vulnerability assessments with the minimum safety factor derived from the three simplified numerical checks results into the association of the building with a low, medium, high or very high vulnerability class (Table 3).

Table 3. Anti-seismic classification [20].

Safety Assessment for the State of Life Preservation Limit (SLV)					A+
	Results of the Numerical Assessment				
Vulnerability judgment	>80%	80%–60%	60%–40%	40%–20%	<20%
Low vulnerability	A	B1	B2	C1	D1
Medium vulnerability	B1	B2	C1	C2	D2
High vulnerability	B2	C1	C2	D1	D3
Very high vulnerability	C1	C2	D1	D2	E

The methodology has been developed within the framework of the POR-ESFR 2007–2013 for the Umbria Region seismic certification/qualification of the existing asset (Seismic C/Q). Currently the guidelines for the seismic C/Q have been formally acquired by the Umbria Region and the Marche Region.

2.6. Non-Destructive Diagnostics

For interventions on buildings, the Italian Standard requires more or less severe safety factors depending on the knowledge level reached about the typology and details and mechanical characteristics of the building studied. In order to reduce the severity imposed by the regulatory coefficients, a diagnostic study should be necessary and the investigations previously described can help not to disperse resources, highlighting critical points on which possible experimental tests should focus.

High knowledge levels require in-depth and invasive diagnostic investigations, often not compatible with the available economic resources and the intrinsic value of the asset. Setting up a non-destructive or minimally invasive testing campaign (sonic and ultrasonic tests, superficial and deep penetrometric tests, georadar, partial plaster removal, small inspections, etc.) focused on the areas with the main problems can help contain costs but also reach a higher level of knowledge, as required by regulations.

2.7. Identification of Vulnerable Elements and the Level of Damage

In the literature, damage classifications for different types of buildings are available: they are defined from the study of the fragility curves that link the damage probability to the intensity of the seismic event. However, these classifications mainly concern the probable response of a given structural typology to a given seismic action [23–25], not the response of a single vulnerable element in a single building. Therefore, the assessment of the vulnerability level of each element of the building must follow other paths. In 2000 for the Marche Region, Francesco Doglioni proposed to associate the forms of specific vulnerability, often found in existing buildings, with six well-defined causes (Table 4) [26]. This classification can be extended to the entire historical heritage at risk and represents a very important starting point on which the identification of specific vulnerabilities and an initial estimate of the propensity to damage of vulnerable elements for an earthquake of a certain magnitude can be based.

The propensity to damage can be assessed by associating weights to each detected criticality connected with one of the specific vulnerabilities listed in Table 4. These weights can be established according to the detected damage level.

Obviously, this step must be subjected to "expert" judgment. In fact, on the basis of an expert judgment, a certainly subjective, but supported by the results of previous investigations, weight can be assigned to each of the vulnerability forms found and a "judgment" can be formulated on their influence on the overall vulnerable behavior of the building. In this way, each of them will be qualified as absent (0), scarcely influential (1), averagely influential (2), significantly influential (3).

More than one of the specific vulnerabilities listed in Table 4 may be present in the building elements. Now, if each element were affected by more than one form of specific vulnerability, it would be interesting to have an immediate graphical feedback of its vulnerabilities [13–16].

The research presented here focuses on the possibility of using the BIM methodology also in this field, evaluating the effectiveness of a three-dimensional building representation with the graphical identification of elements affected by one or more forms of vulnerability and the classification of their level of influence on the total vulnerability of the building through a given range of colors.

Table 4. Classification of specific vulnerability forms in thematic groups attributed to six different possible causes.

Thematic Groups of Specific Vulnerability Forms Attributed to:			
1	Initial methods of construction	Materials and constructive techniques	• Quality of supports and binders • Masonry adhesion and/or cohesion
		Geometric and dimensional masonry characters	• Related to the masonry section: positioning of the outer leaf • Related to the architectural-structural elements: slenderness
		Elements that reduce the masonry cross section	• Presence of fuel pipes • Interruptions due to plants ducts • Eaves/rain pipes within the masonry section
		Structural elements with inadequate section	• Slenderness • Poor thickness • Towering elements
		Peculiar openings distribution	• Misaligned apertures • Apertures close to corners
		Lack of connection between stone elements	• Inadequate interlocking • Lack of adhesion
2	Building transformation processes	Enlargement Super-elevation Closing/opening of apertures	• Non-interlocked • Discontinuous • Of heterogeneous material
		Subtraction of elements or masonry parts	• Demolition of partitions • Opening of large apertures
		Inadequate or dangerous structural solutions	• Unsupported walls • Pillars above vaults
3	Lack of structural connections and role of existing protection devices	Wall-wall connections	• Discontinuities in wall node
		Wall-roof connections Wall-floor connections	• Non-compensated local thrust • Lack or poor effectiveness of connections
		Inadequate or damaged connection systems	• Lacking or ineffective ties • Inadequate buttresses

Table 4. *Cont.*

Thematic Groups of Specific Vulnerability Forms Attributed to:			
4	Structural deterioration and lack of maintenance	Masonry decay	• Lack of binder in the joints • Masonry de-cohesion • Presence of water • Decay of stone elements
		Decay of timber elements in the roofs	• Heads rotting • Decay of secondary elements and/or of the tables • General decay • Elements deflection
		Lack of maintenance of masonry and plasters	• Exposed outer leaf • Plaster efficiency
		Lack of maintenance of the roof	• Conditions of the covering layer • Inefficient eaves and pipes
5	Previous structural damage not sufficiently repaired	Seismic	• Cracking • Deformations/out of plumb
		Static	• Cracking • Deformations/out of plumb
6	Recent structural refurbishment	Executed with "modern" techniques	• Reinforced concrete beams with inadequate sections and/or detached from the masonry • Non-distributed and nonhomogeneous injections • Reinforced plaster non suitably connected with the masonry

3. Building Information Modeling Applied to Seismic Vulnerability

The 3D design in BIM environment is able to support multidisciplinary collaborative processes useful in the management and control of the building process; in fact, the BIM modeling allows optimization of the planning, implementation, and management of the entire building complex from the architectural, structural, and plant engineering point of view. Each element of the building, created in BIM environment, has parametric characteristics that can be modified: at each project implementation, with additions or changes, the entire design system is updated according to the changes made; this model potentiality allows a flexible management of the entire construction process from design, to building, to maintenance.

BIM modeling can also be used in the interventions on existing assets. In this case, the graphic reproduction of an existing building requires input from a base track; this base can be provided through a points cloud obtained from a laser scanner survey, or a geometric survey file resulting from computer aided design (CAD) programs.

The 3D model, created as a geometric survey of the building, is then parameterized and reproduced as a BIM model. The BIM modeling allows association of to each building element (component) and all its characteristics (geometry, material, mechanical parameters, etc.) in parametric form. Taking advantage of this BIM environment peculiarity, it seems interesting to associate to each element, in parametric form, with the vulnerability characteristics resulting from previous investigations.

The first step required by the parameterization procedure is creating an association between the specific vulnerabilities found and their assigned weights for each individual element of the building. The purpose of having a graphical resolution of the vulnerabilities requires that each vulnerability and each weight is associated with a field that identifies the type of specific vulnerability and the weight it could have on the overall vulnerability of the building. The graphical representation offered by the BIM modeling is certainly useful to have an immediate picture of the distribution of some vulnerabilities, and their danger level, throughout the whole building. These graphic representations can be obtained through targeted queries, a classic approach to work in a BIM environment.

These queries can concern the same specific vulnerability but with different weight: 0 = absent, 1 = light damage, 2 = moderate damage, 3 = severe damage.

- Query01_Querying the same vulnerability:
- "Presence of flue pipes" classified with weight 2.
- "Presence of flue pipes" classified with weight 3.

In the example shown, query01 gives the position of voids inside both load-bearing and non-load-bearing walls, and the weight associated with this vulnerability according to the recorded damage level.

Or they can concern two different vulnerabilities

- Query02_Querying two vulnerabilities:
- "Uncompensated localized roof thrust" ranked with weight 3;
- "Unconnected building expansion" classified with weight 2.

Query02, thus defined, gives the critical and significant points for the possible detachment or partial overturning of the masonry structure.

Another possible query is for three or more vulnerabilities.

- Query03_Querying three or more vulnerabilities:
- "No or poor effectiveness of wall-roof connection" classified with weight 2.
- "Missing or ineffective tie rods" classified with weight 2.
- "Condition of the roof covering" classified with weight 1.

This type of query, for instance, is useful for understanding the state of connection and degradation of the roof structures with respect to the underlying masonry.

The parametric modeling and the performed queries presented here seem to overcome the difficulties of a synthetic and global reading of the vulnerabilities present in a building. In fact, the classification of vulnerabilities is generally reported on spreadsheets and tables that are not easy or quick to interpret, especially for non-professionals, as the property owners can be. A graphic representation certainly helps the owner to better understand the problems of his building and to take informed decisions on the interventions to be carried out.

4. Application of the Proposed Procedure to the Case Study of Cuccagna Farmhouse in Milan (North Italy)

The methodology presented in this work has been applied in a project of redevelopment and anti-seismic safety of a historical asset in the city of Milan, Italy.

In 2017, the pilot project PRE.CU.R.S.OR. was the winner of a grant from the Fondazione Cariplo di Milano in the field of Art and Culture-Cultural Heritage at Risk, a funding line that supports preservation interventions on assets protected by law. The involved asset is Cascina Cuccagna, one of the rural buildings still existing in the city of Milan (Figure 3). Until about the middle of the 20th century, farmsteads represented the place of life and productivity for most of the inhabitants of the Lombard plain. Still today, they are a remarkable architectural resource and often, as in the case of Cascina Cuccagna, they are used with different functions from the original ones.

(a)

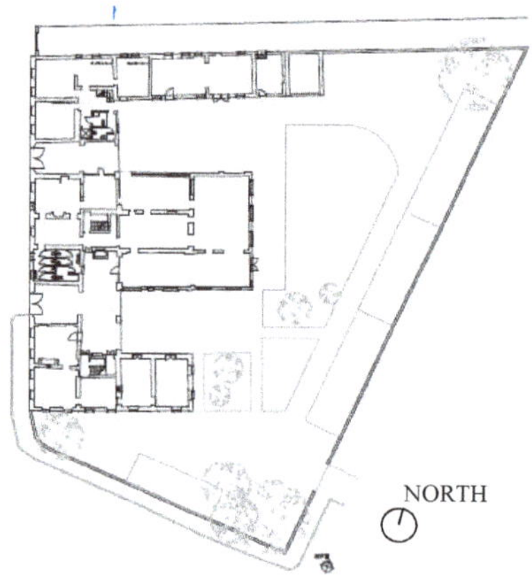

(b)

Figure 3. Cascina Cuccagna, (**a**) overall view and (**b**) ground floor plan.

Cascina Cuccagna represents an ideal scenario for a replicable "pilot project" of progressive improvement of the earthquake resistance conditions on historical buildings. In fact, from the construction point of view, its model is highly repetitive in Lombardy and more generally in northern Italy. It is located in a context of medium-low seismicity that allows testing of investigation and intervention techniques in safe conditions. Cascina Cuccagna is also the site of many cultural, social, and economic activities and a good place for the culture dissemination of risk protection through the direct observation of the actions to be taken, their invasiveness and interference with the users' lives, the implementation timing, the possibility of extending them over time, and the related costs.

4.1. Historical Investigation, Survey, Redrawing, and Crack Pattern

Based on the historical research, the phases of expansion and modification of the original building over the years could be reconstructed (Figure 4).

The contact surfaces between portions built in different periods are delicate and require attention because, often, the extensions are made without sufficient connection to adjacent constructions. Some of these surfaces have also been highlighted by redrawing the crack pattern (Figure 5): the red cracks in Figure 5 are an example of this. The crack pattern survey performed room by room (Figure 6) has highlighted other criticalities where a more in-depth investigation is required. Instead, the building redrawing in a BIM environment from the laser scanner survey points cloud has highlighted some shortcomings at the deformation level: floor sags and possible lack of plomb (Figure 7).

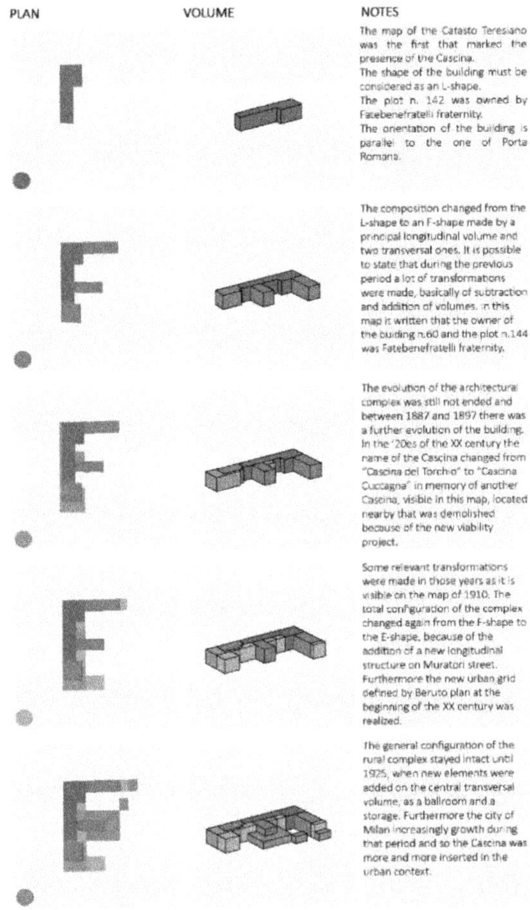

Figure 4. Historic survey of construction phases.

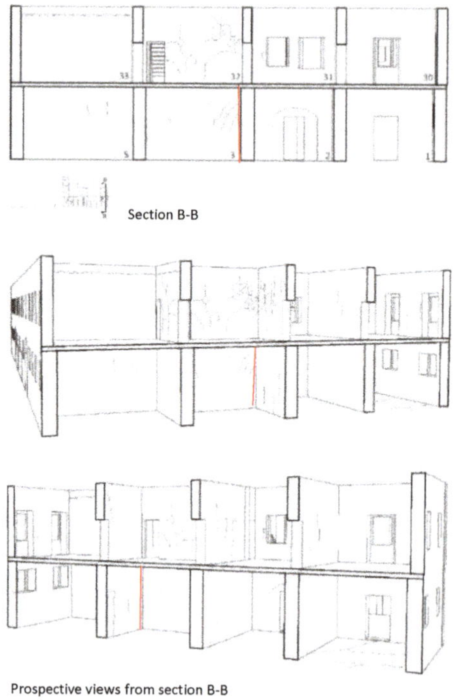

Figure 5. Example of crack pattern (in red a crack probably due to lack of connection between two blocks built in different construction periods).

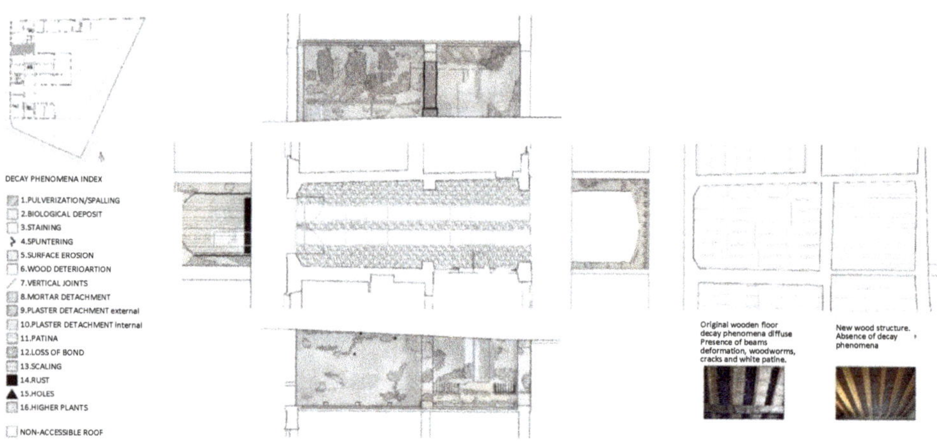

Figure 6. Crack pattern representation for each room of the building.

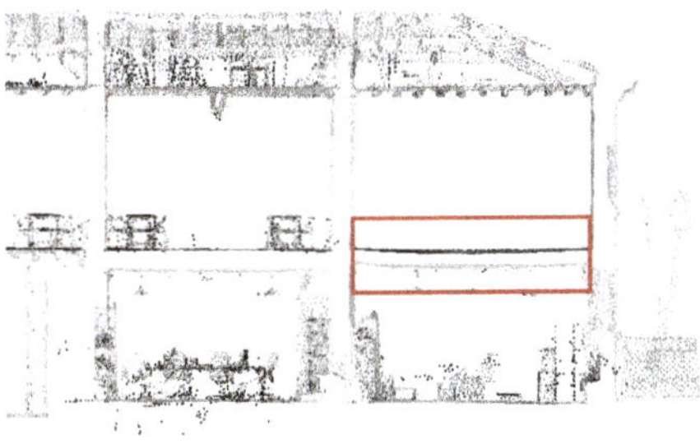

(a)

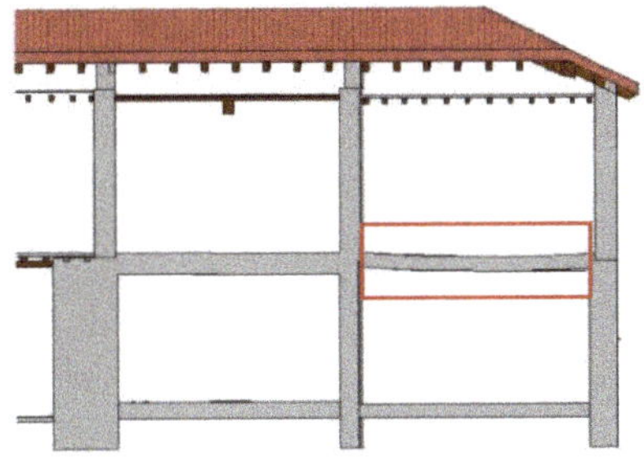

(b)

Figure 7. Example of survey of a flexural anomaly in a floor beam (**a**) laser scanner survey; (**b**) section redrawing.

The construction of the 3D model, again in a BIM environment (Figure 8), was the next step in planning the management of the whole structural system in the phase of identification of specific vulnerabilities and in the phase of design and management of seismic consolidation and adaptation interventions.

(a)

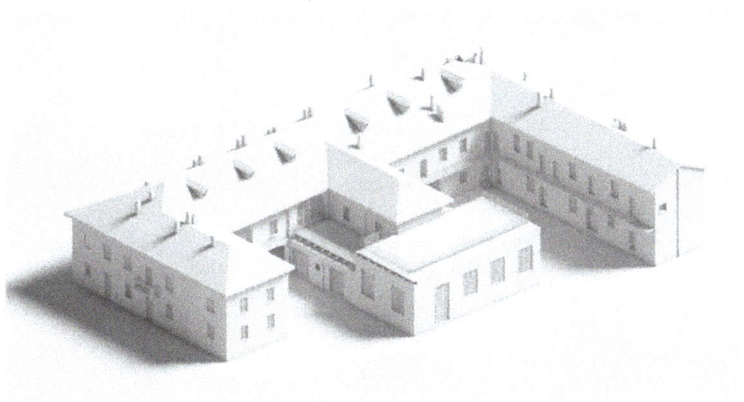

(b)

Figure 8. Reconstruction of the parametric 3D model and photorealistic representation of the white-model render: (**a**) point cloud, (**b**) re-drawing.

4.2. Mechanical Characterization of Materials and Rapid Vulnerability Analysis

The survey on the masonry quality has shown masonry with variable quality, depending above all on the time of construction (Figure 4). In this case, the calculation of a safety factor SF (%) (Equation (2)) is not required, but we can assume that: for class A masonry, SF can be considered greater than 80% of the value required by law; for class B masonry, it can be considered in the range of 50%–75%; while for class C masonry, it can be considered between 25% and 50% of the value required by law. The walls tested, except for a couple of cases, show a medium/high behavior to the transmission of vertical loads, and some criticalities for the transmission of out of plane loads on the most deteriorated panels.

The rapid vulnerability assessment showed a good overall behavior for the whole building (SF_{VG} > 80%) and also a good behavior of the wall leaves which, in case of an earthquake, could be subject to tilting mechanisms (SF_{VL} > 80%). On the other hand, some roof and inter-storey floors showed problems in transmitting vertical loads (60% < SF_{VS} < 40%).

The presence of non-structural elements that could cause vulnerability, such as flue pipes or balconies and walkways, together with poor floor-masonry anchoring, led to a vulnerability level of type B2 (Table 3).

The surveys, crack pattern observation and the rapid evaluations which were carried out have indicated some important shortcomings, worthy of diagnostic investigation. Therefore, the following were carried out:

- Geo-radar surveys in some rooms of the Cascina to understand the direction of floor structural elements when covered with a thatched ceiling, to verify the wall thickness as well as the foundations depth and size. The foundations investigation has also confirmed the probable difference in the construction period of some parts of the building [27,28];
- Sonic investigations on some walls, in those areas where the crack pattern showed fissures of doubtful interpretation, in order to investigate the panel homogeneity and identify possible detachments of internal leaves and/or surface plasters;
- Superficial and deep penetrometric tests for mortars characterization;
- Bricks characterization based on the behavior detected on bricks preserved on site and probably contemporary to the ones in use. The results obtained were compared with values available in the literature for materials of the same type and period [29–31];
- Penetrometric tests (©Resistograph) to characterize the timber floors and roofs [32–34].

Normally, a focused diagnostic phase such as the one described above affects less than 6% the cost of the building's safety intervention.

4.3. Identification of Specific Vulnerabilities

From the rapid evaluations and the experimental tests carried out, each vertical and horizontal structural element and all the non-structural elements could be classified as indicated in Table 4. Each type of specific vulnerability found on individual elements was assigned a weight, according to the damage level obtained in the previous investigations: absent damage (0), scarcely influential (1), averagely influential (2), significantly influential (3). The result of this operation is presented in a matrix with a correspondence between structural elements and specific vulnerabilities, useful to characterize the elements in a BIM environment. In the commercial software that was used, in addition to the geometric, mechanical, and material characteristics, each structural element could be associated with the table of specific vulnerabilities and weights assigned to each item referred to that element. (Figure 9).

4.4. Queries on Vulnerability Levels Carried Out in the BIM Environment

Once the 3D model of the building had been prepared in a BIM environment and the matrix of the correspondence between structural elements and specific vulnerability had been constructed, the queries were carried out.

After associating the vulnerabilities and their respective weights to the building elements, a colored field (selected by the user for each query) was connected with each weight and queries were carried out about the presence or absence of vulnerability elements and their influence on the assessment of the overall vulnerability of the building (Figure 9).

Different types of queries can be performed. The system can be queried about the presence of a certain type of specific vulnerability throughout the building and its degree of incidence on a given element or on the overall vulnerability of the system (Figure 10). The presence of two or more specific vulnerabilities with different degrees of incidence on the overall vulnerability can also be visualized (Figures 10 and 11).

An interesting aspect is the possibility to know which elements are particularly vulnerable because they are affected by two or more criticalities, each measured by its weight on the global vulnerability. By carrying out a single query, the BIM system is able to respond to the question by highlighting the elements that present all (and only) the queried criticalities and with the weights investigated. The graphic result is the coloring of the elements affected by these vulnerabilities (Figures 12 and 13).

If effective structural improvements had been made, like an improvement of the connection between the walls and the thrusting roof, the level of damage associated with the elements will be decreased; so, by making new queries, the effectiveness of the improvements and their location will be visualized in the model.

Masonry Wall Thickness 63 cm.	
	Damage level
quality of supports and binders	2.0000
masonry adhesion and/or cohesion	2.0000
related to the masonry section: positioning of the outer leaf	
related to the architectural-structural elements: slenderness	
masonry adhesion and/or cohesion	
presence of chimneys	
interruptions due to plants ducts	
eaves/rain pipes within the masonry section	

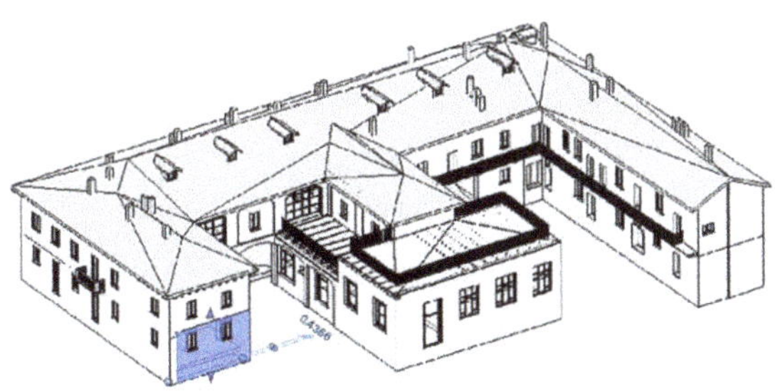

(a)

Figure 9. *Cont.*

Roof Cover
Roof tile

	Damage level
lack of binder in the joints	
masonry de-cohesion	
presence of water	
decay of stone elements	
masonry adhesion and/or cohesion	
heads rotting	
decay of secondary elements and/or of the tables	
general decay	1.0000
exposed outer leaf	1.0000
plaster efficiency	
conditions of the covering layer	
inefficient eaves and pipes	1.0000

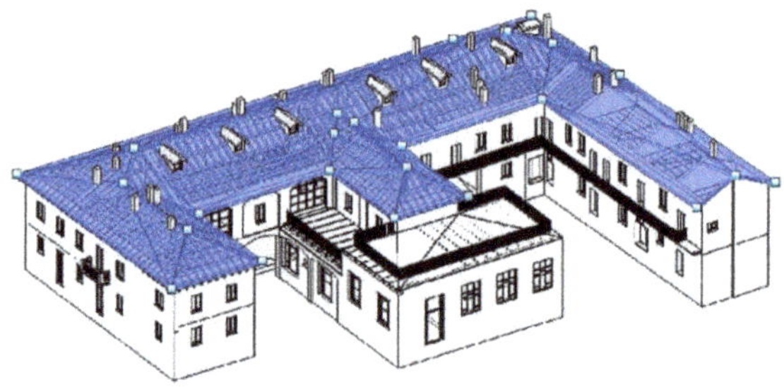

(**b**)

Figure 9. Example of coupling between structural element, specific vulnerability, and associated weights: (**a**) masonry wall thickness; (**b**) roof cover.

CHART_01	
QUERIED VULNERABILITY INDEXES	
Presence of chimneys	
LEGEND	
Label	Rule
☑	presence of chimneys = 2
☑	presence of chimneys = 3

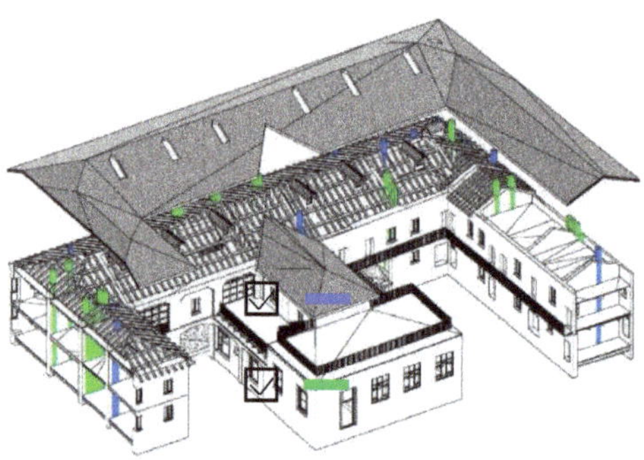

Figure 10. Simple query: presence of flue pipes and their degree of incidence on the vulnerability of the wall panel.

CHART_02	
QUERIED VULNERABILITY INDEXES	
Lack or poor effectiveness of connections walls-roof Lacking or ineffective ties Conditions of the covering layer	
LEGEND	
Label	Rule
☑ (blue)	lack or poor effectiveness of connections walls-roof = 2
☑ (red)	lacking or ineffective ties = 2
☑ (yellow)	conditions of the covering layer = 1

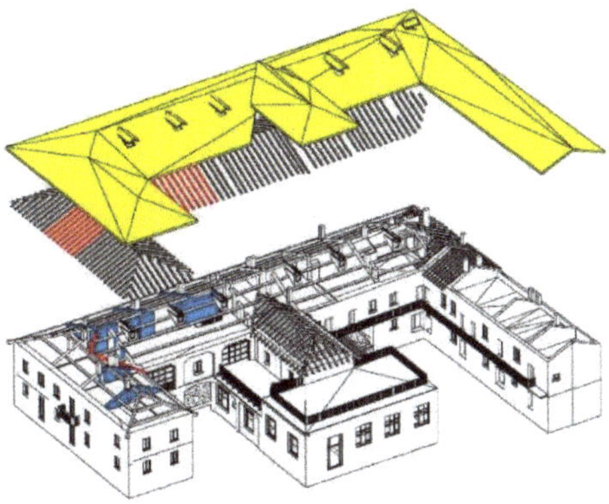

Figure 11. Integrated queries: display of three different types of specific vulnerabilities with an impact of less than 2 on the overall vulnerability.

QUERIED VULNERABILITY INDEXES
General decay Deformations Lack or poor effectiveness of connections
LEGEND
Label   Rule ☑ General decay = 2 ☑ Deformations = 2 ☑ Lack or poor effectiveness of connections = 1

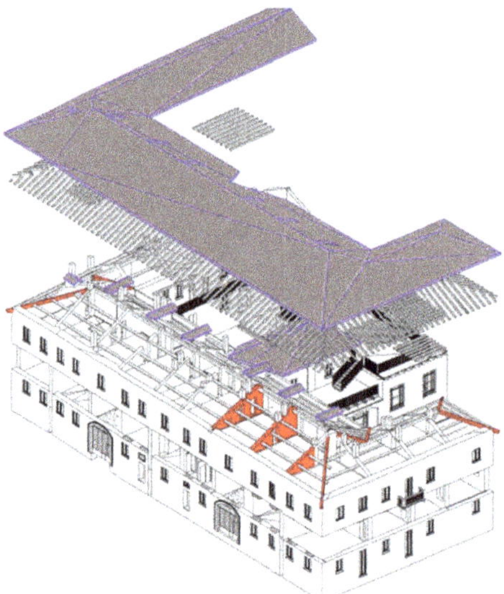

Figure 12. Multiple queries: presence of multiple vulnerabilities with different weights on the same element of the system. The coloring identifies the elements affected by all 3 vulnerabilities in the legend with their weights.

QUERIED VULNERABILITY INDEXES
Lack or poor effectiveness of connections walls-roof Lacking or ineffective ties Conditions of the covering layer
LEGEND
Label Rule ☑ Deformations = 3 ☑ Lack or poor effectiveness of connections = 3

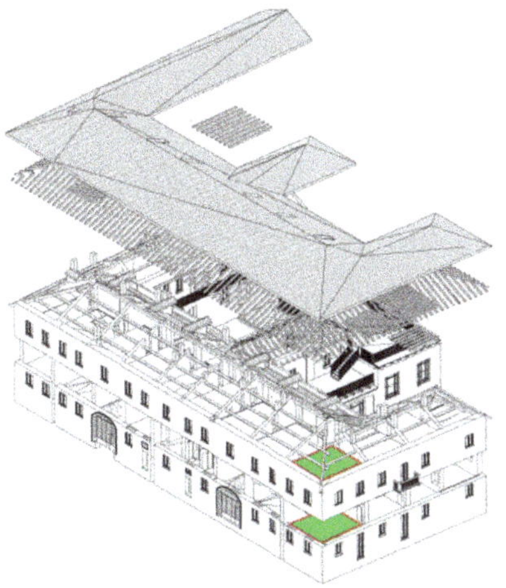

Figure 13. Multiple queries: presence of multiple vulnerabilities with different weights on the same element of the system. The coloring identifies the elements affected by all 2 vulnerabilities in the legend with their weights.

5. Conclusions and Future Developments

The research issue of assessing the seismic vulnerability of existing buildings is still unresolved. For about thirty years, scholars and researchers have been working on quantifying the seismic vulnerability of historical buildings, following different approaches at different levels of detail (rapid, static, dynamic approaches, etc.).

This open-ended problem is raising awareness among the public and owners of the need for implementing safety measures even on buildings not hit by the earthquake in areas with medium-low seismicity.

In this work, we presented the application of a procedure for assessing the seismic vulnerability of an existing building based on survey tools already present in the guidelines of several Italian Regions.

The procedure has proved to be effective and easy to use for a professional. The innovative part of the work carried out consisted of the graphic representation of the results obtained.

Until now, the degree of vulnerability of individual elements or macro elements was only possible in tabular or diagrammatic form, not easy to understand for non-experts. In this work, the use of design in a BIM environment has been experimented. The adopted investigation procedure and the non-destructive diagnostic tests used allowed definition of, for each structural element, a detailed classification of its vulnerabilities and their weight on the overall building vulnerability.

Therefore, in addition to the classic geometric, mechanical, and material characteristics, the use of a BIM model allowed associating to each structural element the presence of vulnerability aspects and their weight. By means of focused queries, specific vulnerabilities present in the building and the reached level of damage (measured as a weight on the total vulnerability) could be highlighted. The graphic representation is very helpful to understand both the location of critical points and the problem severity. The owner becomes aware that the building presents problems and that the interventions have different degrees of urgency but are necessary; on the other hand, the designer has a tool to plan, in terms of time, the maintenance and adaptation interventions and to control ex-post the effectiveness of the interventions carried out.

The procedure presented here served as a basis for planning the consolidation and safety interventions on Cascina Cuccagna in Milan. Critical issues in the floors of the South-West wing and of the South-East wing have been highlighted that require immediate static safety interventions. The floors in the corner area between the South-East and the South-West wings also turned out to be poorly connected to the perimeter masonry and require seismic adaptation. In addition, historical analysis and geo-radar survey of the foundations have shown that the South-West wing is posterior to the South-East wing; redrawing of the cracked pattern has shown a possible poor connection between the perimeter walls of the two wings. All these aspects have been reported in the BIM environment and expressed in graphic form. Thus, the designers were able to plan the most urgent interventions.

Of course, designing in a BIM environment is not without difficulties. First of all, every element of the project must be precisely defined; if this gives great advantages to managing the construction site, certainly it is a not easy constraint in the architectural design phase. In fact, if the architectural design criteria do not follow the BIM environment logic from the very beginning, then the subsequent modeling in a BIM environment is not so easy. Difficulties have also been met due to the compatibility between CAD graphics and structural modeling programs, used in professional practice, with modeling in a BIM environment; this affects the easy transfer of project files in this environment.

These problems are mitigated when the application deals with existing assets, where geometries and materials are given and the implementation in a BIM environment is easier. However, it is necessary to underline that working on existing buildings also requires modeling of structural sections often out of plan, variable, or deformed geometry not easy to shape with the simple elements present in the software. Therefore, simplifications or devices are often used to facilitate the three-dimensional construction of the building and the interchange with third-party software (structural, energy, computational, etc.). This aspect must be solved as well as the improvement of graphic resolution; the architectural side is still very poor and forces professionals to produce two or more types of design tables. Another aspect in which the current BIM modeling is poor is its software compatibility to facilitate professional use.

As for the procedure introduced here, the authors believe that the proposed procedure can be replicated with the same degree of effectiveness on other existing buildings and represents a real way of involving the client in the safety processing of the existing historical heritage. However, in order to improve the input of collected data and the graphic output, it will still require application efforts concerning both different vulnerabilities and monitoring of the effectiveness of the implemented actions for static and seismic adaptation.

Author Contributions: Conceptualization, E.G.; Methodology, E.G. and A.A.; Structural evaluation, Writing, E.G.; Restoration, Reviewing and Editing, A.A.; Methodology, Validation, F.M.; Laser scanner survey, BIM application, F.V. All authors have read and agreed to the published version of the manuscript.

Funding: The following research is funded by the CARIPLO Foundation, Milan, Italy, in the context of the 2017 Art and Culture-Cultural Heritage at Risk Calls, Project PRE.CU.R.S.OR.

Acknowledgments: Thanks to the ACCC association of Cascina Cuccagna, the design team of Hydea s.p.a. and Eng. Luciano Ardito for the collaboration undertaken in this project.

Conflicts of Interest: The authors declare no conflict of interest.

References

1. Dell'Orto, C.; Guzzetti, F.; Maroldi, F.; Molina, C. G.I.S. for the post-earthquake damage assessment. The case of the Fontecchio historic heritage. In Proceedings of the National Conference Sicurezza e Conservazione nel Recupero dei Beni Culturali Colpiti da Sisma. Strategie e Tecniche di Ricostruzione ad un Anno dal Terremoto Abruzzese, Venice, Italy, 8–9 April 2010; CD-ROM. pp. 1–10. (In Italian).

2. Dell'Orto, C.; Maroldi, F. Post-Earthquake damage evaluation and reconstruction strategies: G.I.S. on Fontecchio. In Proceedings of the 14th National Conference ASITA, Brescia, Italy, 9–12 November 2010; CD-ROM. pp. 835–840. (In Italian).
3. Paudel, B.; Pradhan, P.K. Seismic Vulnerability Assessment of Buildings in GIS Environment: Dhankuta Municipality, Nepal. *Geogr. J. Nepal* **2011**, *9*, 1–12. [CrossRef]
4. Ricci, P.; Verderame, G.M.; Manfredi, G.; Pollino, M.; Borfecchia, F.; De Cecco, L.; Martini, S.; Pascale, C.; Ristoratore, E.; James, V. Seismic Vulnerability Assessment Using Field Survey and Remote Sensing Techniques. In Proceedings of the International Conference on Computational Science and its Application, ICCSA 2011, Santander, Spain, 20–23 June 2011; Murgante, B., Ed.; Springer: Berlin/Heidelberg, Germany, 2011. Part II, LNCS 6783. pp. 109–124. [CrossRef]
5. Ahmed, M.M.; Jahan, I.; Alam, J. Earthquake Vulnerability Assessment of Existing Buildings in Cox's-Bazar using Field Survey & GIS. *IJERT Int. J. Eng. Res. Technol.* **2014**, *3*, 1147–1156. Available online: https://www.ijert.org/research/earthquake-vulnerability-assessment-of-existing-buildings-in-coxs-bazar-using-field-survey-gis-IJERTV3IS080729.pdf (accessed on 8 July 2014).
6. Rahman, N.; Ansary, M.A.; Islam, I. GIS based mapping of vulnerability to earthquake and fire hazard in Dhaka city, Bangladesh. *Int. J. Disaster Risk Reduct.* **2015**, *13*, 291–300. [CrossRef]
7. Rezaie, F.; Panahi, M. GIS modeling of seismic vulnerability of residential fabrics considering geotechnical, structural, social and physical distance indicators in Tehran using multi-criteria decision-making techniques. *Nat. Hazards Earth Syst. Sci.* **2015**, *15*, 461–474. [CrossRef]
8. Taffarel, S.; Marson, C.; Valotto, C.; Rovevrato, M.; Munari, M.; Da Porto, F.; Modena, C. Seismic vulnerability maps of Timisoara historical centre based on fragility curves. In Proceedings of the 10th International Conference on Structural Analysis of Historical Constructions, SAHC2016, Leuven, Belgium, 13–15 September 2016; Von Balen, K., Verstrynge, E., Eds.; CRC Press: Boca Raton, FL, USA, 2016.
9. Taffarel, S.; Caliman, M.; Valluzzi, M.R.; Modena, C. Seismic vulnerability assessment of clustered historical centers: Fragility curves based on local collapse mechanisms analyses. In Proceedings of the IB2MAC, 16th International Brick and Block Masonry Conference, Trends, Innovations and Challenges, Padua, Italy, 26–30 June 2016; Modena, C., da Porto, F., Valluzzi, M.R., Eds.; CRC Press, Taylor & Francis Group: London, UK, 2016. A Balkema Book, electronic support. pp. 2499–2506, ISBN 9781138029996. [CrossRef]
10. Sadrykia, M.; Delavar, M.R.; Zare, M. A GIS-Based Fuzzy Decision Making Model for Seismic Vulnerability Assessment in Areas with Incomplete Data. *ISPRS Int. J. Geo-Inf.* **2017**, *6*, 119. [CrossRef]
11. Taffarel, S. Metodi Speditivi Per la Valutazione Della Vulnerabilità Sismica Del Costruito Storico: Approccio All'incertezza Nelle Forme di Aggregazione Complessa a Diversa Scala. Ph.D. Thesis, dBC Department, University of Padua, Padua, Italy, 2017. (In Italian).
12. Coccia, D.E. Interoperabilità BIM-GIS Nella Classificazione Della Vulnerabilità Sismica Degli Edifici. Master's Thesis, Politecnico di Torino, I School of Engineering, Turin, Italy, 2017. (In Italian).
13. Welch, D.P.; Sullivan, T.; Filiatrault, A. Potential of Building Information Modelling for seismic risk mitigation in buildings. *Bull. New Zealand Soc. Earthq. Eng.* **2014**, *47*, 253–263. Available online: https://www.nzsee.org.nz/library/nzsee-quarterly-bulletin/vol-41-50/ (accessed on 8 July 2014). [CrossRef]
14. Christodoulou, S.E.; Vamvatsikos, D.; Georgiou, C. A BIM-Based Framework for Forecasting and Visualizing Seismic Damage, Cost and Time to Repair. In *eWork and eBusiness in Architecture, Engineering and Construction*; Menzel, Scherer, R., Eds.; Taylor & Francis Group: London, UK, 2010; pp. 33–38. ISBN 978-0-415-60507-6.
15. Alirezaei, M.; Noori, M.; Tatari, O.; Mackie, K.R.; Elgamal, A. BIM-Based Damage Estimation of Buildings under Earthquake Loading Condition. *Procedia Eng.* **2016**, *145*, 1051–1058. [CrossRef]
16. Xu, Z.; Lu, X.; Zeng, X.; Xu, Y.; Li, Y. Seismic loss assessment for buildings with various –LOD BIM data. *Adv. Eng. Inform.* **2019**, *39*, 112–126. [CrossRef]
17. Borri, A.; De Maria, A.; Casaglia, S. The EAL-M method for the seismic classification of the existing masonry buildings: A comparison between different methods and preliminary evaluations of other typologies. *Progett. Sismica* **2014**, *5*, 11–29. Available online: http://dx.medra.org/10.7414/PS.5.2.11-29 (accessed on 8 July 2014). (In Italian).
18. Borri, A.; De Maria, A. Masonry Quality Index (MQI): Correlation with the mechanical characteristics and knowledge levels. *Progett. Sismica* **2015**, *6*, 45–63. (In Italian) [CrossRef]
19. Borri, A.; Corradi, M.; Castori, G.; De Maria, A. A method for the analysis and classification of historic masonry. *Bull. Earthq. Eng.* **2015**, *13*, 2647–2665. [CrossRef]

20. Borri, A.; De Maria, A. La Classificazione Sismica: Un Protocollo Metodologico Già Operativo Applicabile Agli Edifici Esistenti in Murature, Ingegno-Web. 2016. Available online: https://www.ingenio-web.it/6003-la-classificazione-sismica-un-protocollo-metodologico-gia-operativo-applicabile-agli-edifici-esistenti-in-muratura (accessed on 8 July 2016).
21. Binda, L.; Saisi, A.; Tiraboschi, C. Investigation procedures for the diagnosis of historic masonries. *J. Constr. Build. Mater.* **2000**, *14*, 199–233. Available online: https://www.sciencedirect.com/journal/construction-and-building-materials/vol/14/issue/4 (accessed on 8 July 2000). [CrossRef]
22. Binda, L.; Borri, A.; Cardani, G.; Doglioni, F. *Scheda Qualità Muraria: Relazione Finale e Linee Guida per la Compilazione Della Scheda di Valutazione Della Qualità Muraria*; Report ReLUIS 2005–2008; DPC Department Of Civil Protection: Rome, Italy, 2009. (In Italian)
23. Erberik, M.A. Generation of fragility curves for Turkish masonry buildings considering in-plane failure modes. *Earthq. Eng. Struct. Dyn.* **2008**, *37*, 387–405. [CrossRef]
24. Rota, M.; Penna, A.; Magenes, G. A methodology for deriving analytical fragility curves for masonry buildings based on stochastic nonlinear analyses. *Eng. Struct.* **2010**, *32*, 1312–1323. [CrossRef]
25. Pagnini, L.C.; Vicente, R.; Lagomarsino, S.; Varum, H. A mechanical model for the seismic vulnerability assessment of old masonry buildings. *Earthq. Struct.* **2011**, *2*, 25–42. [CrossRef]
26. Doglioni, F. *Codice di Pratica (Linee Guida) per la Progettazione Degli Interventi di Riparazione, Miglioramento Sismico e Restauro dei Beni Architettonici Danneggiati dal Terremoto Umbro-Marchigiano del 1997*; BUR, 29 settembre 2000, bollettino ufficiale della regione Marche, anno XXXI • n. ed. s. 15; Regione Marche: Ancona, Italy, 2000; Available online: http://www.ahrcos.it/newsite/terremoto/Codice_di_pratica_terremoto_umbromarchigiano97.pdf (accessed on 8 July 2000). (In Italian)
27. Binda, L.; Zanzi, L.; Lualdi, M.; Condoleo, P. The use of georadar to assess damage to a masonry Bell Tower in Cremona, Italy. *NDT E Int.* **2005**, *38*, 171–179. [CrossRef]
28. Binda, L.; Lualdi, M.; Saisi, A. Investigation strategies for the diagnosis of historic structures: On-site tests on Avio Castle, Italy, and Pišece Castle, Slovenia. *Can. J. Civ. Eng.* **2008**, *35*, 555–566. [CrossRef]
29. Candela, M.; Cattari, S.; Lagomarsino, S.; Fonti, R. Prove in situ per la caratterizzazione della muratura in pietrame grezzo alle azioni nel piano e fuori dal piano. In Proceedings of the Conference of the Italian National Association of Earthquake Engineering (ANIDIS), Padua, Italy, 30 June–4 July 2013; pp. 1–10. (In Italian).
30. Vintzileou, E. Testing Historic Masonry Elements and/or Building Models. In *Perspectives on European Earthquake Engineering and Seismology*; Ansal, A., Ed.; Springer International Publishing: Cham, Switzerland, 2014; Volume 1, Chapter 8. [CrossRef]
31. Marastoni, D.; Benedetti, A.; Pelà, L.; Pignagnoli, G. Torque Penetrometric Test for the in-situ characterisation of historical mortars: Fracture mechanics interpretation and experimental validation. *Constr. Build. Mater.* **2017**, *157*, 509–520. [CrossRef]
32. Ilharco, T.; Guedes, J.M.; Arêde, A.; Pauepério, E.; Costa, A.G. Analysis and diagnosis of timber structures in Porto Historical Centre. In Proceedings of the 6th International Conference on Structural Analysis of Historic Construction, Bath, UK, 2–4 July 2008; D'Ayala, D., Fodde, E., Eds.; Taylor and Francis Group: London, UK, 2008; pp. 653–661, ISBN 978-0-415-46872-5.
33. Piazza, M.P.; Riggio, M.R.; Brentari, G.B. Strengthening and control methods for old timber trusses: The queen-post truss of the Trento theatre. In Proceedings of the 4th International Conference on Structural Analysis of Historic Construction, Padua, Italy, 10–13 November 2004; Modena, C., Lourenço, P., Roca, P., Eds.; Taylor and Francis Group: London, UK, 2005; pp. 957–965.
34. Kloiber, M.; Tippner, J.; Praus, L.; Hrivinák, J. Experimental Verification of a New Tool for Wood Mechanical resistance measurement. *Wood Res.* **2012**, *57*, 383–398. Available online: http://www.woodresearch.sk/wr/201203/05.pdf (accessed on 8 July 2012).

applied sciences

MDPI

Article

Simulation of Three Constitutive Behaviors Based on Nonlinear Ultrasound

Zaifu Zhan, Shen Wang *, Fuping Wang, Songling Huang, Wei Zhao and Zhe Wang

State Key Laboratory of Power System, Department of Electrical Engineering, Tsinghua University, Beijing 100084, China; zhanzf18@mails.tsinghua.edu.cn (Z.Z.); wangfuping97@mails.tsinghua.edu.cn (F.W.); huangsling@tsinghua.edu.cn (S.H.); zhaowei@mail.tsinghua.edu.cn (W.Z.); z-w16@mails.tsinghua.edu.cn (Z.W.)

* Correspondence: wangshen@mail.tsinghua.edu.cn

Received: 12 February 2020; Accepted: 6 March 2020; Published: 13 March 2020

check for updates

Abstract: Nonlinear ultrasound has attracted more and more attention. In classical acoustic nonlinear theory, the source of nonlinearity is the change of constitutive relation of materials. Structure response that distorts after a single tone ultrasound wave is important to detect imperfection. This is rarely found in current simulations. The current simulation always introduces defects which do not match to the classical acoustic nonlinear theory. In this manuscript, the recurrence expressions of three kinds of imperfect materials for subroutine are given. The verifying simulation model that is used for verifying recurrence equations and wave propagation model that are used for analysing the process of ultrasonic propagation are established. The results show that the two constitutive models are effective in the verifying simulation and the hysteresis material has some special characteristics. Finally, ultrasonic propagation in two types of materials produce the expected harmonics, which build foundations for simulations of nonlinear ultrasound.

Keywords: classical acoustic nonlinearity; constitutive behavior; ultrasonic testing

1. Introduction

Under cyclic loads or harsh environments, no matter what materials, parts and complete structures will undergo material properties degradation. Continuous degradation leads to predictable and unpredictable problems, sometimes with serious consequences. Therefore, it is necessary to do health testing for structures or materials in practical engineering [1–5].

The early degradation of the mechanical properties takes up most of the life of the structures. Generally in metals, the damage first appears in the form of dislocation substructures such as veins and persistent slip bands (PSBs), and these accumulate at grain boundaries to produce strain localization and then further produce micro-cracks. This process accounts for 80–90% [6,7] of the life of the whole structure when the metal material structure is bearing a cyclic load or harsh environment. When the cracks show, the life the of material is almost over, so the testing of the early properties of the materials is particularly important.

With the development of ultrasonic non-destructive testing technology, linear ultrasound cannot satisfy the current testing needs, because linear ultrasound technology has no high sensitivity in the early stage of deterioration of material mechanical properties [8]. Studies show that the dislocation substructures and resulting microplastic deformation do not cause large changes in the macroscopic properties such as elastic modulus, sound speed and attenuation of a material [9,10], thus the changes in the linear ultrasonic quantities are not large enough to be accurately measured with conventional linear ultrasound.

The testing method of nonlinear ultrasound has attracted considerable attention in recent years [3,11–16]. Ultrasonic waves are highly sensitive to material damage, propagate in all directions, spread quickly and have high penetration into materials [17]. The nonlinear phenomenon manifests itself with second, third and even more harmonic generation: when a sinusoidal ultrasonic wave of finite amplitude at the f frequency crosses a nonlinear elastic material, harmonics at the $2f$,$3f$ (et al.) frequency are generated [18]. In refs [8–10], experimental results and physical models show that the harmonic amplitude is directly related to material damage. Therefore, it can be expected that the degree of material degradation can be evaluated by measuring the nonlinearity of the ultrasonic wave propagated through the target material.

The sources of nonlinearity can be divided into classical acoustic nonlinearity and contact acoustic nonlinearity (CAN). CAN manifests the phenomenon of nonlinearity by introducing a micrometer- or even nanometer-level small crack, which has been done in many simulations [17,19–23] and experiments [17,24–26]. However, it is known that when the material is under a load or in a severe environment, not only a small area of the material is damaged (e.g., micro-cracks), but also every under-service part of a structure is damaged to different degrees, which is rarely discovered in the current simulation. It is difficult to decide where to set micro-cracks in simulation for the structure under uniform load, but the material or structure must be damaged in service, so it is necessary to study the degradation of the mechanical properties of the material or structure.

The work of this manuscript is based on classical acoustic nonlinearity. When the material is undamaged, it complies with the linear constitutive relation. The wave propagation in the material is linear. However, when the material is damaged, the mechanical constitutive behavior no longer obeys the linear relation and the relation is complex and nonlinear [16,27,28]. As a result, the distortion of the wave form appears and the harmonics are derived because the constitutive relation is no longer linear.

Different materials exhibit different constitutive behaviors when damaged. The coefficient of material constitutive relation is also different for different damage degrees. In current simulation research, no one has carefully studied the constitutive relationship of materials. In addition, few researchers has verified the constitutive behavior of material in the simulation. Almost all simulations show the nonlinear results directly, but it is not known whether the nonlinearity is caused by the nonlinear constitutive relationship. In this manuscript, subroutines are written to defined three constitutive relationships which are first and second order perturbation of classic nonlinearity and nonlinear hysteresis respectively. Ref [28] compares linear wave propagation of a monofrequency signal with the results for the various nonlinear and hysteretic contributions. The results of the comparison are shown in Figure 1.

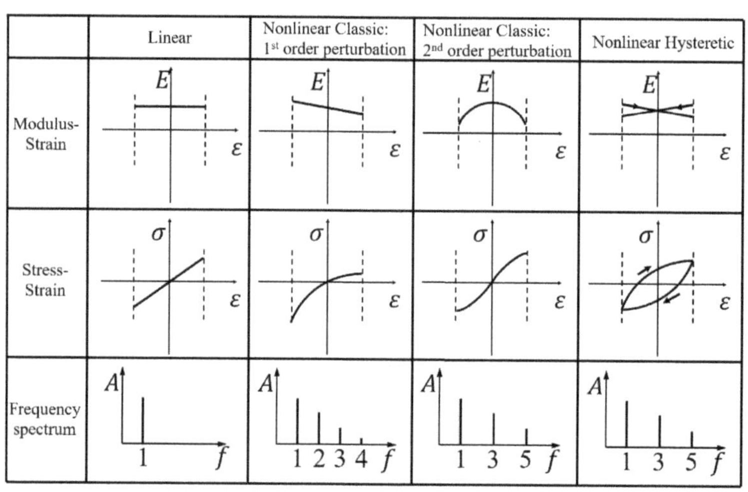

Figure 1. Schematic overview of the nonlinear contributions to the constitutive equation and its implication for dynamic one-dimensional wave propagation of a finite-amplitude monofrequency signal.

The remainder of this manuscript is organized as follows: Section 2 introduces not only the basic theoretical background of nonlinear ultrasound and higher harmonics generation in theory but also three constitutive behaviors which are written by a subroutine in simulations. In Section 3, a finite element model of the propagation of nonlinear longitudinal waves is described in detail. finite element method (FEM) simulation results are presented and discussed in Section 4, and conclusions are drawn and future studies proposed in Section 5.

2. Classical Nonlinear Elasticity

In this section, the derivation of nonlinear wave equations in classical nonlinearity will be reviewed, then we will introduce the constitutive relation of general materials. The general constitutive relation can be decomposed to three basic constitutive relations.

In order to save computing resources, the stress is calculated by strain thorough recurrence expression of the stress-strain relationship in simulations. Therefore, the recurrence relation of each constitutive relation will be given in the follow subsection.

In order to unify the symbols in equations and make it easy for readers, the parameters involved in this paper and their meanings are shown in Table 1.

Table 1. Main symbols used in the equation.

Free energy	F
Lamé coefficients	λ , μ
Strain tensor	ε_{ik}
Displacement component	u_i
Cartesian coordinate	x_i
Quadratic nonlinear perturbation coefficient	β
Cubic nonlinear perturbation coefficient	ϵ
Hysteresis	α
Simplified first perturbation coefficient	$C_1(=-\frac{1}{2}E_0\beta)$
Simplified second perturbation coefficient	$C_2(=-\frac{1}{3}E_0\epsilon)$
Simplified nonlinear hysteretic coefficient	$C_3(=\frac{1}{2}E_0\alpha)$
Young's modulus	E
Kronecker symbol	δ_{ij}
Strain rate	$\dot{\varepsilon}=d\varepsilon/dt$

2.1. Nonlinear Wave Equation

This subsection will begin with Hooke's law and simply study the source of classical acoustic nonlinearity through theoretical analysis.

This is the general expression for the free energy of a deformed isotropic body [27]:

$$F=F_0+\frac{1}{2}\lambda\varepsilon_{ii}^2+\mu\varepsilon_{ik}^2 \quad (1)$$

where, ε_{ik} is the strain tensor whose linear expression is [27]:

$$\varepsilon_{ik}=\frac{1}{2}(\frac{\partial u_i}{\partial x_k}+\frac{\partial u_k}{\partial x_i}) \quad (2)$$

When nonlinearity is considered, the strain tensor should be expanded [27]:

$$\varepsilon_{ik}=\frac{1}{2}(\frac{\partial u_i}{\partial x_k}+\frac{\partial u_k}{\partial x_i}+\frac{\partial u_l}{\partial x_k}\frac{\partial u_l}{\partial x_i}) \quad (3)$$

If only a one-dimensional situation is considered, the linear wave equation:

$$\frac{\partial^2 u}{\partial t^2} - c_0 \frac{\partial^2 u}{\partial x^2} = 0 \tag{4}$$

should be amended to:

$$\frac{\partial^2 u}{\partial t^2} - c_0 (1 - \beta \frac{\partial u}{\partial x} + \epsilon (\frac{\partial u}{\partial x})^2) \frac{\partial^2 u}{\partial x^2} = 0 \tag{5}$$

There is another simpler theoretical explanation that when the material is undamaged, the relationship between stress and strain is linear:

$$\sigma = E\varepsilon \tag{6}$$

When the material is damaged, however, the relationship between stress and strain is no longer linear:

$$\sigma = E\varepsilon(1 + \beta\varepsilon + \cdots) \tag{7}$$

Substituting this relation into the one-directional wave equation:

$$\rho \frac{\partial^2 u}{\partial t^2} = \frac{\partial \sigma}{\partial x} \tag{8}$$

Then the nonlinear wave equation is obtained:

$$\rho \frac{\partial^2 u}{\partial t^2} = A \frac{\partial^2 u}{\partial x^2} + B \frac{\partial u}{\partial x} \frac{\partial^2 u}{\partial x^2} + \cdots \tag{9}$$

Cantrell [7,29] et al. also established a similar second nonlinear ultrasonic wave equation:

$$\rho \frac{\partial^2 u}{\partial t^2} = K_2 \frac{\partial^2 u}{\partial x^2} + (3K_2 + K_3) \frac{\partial u}{\partial x} \frac{\partial^2 u}{\partial x^2} \tag{10}$$

Therefore, it is found that when the material constitutive relation changes, the wave equation is no longer linear. Based on this theory, when the material is damaged, the input of the system is monochromatic, and the output must contain harmonics.

The derivation of wave equation above does not consider the nonlinear hysteresis for simplicity.

2.2. General Constitutive Relationship

To first approximation, the one-dimensional constitutive relation between the stress σ and the strain ε used in simulations of the dynamic behavior of solids can be expressed as follows [28]:

$$\sigma = \int E(\varepsilon, \dot{\varepsilon}) d\varepsilon \tag{11}$$

with E the nonlinear and hysteretic modulus given by [28]

$$E(\varepsilon, \dot{\varepsilon}) = E_0\{1 - \beta\varepsilon - \epsilon\varepsilon^2 - \alpha[\Delta\varepsilon + \varepsilon(t) sign(\dot{\varepsilon})] + \cdots\} \tag{12}$$

where E_0 is the linear modulus, $\Delta\varepsilon$ is the local strain amplitude over the previous period, $sign(\dot{\varepsilon}) = 1$ if $\dot{\varepsilon} > 0$ and $sign(\dot{\varepsilon}) = -1$ if $\dot{\varepsilon} < 0$.

In simulation, the component of constitutive relation is needed:

$$\sigma_{ij} = \lambda \delta_{ij} \varepsilon_{kk} + 2\mu \varepsilon_{ij} \tag{13}$$

where δ_{ij} is the Kronecker Symbol, $\delta_{ij} = 1$ if $i = j$ and $\delta_{ij} = 0$ if $i \neq j$.

2.3. Nonlinear Classic: 1st Perturbation (β)

If only the first perturbation is considered, the nonlinear modulus will be simplified to:

$$E(\varepsilon, \dot{\varepsilon}) = E_0(1 - \beta\varepsilon) \tag{14}$$

The one-dimensional constitutive relation between the stress σ and the strain ε should be:

$$\sigma = \int E(\varepsilon, \dot{\varepsilon}) d\varepsilon = E_0\varepsilon - \frac{1}{2}\beta E_0 \varepsilon^2 \tag{15}$$

In the simulations, the code of the recursive constitutive equations are programmed as follows:

$$\begin{cases} \Delta\varepsilon_{ij}^{m+1} = \varepsilon_{ij}^{m+1} - \varepsilon_{ij}^{m} \\ \lambda_{ij}^{m+1} = C_1 \dfrac{\varepsilon_{ij}^{m+1} + \varepsilon_{ij}^{m}}{2} + \lambda_0 \\ \mu_{ij}^{m+1} = C_1 \dfrac{\varepsilon_{ij}^{m+1} + \varepsilon_{ij}^{m}}{2} + \mu_0 \\ \sigma_{ij}^{m+1} = \sigma_{ij}^{m} + \delta_{ij}(\lambda_{11}^{m+1}\Delta\varepsilon_{11}^{m+1} + \lambda_{22}^{m+1}\Delta\varepsilon_{22}^{m+1} + \lambda_{33}^{m+1}\Delta\varepsilon_{33}^{m+1}) + 2\mu_{ij}^{m+1}\Delta\varepsilon_{ij}^{m+1} \end{cases} \tag{16}$$

where, λ_0 and μ_0 are Lamé coefficients when the material is linear.

2.4. Nonlinear Classic: 2nd Perturbation (ϵ)

If only the second perturbation is considered, the nonlinear modulus will be simplified to:

$$E(\varepsilon, \dot{\varepsilon}) = E_0(1 - \epsilon\varepsilon^2) \tag{17}$$

The one-dimensional constitutive relation between the stress σ and the strain ε should be:

$$\sigma = \int E(\varepsilon, \dot{\varepsilon}) d\varepsilon = E_0\varepsilon - \frac{1}{3}\epsilon E_0 \varepsilon^3 \tag{18}$$

In the simulations, the code of the recursive constitutive equations are programmed as follows:

$$\begin{cases} \Delta\varepsilon_{ij}^{m+1} = \varepsilon_{ij}^{m+1} - \varepsilon_{ij}^{m} \\ \lambda_{ij}^{m+1} = C_2 (\dfrac{\varepsilon_{ij}^{m+1} + \varepsilon_{ij}^{m}}{2})^2 + \lambda_0 \\ \mu_{ij}^{m+1} = C_2 (\dfrac{\varepsilon_{ij}^{m+1} + \varepsilon_{ij}^{m}}{2})^2 + \mu_0 \\ \sigma_{ij}^{m+1} = \sigma_{ij}^{m} + \delta_{ij}(\lambda_{11}^{m+1}\Delta\varepsilon_{11}^{m+1} + \lambda_{22}^{m+1}\Delta\varepsilon_{22}^{m+1} + \lambda_{33}^{m+1}\Delta\varepsilon_{33}^{m+1}) + 2\mu_{ij}^{m+1}\Delta\varepsilon_{ij}^{m+1} \end{cases} \tag{19}$$

2.5. Nonlinear Hysteretic (α)

If only the hysteresis of material is considered, the nonlinear modulus will be simplified to:

$$E(\varepsilon, \dot{\varepsilon}) = E_0\{1 - \alpha[\Delta\varepsilon + \varepsilon(t) sign(\dot{\varepsilon})]\} \tag{20}$$

In simulations, increment step and mesh elements are both set very small. Therefore, eliminating the $\Delta\varepsilon$ can be simple and satisfy the accuracy at the same time. The one-dimensional constitutive relation between the stress σ and the strain ε should be:

$$\sigma = \int E(\varepsilon, \dot{\varepsilon}) d\varepsilon = \begin{cases} E_0\varepsilon - \frac{1}{2}E_0\alpha\varepsilon^2, & \dot{\varepsilon} > 0 \\ E_0\varepsilon + \frac{1}{2}E_0\alpha\varepsilon^2, & \dot{\varepsilon} < 0 \end{cases} \quad (21)$$

In the simulations, the code of recursive constitutive equations are programmed as follows:

$$\begin{cases} \Delta\varepsilon_{ij}^{m+1} = \varepsilon_{ij}^{m+1} - \varepsilon_{ij}^{m} \\ \lambda_{ij}^{m+1} = c_3' C_3 \dfrac{\varepsilon_{ij}^{m+1} + \varepsilon_{ij}^{m}}{2} + \lambda_0 \quad (c_3' = 1, if \Delta\varepsilon_{ij}^{m+1} < 0 \quad and \quad c_3' = -1, if \Delta\varepsilon_{ij}^{m+1} > 0) \\ \mu_{ij}^{m+1} = c_3' C_3 \dfrac{\varepsilon_{ij}^{m+1} + \varepsilon_{ij}^{m}}{2} + \mu_0 \\ \sigma_{ij}^{m+1} = \sigma_{ij}^{m} + \delta_{ij}(\lambda_{11}^{m+1}\Delta\varepsilon_{11}^{m+1} + \lambda_{22}^{m+1}\Delta\varepsilon_{22}^{m+1} + \lambda_{33}^{m+1}\Delta\varepsilon_{33}^{m+1}) + 2\mu_{ij}^{m+1}\Delta\varepsilon_{ij}^{m+1} \end{cases} \quad (22)$$

3. Numerical Simulation

Two two-dimensional (2D) finite element method (FEM) models were built with Abaqus/CAE and dynamic simulations were conducted with Abaqus/Explicit software.

The Explicit solver of Abaqus specializes in solving complex problems. It does not solve the coupled equation of systems or generate the total stiffness matrix in each increment step, but recurs the result in a small increment step in the time domain. It only calculates the total energy in the simulation to determine whether there is divergence in each increment step.

What is more, the solver provides an interface—VUMAT, which is used to define the mechanical constitutive behavior of a material thorough FORTRAN language. Recursive expressions mentioned in Section 2 are achieved by subroutine.

There are two 2D FEM models were build. The first model was built for verifying the correctness of the codes. The propagation of the longitudinal wave was simulated in the second model.

3.1. Verifying the Model

In order to verify the reliability of the program, the following model was established.

The 2D verifying model was a square with 10 mm side length as shown in Figure 2. The boundary condition set on the left side was encastré which means that the left edge could not move in the x and y directions during the whole process. Displacement in the y direction of the top and bottom edges was set to 0 so it could not move in the y direction. In the right side of the model, the displacement boundary condition was used to produce deformation so as to analyze the relation between stress and strain. The displacement signal in the right side of the verifying model was a triangular wave which is shown in Figure 3. The amplitude of the boundary condition was set to 0.25 mm.

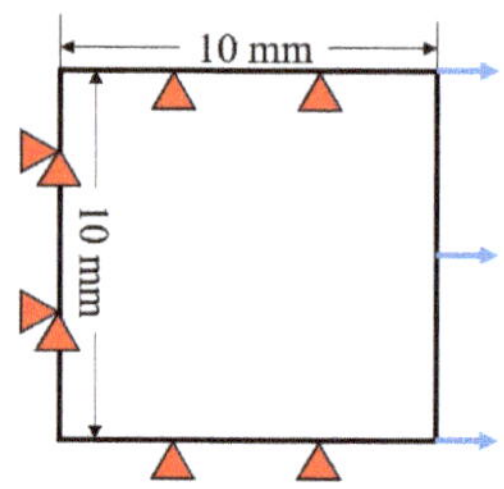

Figure 2. Geometry of the verifying model for simulations.

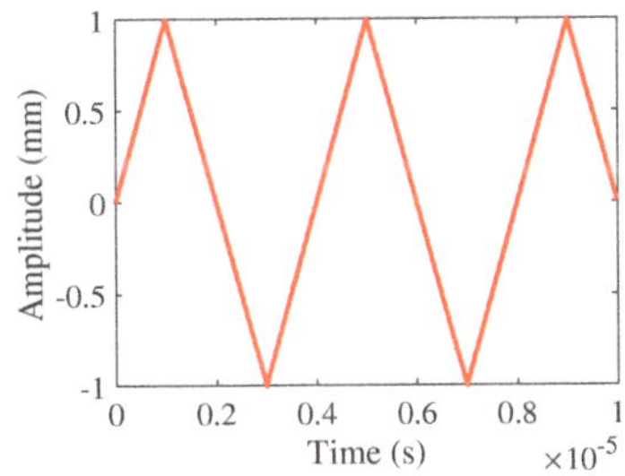

Figure 3. Displacement signal for the right side boundary condition of the verifying model.

We chose steel as the simulation material because it is widely used. The material properties are shown in Table 2. In this manuscript, all materials in the simulation were steel.

Table 2. Material properties in simulations.

Material Type	Mass Density	Young's Modulus	Poisson's Ratio
Steel	7800 kg/m^3	2×11 Pa	0.3

Abaqus software contains a wealth of mesh elements libraries. We chose CPE4R (i.e., linear reduction integral plane strain type) as the element type. It should be noted that the linear reduced integral refers to the linear interpolation used by the mesh elements in the calculation, which has nothing to do with the nonlinear source mentioned in this manuscript, but only for saving calculation resources and meeting the algorithm of Abaqus/Explicit solver.

The most important consideration is that only one mesh element was formed in the verifying model which means the whole model was an element. It can be clearly known that the constitutive relation does change in simulations by observing the stress and strain of an element. The whole simulation can only be correct when the constitutive behavior of one element is correct.

3.2. Wave Propagation Model

After the code written for the different constitutive behavior was verified, the wave propagation in a solid was simulated to study different nonlinear phenomena from different materials whose constitutive behavior was different.

The 2D model was 40 mm long and 20 mm thick as shown in Figure 4 The material properties are in Table 2. The boundary conditions of four sides in the model were free boundary conditions which means the stress on the boundary was zero.

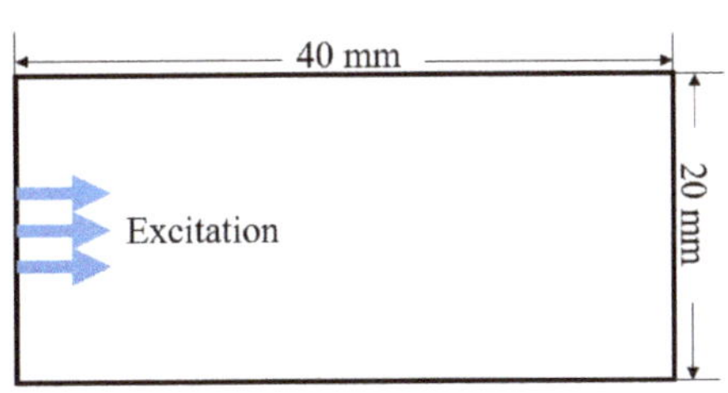

Figure 4. The 2D wave propagation model and the place of excitation.

The excitation was applied in the left edge of model within the width of 6 mm and the excitation amplitude was 90 nm. The excitation signal was a five cycle tone-burst, as shown in Figure 5. The window function was used to prevent spectrum leakage. The excitation signal adopted a frequency of 3×10^6 Hz, and its expression was as follows:

$$y = sin(2\pi ft)(0.08 + 0.46(1 - cos(2\pi ft/5))) \tag{23}$$

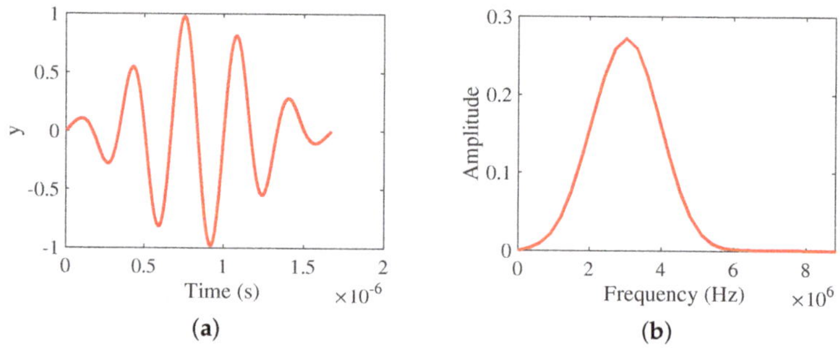

Figure 5. (**a**) Temporal waveform of the excited tone burst signal. (**b**) Frequency spectrum of the excited tone burst signal.

Mesh generation is important in FEM. The mesh element size should be much smaller than the wavelength, otherwise there is no wave propagation phenomenon. The accuracy requires that the smallest wavelength λ_{min} must be correctly sampled to describe the propagating mode in the frequency range for analysis. The denser the mesh, the more accurate the result. However, more density means more computing resource usage. The mesh element size should balance the accuracy of results with computing resources usage. Therefore, the spatial discretization of δx_1 and δx_2 of each element of the mesh satisfies the condition [30]:

$$\frac{\lambda_{min}}{max(\delta x_1, \delta x_2)} > 10 \tag{24}$$

Due to the need to measure harmonics, the $max(\delta x_1, \delta x_2)$ of mesh elements was set to 0.05 mm. The type of mesh element was CPE4R which specializes in plane strain problems.

Another stability condition is that the time step must be chosen so that no wave can propagate across one mesh spacing in less than one time step. Typically, the time discretization must satisfy the condition [30]:

$$\Delta t \leq 0.7 \frac{\min(\delta x_1, \delta x_2)}{c_l} \tag{25}$$

where c_l is the velocity of the longitudinal wave. Therefore, in order to be precise, the step time was set as 1×10^{-9} s, and the total simulation time was 8×10^{-6} s.

4. Simulation Result and Discussion

This section firstly presents and discusses the simulation results generated from three group verifying models. Simulation results from models of first and second order perturbation of classic nonlinearity agreed with the nonlinear theory. However, the results of nonlinear hysteresis had some differences from the ideal situation. Some analysis will be made based on the results. Finally, two stable codes of classic nonlinearity were used in the simulation of wave propagation. The growth of harmonics in different materials was obtained by multiple simulations.

4.1. Verification Simulation

4.1.1. Nonlinear Classic: 1st Perturbation (β)

When the first order perturbation was added to the verifying model, three deformation results of model are shown in Figure 6.

The solid line represents the deformation when the displacement was zero. The thick dotted line and thin dotted line are deformation of maximum tension and compression respectively.

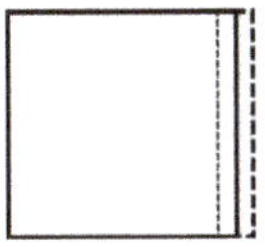

Figure 6. Deformation of verifying model.

From the deformation in Figure 6, it can be seen that verifying model only deformed in the x direction. Owing to the whole model being just one element and none of the other three sides can move, this element only produced strain in x direction, which was ε_{11}.

In Figure 7, two groups of simulations show that the constitutive relation was changed by the subroutine. Figure 7a,b is the change of strain in the simulation process. They were exactly the same because of the boundary condition. The stress of the verifying model was very different as shown in Figure 7c,d. We can see that stress-time curve did distort when $C_1 = -4 \times 10^6$. It is obvious that the upper part and the lower part of the stress were no longer symmetrical. Although it can be seen that the stress changed obviously, it was not intuitive enough. Every point of stress and strain at every moment can be found and connected, which is shown in Figure 7e,f. It is clear that the stress-strain relationship was no longer linear. Therefore, it can be confirmed that the constitutive behavior of the material was indeed changed to what we want.

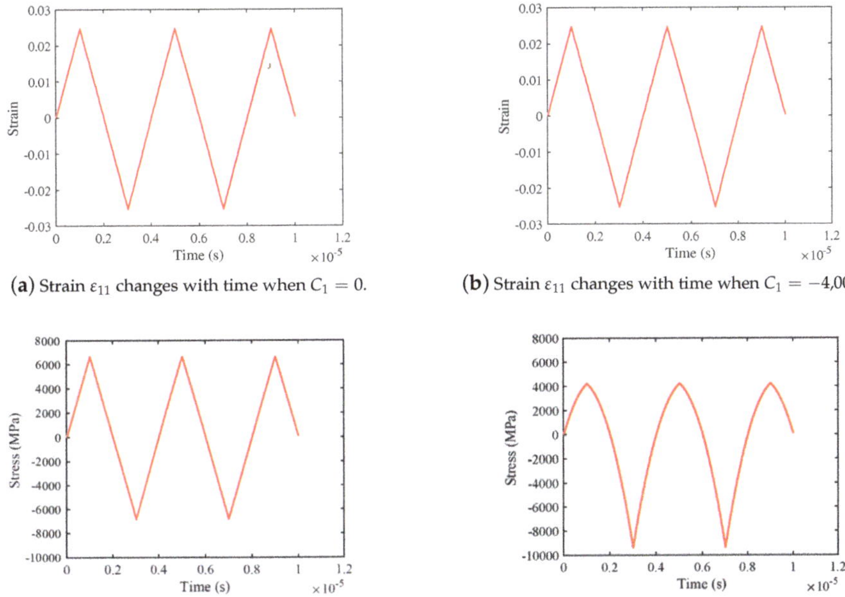

(a) Strain ε_{11} changes with time when $C_1 = 0$. **(b)** Strain ε_{11} changes with time when $C_1 = -4{,}000{,}000$.

(c) Stress σ_{11} changes with time when $C_1 = 0$. **(d)** Stress σ_{11} changes with time when $C_1 = -4{,}000{,}000$.

Figure 7. *Cont.*

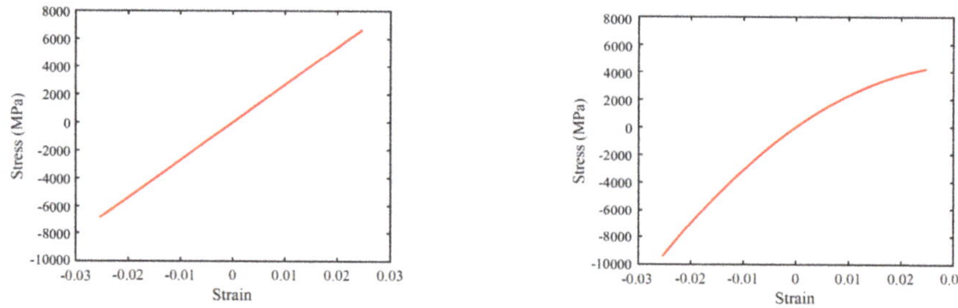

(e) Relation of stress and strain when $C_1 = 0$.

(f) Relation of stress and strain when $C_1 = -4{,}000{,}000$.

Figure 7. Comparative analysis of strain and stress when C_1 is equal to 0 and $-4{,}000{,}000$ respectively.

A series of stress-strain curves could be obtained by changing the value of C_1 as shown in Figure 8. It can be seen that the degree of stress-strain nonlinearity increased with the increase of the first-order perturbation. In the following simulation, several values of perturbation were selected to simulate the propagation of ultrasound.

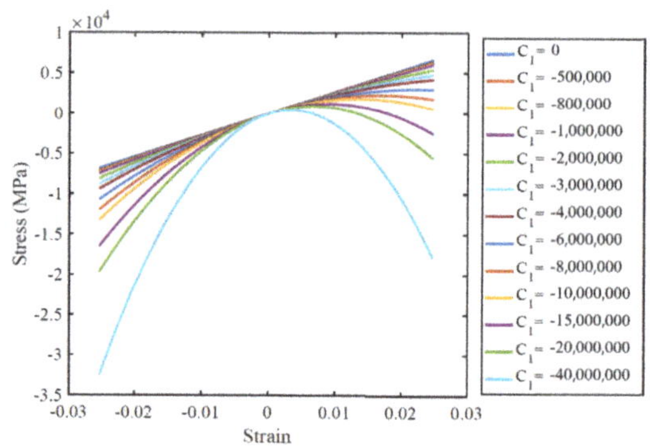

Figure 8. Stress-strain curves under different first order perturbations.

Simulation results of Figures 7 and 8 were consistent with the theory of Ref [28], which confirms the correctness of code of the subroutine.

4.1.2. Nonlinear Classic: 2nd Perturbation (ϵ)

When only adding the second order perturbation to the verifying model, three comparison results of verifying model are shown in Figure 9 like the last subsection. In the same way, the comparison between two groups of simulations verified the code of the second perturbation.

It is clear that the constitutive relationship changed by comparison in Figure 9. Figure 9a,b is the change of strain in the simulation process. The stress of the verifying model was very different as shown in Figure 9c,d when C_2 was different. It is certain that stress-time curve distorted when $C_2 = -3 \times 10^8$, which led to the change of constitutive behavior as shown in Figure 9e,f.

In addition, lots of simulations were conducted and the curve of relation between strain and stress were obtained as shown in Figure 10. Figure 10 shows that the constitutive relation became more and more complicated with the change of the perturbation coefficient C_2. Therefore, it can be confirmed that the constitutive behavior of the material was indeed changed to meet the expected goal.

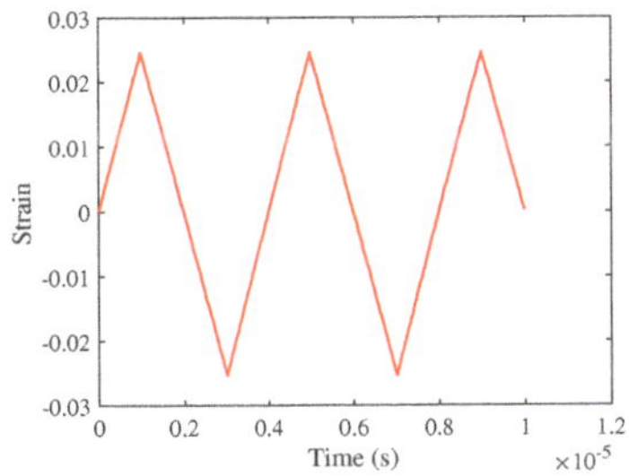

(**a**) Strain ε_{11} changes with time when $C_2 = 0$.

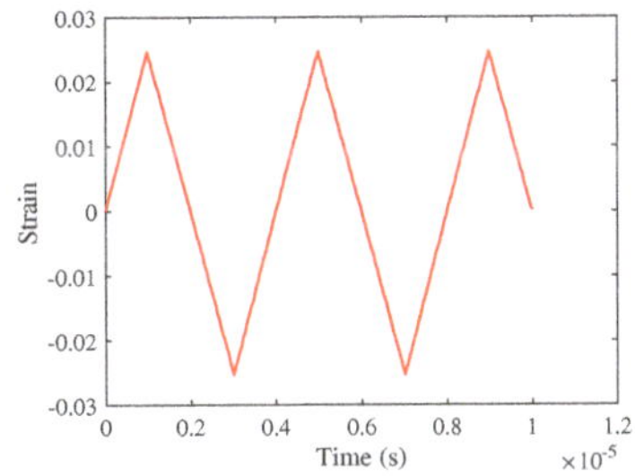

(**b**) Strain ε_{11} changes with time when $C_2 = -3 \times 10^8$.

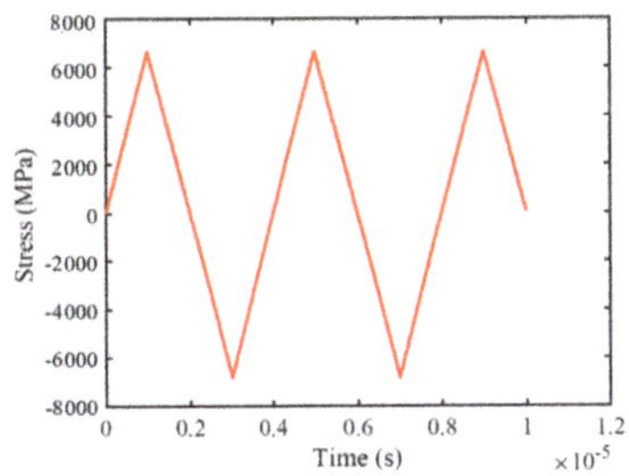

(**c**) Stress σ_{11} changes with time when $C_2 = 0$.

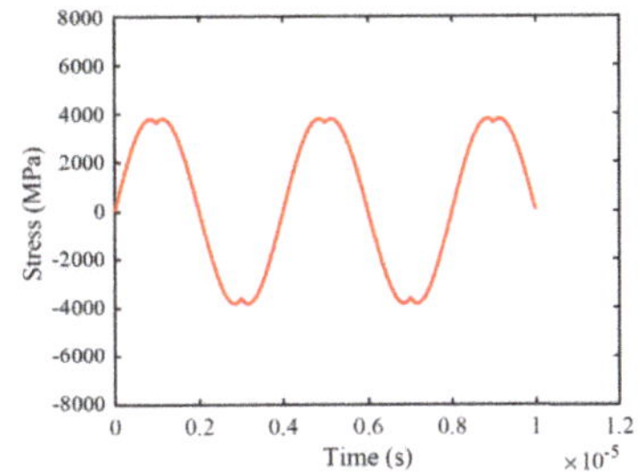

(**d**) Stress σ_{11} changes with time when $C_2 = -3 \times 10^8$.

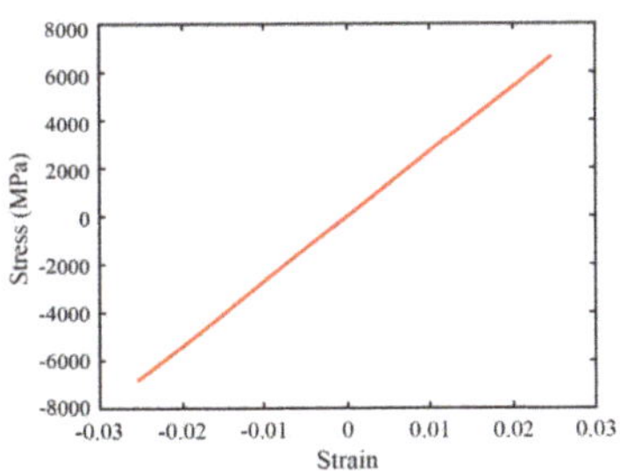

(**e**) Relation of stress and strain when $C_2 = 0$.

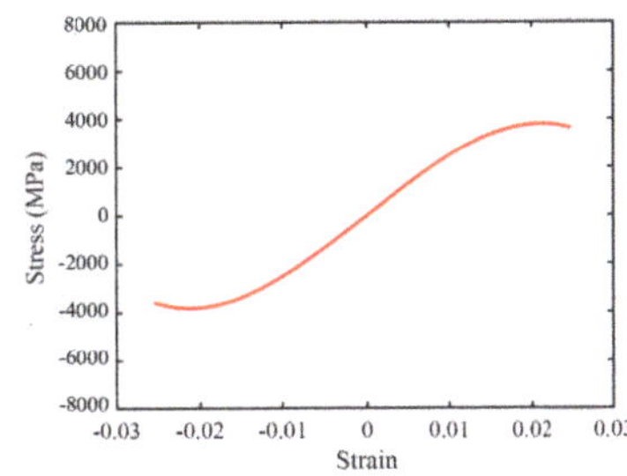

(**f**) Relation of stress and strain when $C_2 = -3 \times 10^8$.

Figure 9. Comparative analysis of strain and stress when C_2 is equal to 0 and -3×10^8 respectively.

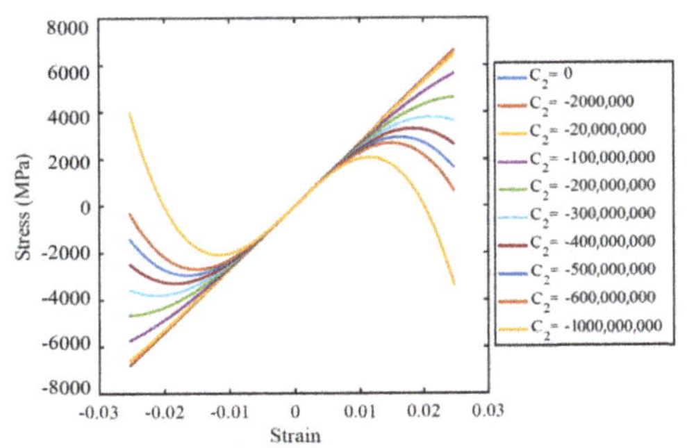

Figure 10. Stress-strain curves under different second order perturbations.

In Figure 9d, waveform distortion was more serious by comparison with Figure 7d. However, the stress upper and lower waveform was symmetrical in this simulation. This phenomenon also explains why the second order perturbation only produced odd harmonics which was exhibited in Figure 1 of the Introduction section. This was the main difference between the first order perturbation and the second order perturbation. It is a reminder that the properties of materials can be judged by a stress curve. For example, in Ref [21], a strain curve is used to measure the damage of materials.

Figure 1 is the result of theory from Ref [20] which is consistent with the result of simulation as shown in Figures 9 and 10, which confirmed the correctness of the code of the subroutine.

4.1.3. Nonlinear Hysteresis (α)

In th everifying model, the nonlinear hysteresis was only added to a constitutive relationship. The modified constitutive relationship is Equation (18).

Figure 11a is the result of stress of verifying model when $C_3 = 1 \times 10^7$. The distortion of stress curve was more severe compared with the previous two results. It can be found that, in the process of tension and compression, the stress curve showed an upward trend. Similarly, the whole curve of stress-strain relationship rose in the process of deformation which is shown in Figure 11b. This is because of the recursive equations and the curve must start at the origin point (0,0).

Simulations were conducted and the curves of relation between strain and stress are shown in Figure 12.

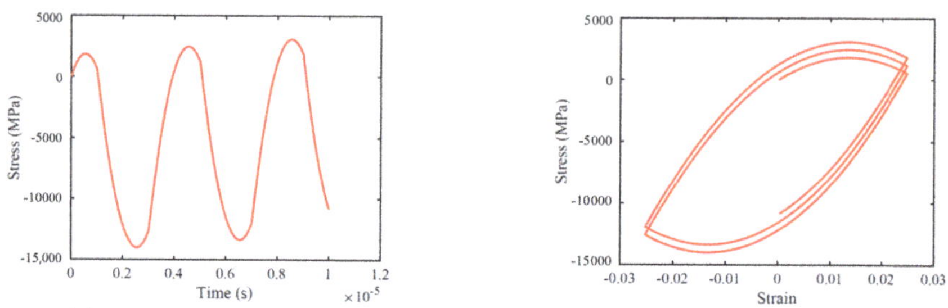

(**a**) Stress ε_{11} changes with time. (**b**) Relation of stress and strain.

Figure 11. Simulation result of hysteretic material in the verifying model when $C_3 = 1 \times 10^7$.

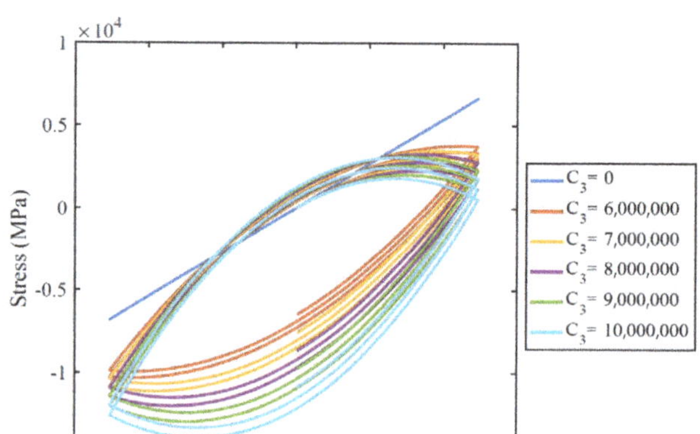

Figure 12. Stress-strain curves for different hysteresis.

The result in Figures 11 and 12 also means that repeated tension and compression made the materials reach the ideal hysteretic properties in the degradation process. In the simulation, if the target is to study the constitutive relation of hysteretic materials, the best way is the look-up table method. However, this method is limited because different damage degrees correspond to different constitutive curves. The constitutive formula of hysteretic materials has been defined by Nazarov and Ostrovsky [31] in 1988 but there are several coefficients involved in the formula. These coefficients need to be measured by actual experiments. Therefore, it is difficult to involve several coefficients in the simulation before conducting experiments.

4.2. Wave Propagation Simulation

4.2.1. Nonlinear Classic: 1st Perturbation (β)

The models were built as described in Section 3.2. A group of simulations were conducted by changing the value of the simplified first perturbation coefficient C_1. C_1 was respectively set to 0, −2000, −20,000, −200,000, and −2,000,000. When C_1 equals 0, the material was not damaged.

The propagation of wave in simulation was exhibited by the stress nephogram which is more intuitive as shown in Figure 13. From this figure, it can be seen that the excitation produced not only longitudinal waves but also transverse waves, and the longitudinal wave beam spread in propagation. As a result, some longitudinal waves reflected in the upper and lower surfaces. It was difficult to find the difference from the stress nephogram after material constitutive relation changed. Therefore, the stress nephogram was not used to compare the result.

The resulting signal is shown in Figure 14 when $C_1 = -2{,}000{,}000$. The second and third harmonics are shown very clearly. Due to the phenomenon of diffraction in the process of propagation, however, the harmonic magnitude can be not directly compared because of diffraction in the process of wave propagation.

In order to characterize the accumulation of harmonics better in the process of ultrasonic propagation, the nonlinear parameter β' was introduced as nonlinear ultrasonic detection coefficients. The expression of β' is as follows:

$$\beta' = \frac{A_2}{A_1^2} \tag{26}$$

where, A_i is the amplitude of the ith harmonic.

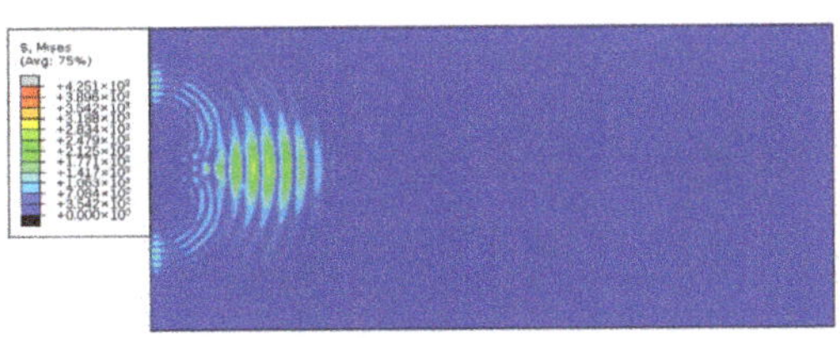

(a) t = 2.001 × 10^−6^ (s)

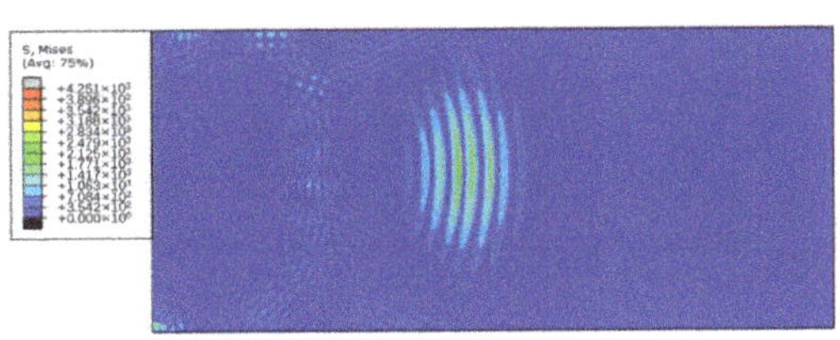

(b) t = 4.001 × 10^−6^ (s)

Figure 13. *Cont.*

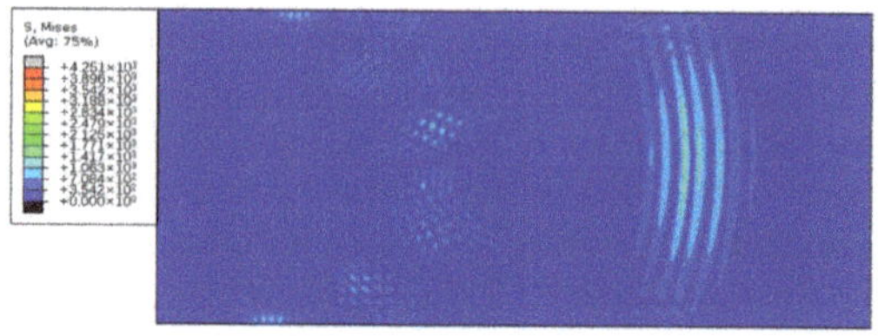

(c) t = 6.001 × 10^{-6} (s)

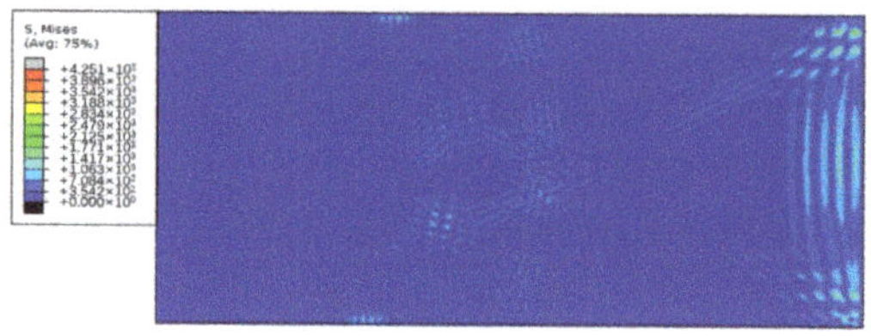

(d) t = 8 × 10^{-6} (s)

Figure 13. Stress nephogram at different times during ultrasonic propagation.

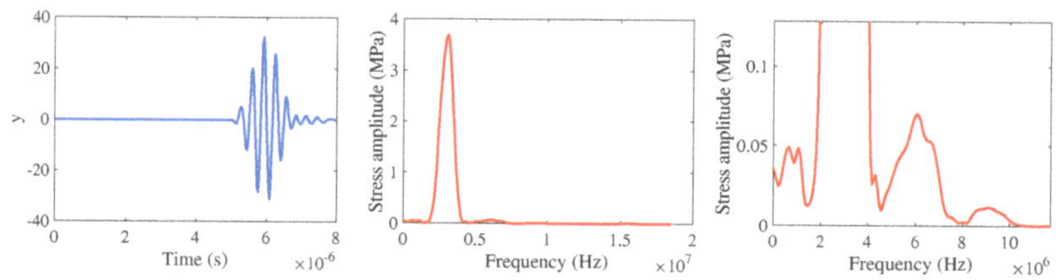

(a) (b) (c)

Figure 14. Simulation result of the first order perturbation when $C_1 = -2{,}000{,}000$: (**a**) Time domain signal. (**b**) Spectrum of signal. (**c**) Harmonic of signal.

Figure 15a shows that, under different perturbation coefficients, the nonlinear ultrasonic detection coefficient β' changed with wave propagation distance. When the perturbation coefficient C_1 was larger than a certain value, the coefficient β' increased with the propagation distance. Figure 15b is the result of Ref [17] which also used the subroutine to define the material. However, it did not give a definite value of C_1. The correctness of simulation result was verified by contrast, which also verified the code that is clearly described in Section 2.

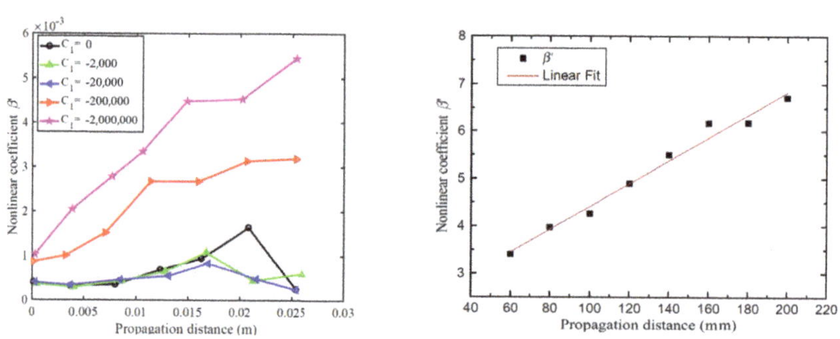

(a) (b)

Figure 15. Coefficient β' changes with propagation distance. (**a**) result of wave propagation model. (**b**) result from Ref [17].

4.2.2. Nonlinear Classic: 2nd Perturbation (ϵ)

A group of simulations in which the second perturbation coefficient C_2 was different were conducted by Abaqus software.

The resulting signal is shown in the Figure 16 when $C_2 = -3 \times 10^7$. It can be obviously seen that the third and fifth harmonics were generated by the second perturbation.

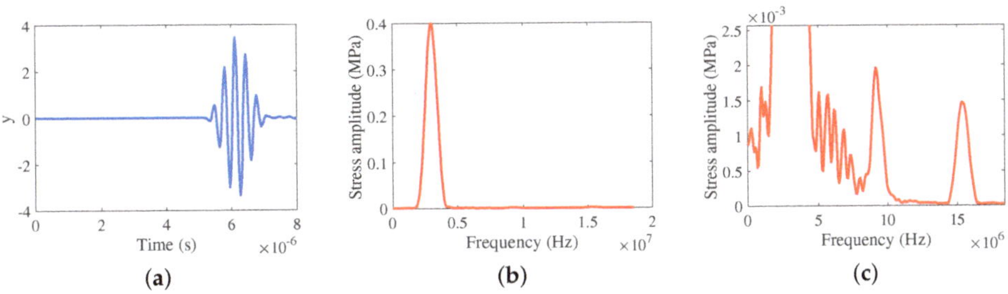

Figure 16. Simulation result of second order perturbation when $C_2 = -3 \times 10^7$: (**a**) Time domain signal. (**b**) Spectrum of signal. (**c**) Harmonic of signal.

Current finite element simulation software requires that the deformation of mesh elements in simulation should not be too large or it can not guarantee the correctness of result. Therefore, the mesh elements should not be too small. On the other hand, if the objective is to measure the third or fifth harmonics, mesh elements must be small enough as described in Equation (24). There is a contradiction between the two theories, so there was no need to compare the amplitude of the third and fifth harmonics in wave propagation.

5. Conclusions

In this manuscript, three types of constitutive relations are reviewed and their subroutines with recursive expression are written. Two kinds of simulation models are established. One is to verify the correctness of subroutines, and the other is to conduct ultrasonic simulation. From the result of the verifying model, the subroutines of the first and second order perturbations satisfy the nonlinear theory. Hysteretic material simulation shows the deterioration process of materials, but it cannot be directly used to simulate the damaged materials and some analyses are made on it. Finally, the simulation of the first and second order perturbation shows that the harmonics appear. Especially, nonlinear coefficient β' increases with propagation distance in first order perturbation simulations and odd harmonics of the second order perturbation simulations are obvious. The work in this manuscript provides a new idea for the simulation of nonlinear ultrasonic testing.

There is still some follow-up work to be studied. Firstly, although recurrence equations can improve the calculation speed, it is not suitable to study the hysteresis material directly. Secondly, the current FE simulation software cannot guarantee the accuracy when the mesh element deformation is too large. Thirdly, the recurrence equations are not verified in three-dimensional simulation.

Author Contributions: Conceptualization, Z.Z., S.W. and F.W.; data curation, Z.Z.; formal analysis, Z.Z.; funding acquisition, F.W. and S.H.; investigation, Z.Z. and S.W.; methodology, Z.Z.; project administration, S.W. and F.W.; resources, Z.Z., S.W. and F.W.; supervision, S.W. and F.W.; validation, Z.Z.; writing—original draft, Z.Z.; writing—review and editing, Z.Z., S.W., F.W., S.H., W.Z. and Z.W. All authors have read and agreed to the published version of the manuscript.

Funding: This research was funded by National Key Research and Development Program grant number 2018YFC0809002.

Acknowledgments: The author greatly appreciates the help of translator Mengyuan Cui.

Conflicts of Interest: The authors declare no conflicts of interest.

References

1. Farrar, C.R.; Worden, K. An introduction to structural health monitoring. *Philos. Trans. A Math. Phys. Eng. Sci.* **2007**, *365*, 303–315. [CrossRef] [PubMed]
2. Marcantonio, V.; Monarca, D.; Colantoni, A.; Cecchini, M. Ultrasonic waves for materials evaluation in fatigue, thermal and corrosion damage: A review. *Mech. Syst. Signal Process.* **2019**, *120*, 32–42. [CrossRef]
3. Blanloeuil, P.; Rose, L.; Veidt, M.; Wang, C.H. Time reversal invariance for a nonlinear scatterer exhibiting contact acoustic nonlinearity. *J. Sound Vib.* **2019**, *417*, 413–431. [CrossRef]
4. Aleshin, V.; Delrue, S.; Trifonov, A.; Matar, O.B.; Abeele, K.V.D. Two dimensional modeling of elastic wave propagation in solids containing cracks with rough surfaces and friction—Part I: Theoretical background. *Ultrasonics* **2017**, *82*, 11–17. [CrossRef]
5. Kim, J.; Park, S.J.; Yim, H. Nonlinear Resonance Vibration Assessment to Evaluate the Freezing and Thawing Resistance of Concrete. *Materials* **2019**, *12*, 325. [CrossRef]
6. Meyendorf, N.G.; Rösner, H.; Kramb, V.; Sathish, S. Thermo-acoustic fatigue characterization. *Ultrasonics* **2002**, *40*, 427–434. [CrossRef]
7. Cantrell, J.H. Substructural organization, dislocation plasticity and harmonic generation in cyclically stressed wavy slip metals. *Proc. R. Soc. A Math. Phys. Eng. Sci.* **2004**, *460*, 757–780. [CrossRef]
8. Nagy, P.B. Fatigue damage assessment by nonlinear ultrasonic materials characterization. *Ultrasonics* **1998**, *36*, 375–381. [CrossRef]
9. Herrmann, J.; Kim, J.Y.; Jacobs, L.J.; Qu, J.; Littles, J.W.; Savage, M.F. Assessment of material damage in a nickel-base superalloy using nonlinear Rayleigh surface waves. *J. Appl. Phys.* **2006**, *99*, 124913. [CrossRef]
10. Kim, J.Y. Experimental characterization of fatigue damage in a nickel-base superalloy using nonlinear ultrasonic waves. *J. Acoust. Soc. Am.* **2006**, *120*, 1266–1273. [CrossRef]
11. Raghavan, A. Guided-wave structural health monitoring. *Diss. Abstr. Int.* **2007**, *975*, 91–114. [CrossRef]
12. Solodov, I.Y.; Krohn, N.; Busse, G. CAN: An example of nonclassical acoustic nonlinearity in solids. *Ultrasonics* **2002**, *40*, 621–625. [CrossRef]
13. Zarembo, I.K.; Krasil'Nikov, V.A. Nonlinear Phenomena in the Propagation of Elastic Waves in Solids. *Sov. Phys. Uspekhi* **1970**, *102*, 778. [CrossRef]
14. Jhang, K.Y.; Kim, K.C. Evaluation of material degradation using nonlinear acoustic effect. *Ultrasonics* **1999**, *37*, 39–44. [CrossRef]
15. Zheng, Y.; Maev, R.G.; Solodov, I.Y. Nonlinear acoustic applications for material characterization: A review. *Can. J. Phys.* **1999**, *77*, 927–967. [CrossRef]
16. Broda, D.; Staszewski, W.; Martowicz, A.; Uhl, T.; Silberschmidt, V. Modelling of nonlinear crack–wave interactions for damage detection based on ultrasound—A review. *J. Sound Vib.* **2014**, *333*, 1097–1118. [CrossRef]
17. Hong, M.; Su, Z.; Wang, Q.; Cheng, L.; Qing, X. Modeling nonlinearities of ultrasonic waves for fatigue damage characterization: Theory, simulation, and experimental validation. *Ultrasonics* **2014**, *54*, 770–778. [CrossRef]
18. Ruiz, A.; Ortiz, N.; Medina, A.; Kim, J.Y.; Jacobs, L. Application of ultrasonic methods for early detection of thermal damage in 2205 duplex stainless steel. *NDT E Int.* **2013**, *54*, 19–26. [CrossRef]
19. Fierro, G.M.; Ciampa, F.; Ginzburg, D.; Onder, E.; Meo, M. Nonlinear ultrasound modelling and validation of fatigue damage. *J. Sound Vib.* **2015**, *343*, 121–130. [CrossRef]
20. Ding, X.; Feilong, L.; Youxuan, Z.; Yongmei, X.; Ning, H.; Peng, C.; Mingxi, D. Generation Mechanism of Nonlinear Rayleigh Surface Waves for Randomly Distributed Surface Micro-Cracks. *Materials* **2019**, *11*, 644. [CrossRef]
21. Wan, X.; Zhang, Q.; Xu, G.; Tse, P. Numerical Simulation of Nonlinear Lamb Waves Used in a Thin Plate for Detecting Buried Micro-Cracks. *Sensors* **2014**, *14*, 8528–8546. [CrossRef] [PubMed]
22. Melchora, J.; Parnell, W.; Bochud, N.; Peralta, L.; Rusa, G. Damage prediction via nonlinear ultrasound: A micro-mechanical approach. *Ultrasonics* **2019**, *93*, 145–155. [CrossRef] [PubMed]
23. Šofer, M.; Ferfecki, P.; Fusek, M.; Šofer, P.; Gnatowska, R.; Malujda, I.; Dudziak, M.; Krawiec, P.; Talaśka, K.; Wilczyński, D.A. The numerical analysis of the interaction between the low-order Lamb wave modes and surface breaking cracks. *MATEC Web Conf.* **2019**, *254*. [CrossRef]

24. Cantrell, J.H.; Yost, W.T. Nonlinear ultrasonic characterization of fatigue microstructures. *Int. J. Fatigue* **2001**, *23*, 487–490. [CrossRef]
25. Shui, G.; Wang, Y.; Qu, J. Advances in nodestructive test and evaluation of material degradation using nonlinear ultrasound. *Adv. Mech.* **2005**, *35*, 54–70.
26. Jiao, J.; Lv, H.; He, C.; Wu, B. Fatigue crack evaluation using the non-collinear wave mixing technique. *Smart Mater. Struct.* **2017**, *26*, 65–71. [CrossRef]
27. Landau, L.D.; Pitaevskii, L.P.; Kosevich, A.M.; Lifshitz, E.M. *Theory of Elasticity*; Elsevier Ltd.: Oxford, UK, 1986.
28. Van Den Abeele, K.A.; Johnson, P.A.; Sutin, A. Nonlinear Elastic Wave Spectroscopy (NEWS) Techniques to Discern Material Damage, Part I: Nonlinear Wave Modulation Spectroscopy (NWMS). *Res. Nondestruct. Eval.* **2000**, *12*, 31–42. [CrossRef]
29. Van Den Abeele, K.; Breazeale, M.A. Theoretical model to describe dispersive nonlinear properties of lead zirconate–titanate ceramics. *J. Acoust. Soc. Am.* **1996**, *99*, 1430–1437. [CrossRef]
30. Jingpin, J.; Xiangji, M.; Cunfu, H.; Bin, W. Nonlinear Lamb wave-mixing technique for micro-crack detection in plates. *NDT E Int.* **2016**, *85*, 63–71. [CrossRef]
31. Ostrovsky, L.A.; Johnson, P.A. Dynamic nonlinear elasticity in geomaterials. *Riv. Nuovo Cimento della Soc. Ital. Fis.* **2001**, *24*, 1–46.

applied sciences

MDPI

Article

Bonding Characteristics of Single- and Joggled-Lap CFRP Specimens: Mechanical and Acoustic Investigations

Claudia Barile *, Caterina Casavola, Vincenzo Moramarco, Carmine Pappalettere and Paramsamy Kannan Vimalathithan

Dipartimento di Meccanica, Matematica e Management, Politecnico di Bari, Via E. Orabona 4, 70125 Bari, Italy; casavola@poliba.it (C.C.); vincenzo.moramarco@poliba.it (V.M.); carmine.pappalettere@poliba.it (C.P.); pk.vimalathithan@poliba.it (P.K.V.)

* Correspondence: claudia.barile@poliba.it; Tel.: +39-080-596-3209

Received: 30 January 2020; Accepted: 3 March 2020; Published: 5 March 2020

check for updates

Abstract: Two different configurations of adhesive-bonded carbon fiber-reinforced plastic (CFRP) specimens, joggled lap-joint specimens and single lap-joint specimens, are mechanically tested. The mechanical tests show that the joggled lap specimens have lower strength than the single lap specimens. The damage modes in both the specimens are analysed by the Acoustic Emission descriptors recorded during the mechanical tests. The acoustic data as cumulative counts and cumulative energy show the critical points of failure in both the specimen groups under loading. Moreover, they also show that the damage modes in both the specimens are dissimilar. Finally, the data provided by acoustic emission descriptors are verified by fractographic analysis on the failed surface.

Keywords: acoustic emission; acoustic energy; acoustic counts; CFRP; single lap joint; joggled lap joints; fractography

1. Introduction

The advancements in designing composite materials, particularly carbon fiber-reinforced plastic (CFRP), have increased their usage tremendously in aerospace applications [1]. CFRPs are not only sought for their high specific strength and stiffness, but also for their feasibility in forming different structures through adhesive bonding, fasteners or the combination of both [2,3]. This paves way for easy application of composite panels for repairing, thereby largely reducing both the manufacturing costs and the idle time for the flight. Each of these advancements in CFRPs has equally created complications in predicting their strength and structural integrity, particularly in composite joints [4].

Adhesive bonds or the combination of adhesive and fasteners are preferred over the conventional bolting and riveting of composite structures, as of late [5,6]. The stress concentration around the circumference of the fastener holes and the non-circumferential stress distribution around the rivets/bolts can accelerate structural failure in the composite materials [7–9]. However, in a bonded joint, uniform stress distributions can be obtained. Their improved specific strength can be attributed to peel stress and shear stress of the adhesive used. Nonetheless, the bonded joints have their limitations. During shearing of the adhesive, stress energy is released in three different modes: during the adhesive opening mode, sliding mode and tearing mode [10–12]. Moreover, the specific strength of the composite structure is governed by the adhesive thickness. Monitoring the bonding quality, prediction of damage propagation through the adhesives and identifying low impact damages have become tedious. This compels the researchers to search for different damage characterization and integrity monitoring techniques.

Acoustic emission (AE) has become one of the most sought-after non-destructive evaluation technique in recent years to characterize the damage progression in a composite structure. It is one of the very few passive non-destructive evaluation techniques, which can provide details of the damage progression during its entire loading history. When a material is strained, it releases sudden energy in the form of elastic waves [13–15]. The AE technique records these elastic waves in terms of both signal-based waveforms and parameter-based data such as peak amplitude, counts, energy, duration and so on (which can be collectively called AE descriptors). Since the AE records the elastic waves, this technique is very sensitive, even to transient signals [16–18]. This makes AE one of the most efficient and reliable techniques for characterizing damage progression.

However, for a composite material under loading, it is challenging to associate different AE descriptors recorded with different types of damage mode. Although researchers have been working tirelessly over the past 20 years and have identified innovative ways to associate the recorded AE signals with the damage modes, this still remain in debate [19,20]. Some researchers have associated specific peak amplitudes and peak frequencies to the specific types of damage modes such as matrix cracking, delamination and fiber breakage [21,22]. Others are not inclined towards this idea and encompass the necessity of using different signal-based descriptors such as frequency centroid and weighted peak frequency [23,24]. In terms of common ground, numerous researchers alike have agreed upon the reliability of using cumulative counts and cumulative acoustic energy [14,17,21]. The radical change in the cumulative counts/energy with respect to time or displacement can identify the critical damage points and help predict larger damage to the structure.

The question that persists is whether the AE technique can identify the different damage modes in a complicated material such as composite materials. In our previous works, the signal-based AE parameter—the wavelets and parameter based function named as the 'Sentry Function'—have been used for characterizing the damage modes [16,17]. Moreover, numerical and finite element models were also designed for mode I delamination tests [6]. This work differs in a way that the relative change in cumulative counts and energy are directly related to the damage modes. In the present research work, the cumulative counts and cumulative energy distribution have been used to identify the different damage modes in a CFRP bonded with adhesives. For this purpose, two different types of CFRP-bonded specimens are used: joggled-lap shear (JLS) and single-lap shear (SLS). The fracture analysis of the specimens is performed by using an optical microscope. The main objective of this study is to understand the critical points of failure and the differences in the damage modes in both JLS and SLS specimens using the AE technique.

2. Materials and Methods

2.1. Materials

For preparing the specimens, a prepreg fabric laminate with ER450 epoxy matrix (SAATI CIT CC206 ER450 43%) was used. The carbon fiber configuration is stitched with layers of fibers overlapping one other. The nominal ply thickness is 0.244 mm. The composite laminates for both JLS and SLS specimens (Figure 1) were cured by the autoclave method. Other details such as number of plies and geometry of the adherend can be found in Tables 1 and 2, respectively. The specimens were prepared by following the ASTM D5868–Standard Test Method for Lap Shear Adhesion for Fiber-Reinforced Plastic (FRP) Bonding [25]. The adhesive used for this study has a shear strength of 25 MPa and a peel strength of 65 MPa. The thickness of the adhesive coated and the area of the adhesive coated are also provided in Tables 1 and 2.

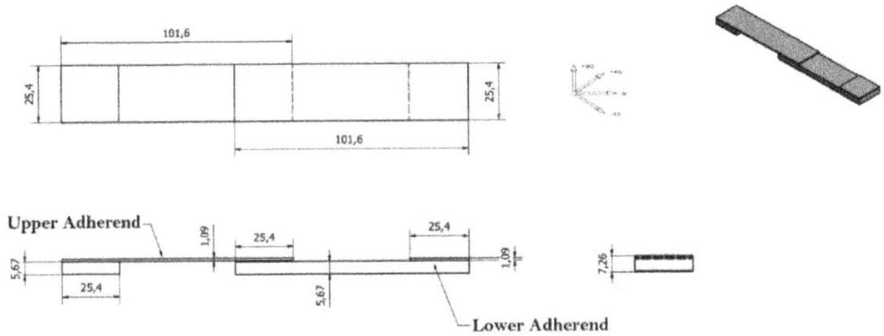

(a)

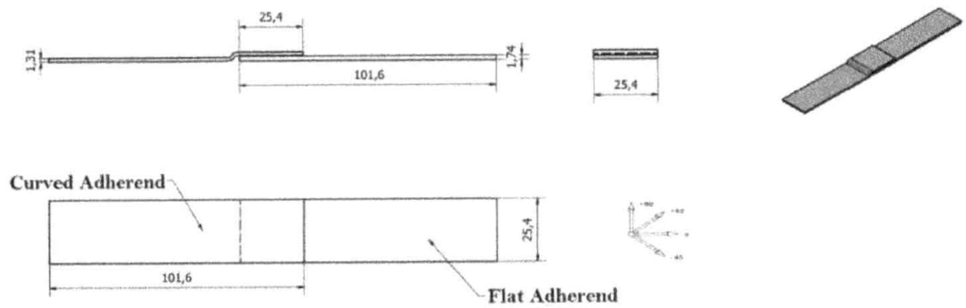

(b)

Figure 1. Single-lap shear (SLS) (**a**) and joggled-lap shear (JLS) (**b**) specimens' nominal dimensions.

Table 1. JLS specimens' geometry and characteristics.

		Flat Adherend		
Length (mm)	**Width (mm)**	**Thickness (mm)**	**No. of Plies**	**Stacking sequence**
101.6 ± 0.12	26.09 ± 0.07	2.0 ± 0.04	8	$[+45/-45]_3/-45/+45$
		Curved Adherend		
Length (mm)	**Width (mm)**	**Thickness (mm)**	**No. of Plies**	**Stacking sequence**
101.6 ± 0.17	26.09 ± 0.05	1.5 ± 0.02	6	$[+45/+45/-45]_S$
		Overlapping Region (Adhesive)		
Length (mm)		**Width (mm)**		**Thickness (mm)**
26 ± 0.21		26.09 ± 0.06		3.67 ± 0.05

Table 2. SLS specimens' geometry and characteristics.

		Upper Adherend		
Length (mm)	**Width (mm)**	**Thickness (mm)**	**No. of Plies**	**Stacking sequence**
101.6 ± 0.11	25.33 ± 0.12	1.3 ± 0.05	5	+45/+45/+45/−45/+45
		Lower Adherend		
Length (mm)	**Width (mm)**	**Thickness (mm)**	**No. of Plies**	**Stacking sequence**
101.6 ± 0.09	25.33 ± 0.14	6.4 ± 0.12	26	$+45/[+45/-45]_{12}/+45$
		Overlapping Region (Adhesive)		
Length (mm)		**Width (mm)**		**Thickness (mm)**
26 ± 0.12		25.33 ± 0.25		8.5 ± 0.11

The adhesive was cured at temperature of 65 °C for 1 hour, then the specimens were tested after 5 days. Three specimens in each type have been studied. The JLS and SLS bonded laminates are provided in Figure 2.

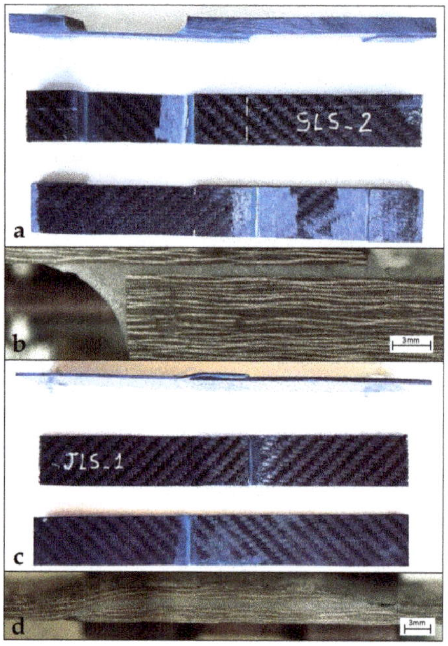

Figure 2. (**a**) SLS specimen, (**b**) joint overlap of SLS specimen, (**c**) JLS specimen, (**d**) joint overlap of JLS specimen.

The reason behind using large variation in thickness between the upper adherend and lower adherend in the SLS specimens (Figure 1a) is to simulate the configuration of the composites used in aerospace structures. SLS configuration, in particular, represents the bonding between the flange of the lower spar of the fuselage and the front bulkhead. JLS configuration (Figure 1b) represents the bonding between the fuselage and the half-wing. Although there is no standard procedure that has been followed for setting the thickness and number of plies, the test procedures were carried out carefully as per the ASTM D5868 standards. Moreover, it has been indicated by many researchers that the thickness of the adherend does not play a significant role in load distribution under displacement-controlled testing conditions [26]. The entire load is distributed through the thickness of the adhesive, which is the crucial factor in this research work.

2.2. Testing Methods

The tension was applied to the specimens under displacement-controlled mode at a speed of 13 mm/min, in accordance with the ASTM D5868 standard and previous works presented in the literature [27,28]. Although the value of the strain rate suggested from the standard method is quite high, it does not significantly influence the mechanical response of the lap joints [29]. The test was carried out in the INSTRON servo-hydraulic testing machine with a maximum load capacity of 100 kN. To record the acoustic activities under loading, two narrow-band general purpose AE sensors were used: R30a (Physical Acoustics, MISTRAS Group, NJ, USA). The sensor has an operating range of 150 kHz to 400 kHz and a resonant frequency of 300 kHz. The reason for selecting the narrow band sensor is because in CFRPs, the recorded AE signals mostly are under the frequency band of 400 kHz. To avoid unnecessary noise, the narrow-band sensor was selected. For the same purpose, the sampling rate was set as 1 MSPS (mega samples per second). The threshold of AE acquisition was set at 35 dB and the recorded signals were amplified by 40 dB through a 2/4/6 AE (Acoustic Emission) preamplifier.

The sensors were placed at 40 mm either side from the centre of the adhesive overlapping region (Figure 3). Silica gel was interposed between the surface of the sensor and the specimens, in order to improve the coupling of the elements. This location was selected based on the length of the overlapping region and to record the AE signals produced within the overlapping region. As indicated in the previous section, the load is carried out only through the adhesives under displacement-controlled mode. For this reason, the AE signals recorded beyond the adhesive overlapping region are discarded. In the fractured specimen, not many damages were observed except in the area covering the overlapping adhesive layer. The surface of the specimen post rupture has been analysed by means of an optical microscope NIKON SMZ800. Moreover, the region of the adhesive and the fractured surfaces have also been studied after the shearing failure.

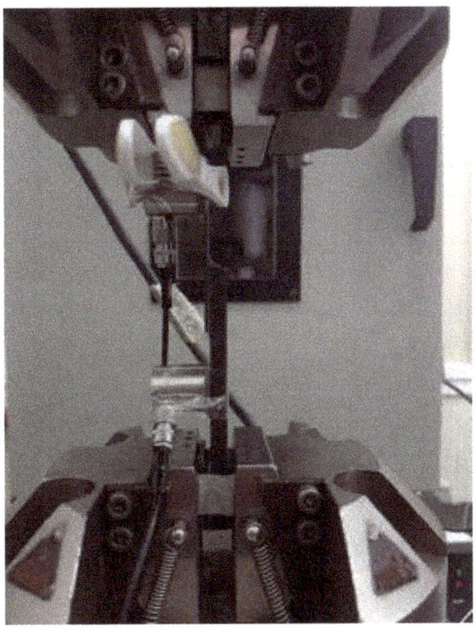

Figure 3. Testing rig with acoustic emission (AE) sensors mounted on the specimen.

It is well known that the distance from the sensor to the acoustic source affects the propagation of the AE signals. This can result in the attenuation of the recorded AE signals. To avoid this attenuation, the AE sensors were calibrated for the attenuation in signal with respect to the distance from the sensor [30]. A pencil break test was performed, which normally produces a peak amplitude of 95 dB to 99 dB at varying distances from the sensor on the SLS and JLS specimens. The table of this calibration was then fed to the PAC PCI 2 data acquisition system, which automatically calibrated the attenuation of the signal.

2.3. Acoustic Emission (AE) Characterization

The recorded acoustic data used for this study are cumulative counts and cumulative energy. Counts are the number of instances of the recorded acoustic signal crosses the set threshold.

The acoustic energy is the total energy of the recorded acoustic event over the time period of the signal t_0 to t_i. If U_i is the transient voltage of the recorded acoustic event, then the acoustic energy E_{AE} is given by Equation (1) [31]:

$$E_{AE} = \int_{t_0}^{t_i} U_i^2(t)dt \tag{1}$$

3. Results and Discussions

3.1. Mechanical Test Results

The bonding characteristics of the JLS and SLS specimens were tested by applying tensile load under displacement-controlled mode as per the ASTM D5868 standard. When the tensile load was applied, the entire load was distributed through the adhesive region. It has been reported by several researchers with solid evidences that the thickness of the adherend does not have a significant role in the distribution of load under shearing [26]. There could also be more than one peak load value, with each of them contributing to the damage characteristics or the bonding characteristics of the specimen.

Figure 4 shows the load vs. time curve for the specimens. In the research work presented, more than one peak can be observed in the load vs. time curve for both specimens. It must be noted that although there are several small peaks observed in both JLS and SLS specimens in Figure 4, the initial peak load is considered as the initial rupture point and the final peak load is considered as the final rupture point. Since these mechanical results provide only a limited amount of information on the damage characteristics, the acoustic emission results have been used in the subsequent sections. Table 3 shows the mechanical characteristics of both JLS and SLS specimens.

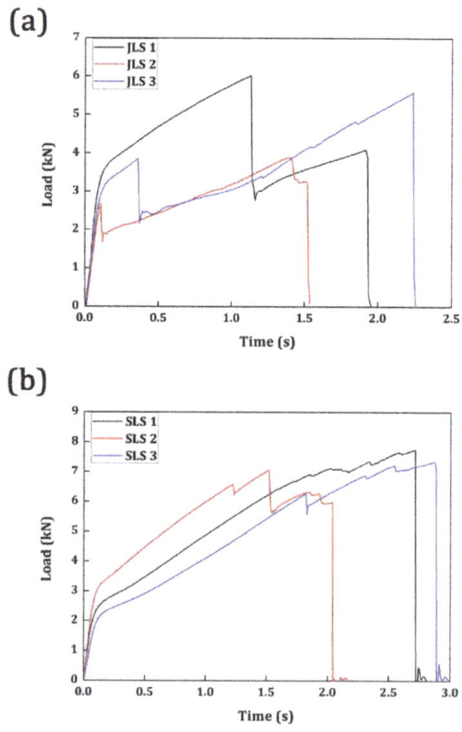

Figure 4. Load vs. time curve: (**a**) JLS Specimens; (**b**) SLS Specimens.

Table 3. Damage characteristics in JLS and SLS specimens.

Specimen	Peak Load (kN)	
	Initial Rupture	Final Rupture
JLS 1	6.02	4.09
JLS 2	2.67	3.89
JLS 3	3.86	5.57
Mean	**4.18**	**4.52**
Std. Dev.	**1.70**	**0.92**
SLS 1	7.05	7.75
SLS 2	7.06	5.96
SLS 3	5.93	7.36
Mean	**6.78**	**7.02**
Std. Dev.	**0.75**	**0.94**

The first major observation from the mechanical test results is the varying peak load values of the same specimen group, both in JLS and SLS. This is common when it comes to the practice of adhesive bonded specimens [27–29] because the bond quality is not only determined by the curing characteristics, the thickness of the adhesive and other mechanical characteristics, but also on the efficiency of the user who handles the adhesives. The adhesives are normally mixed using a hand mixer rather than a mechanical controller owing to its high viscosity and requirements. Thus, it is inevitable that a considerable amount of difference can be observed between the adhesive bonded specimens of the same group. This is the reason why there is a difference in both the peak loads at initial and final ruptures and they are not similar between the same specimen group (refer to Figure 4 and Table 3). Furthermore, under this view, an increase in the number of test specimens would increase the accuracy of the evaluation of the scattering but it not will reduce the latter.

As indicated in the previous sections, the thickness of the adherend does not generally affect the strength of the adhesive bonded shear specimens under displacement-controlled mode. Nonetheless, the thickness of the adhesive carries the majority, if not all the load. In the SLS specimens, owing to their larger thickness than JLS specimens, higher peak loads are accounted for. The average peak load at the initial rupture in SLS specimens is 6.78 kN, while the JLS specimens have an average peak load of 4.18 kN. Similarly, at the final rupture stage, SLS specimens have an average peak load of 7.02 kN, while JLS specimens have 4.52 kN. In SLS specimens, the load is carried along the cross-sectional area of the adhesive region. However, in the JLS specimens, the shear strength of the specimen is also governed by the end curvature effect. This is explained by Taib et. al. [11] in their research work. Single lap joints, when loaded, transfer most of the load in shearing. Joggled lap joints, however, transfer most of the load in peeling. This is also the reason why the joggled lap joints have lower displacement when compared to the single lap joint specimens. The joggled joints suffer lateral deflections rather than shearing, which result in inducing high peel stress on the neutral axis. This results in the crack initiating at the joggled knee of the curvature region. This will result in low peak loads, which are observed in the present research work (see Table 3 and Figure 4). Similar results have been presented by Kishore and Prasad [32], who also addressed the effect of the eccentricity of the joggle in the strength of the joggled lap specimens. It has been addressed by both Taib et. al [32] and Kishore and Prasad [33] that the joggled lap specimens will have lower strength in comparison with the single lap joints. The same has been observed in the presented research work.

3.2. Acoustic Emission Results: Cumulative Counts and Cumulative Energy

The damage characteristics or the mode of failure in the adhesive bonded specimens cannot be explained through the mechanical characteristics. Furthermore, predicting the failure at an early stage is also not easy by studying only the mechanical characteristics. It can be explained by the acoustic emission characteristics. The load vs. time curve of the JLS specimens is plotted over the cumulative counts and the cumulative energy of the AE recorded. It is presented in Figure 5. Similarly, the SLS specimens are presented in Figure 6.

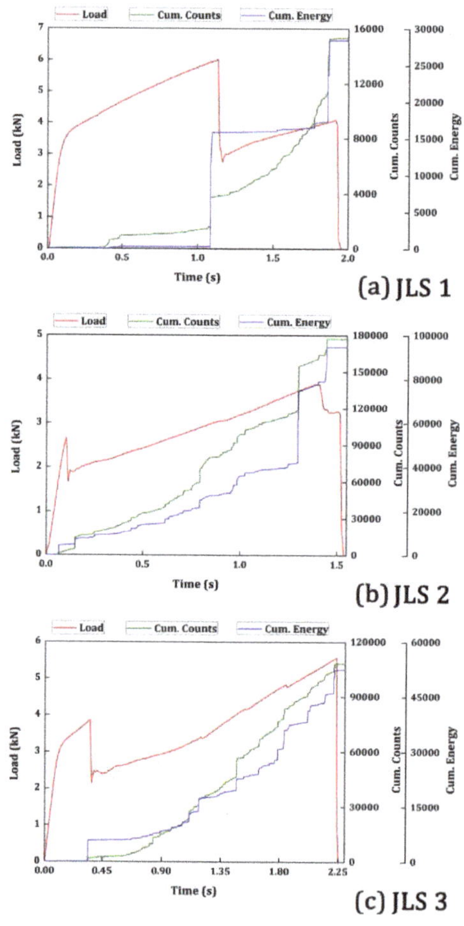

Figure 5. Load, cumulative counts, cumulative energy vs. time: (**a**) JLS 1; (**b**) JLS 2; (**c**) JLS 3.

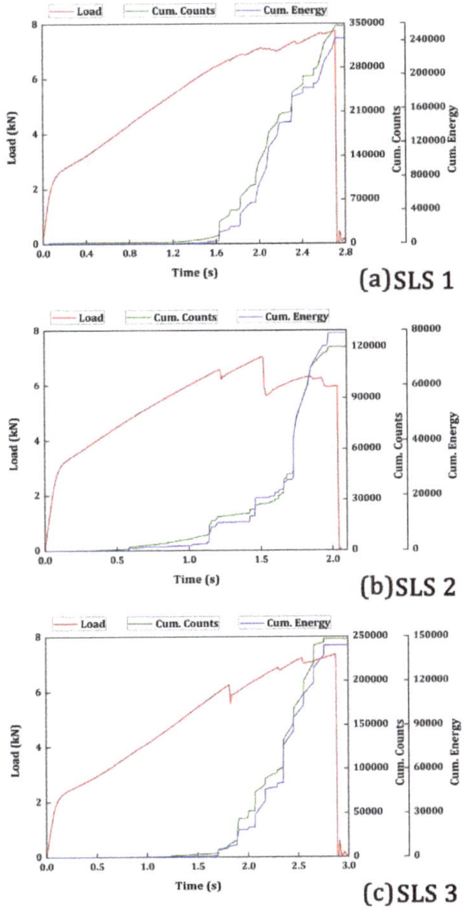

Figure 6. Load, cumulative counts, cumulative energy vs. time: (**a**) SLS 1; (**b**) SLS 2; (**c**) SLS 3.

In Figure 5, it can be observed that whenever there is a load peak in the load vs. time curve in any of the JLS specimens, the cumulative counts and cumulative energy show a steep increase. The initial load drops in all JLS 1, JLS 2 and JLS 3 at 1.1 s, 0.1 s and 0.35 s, respectively, have been associated with the increase in cumulative counts and cumulative energy. Similarly, at the final stages of loading prior to failure 1.8 s, 1.35 s and 2.25 s, respectively, for JLS 1, JLS 2 and JLS 3 are also associated with the increase in cumulative counts and energy. However, there are some instances, particularly around 1.2 s in the JLS 2 specimen, where the cumulative counts and energy increased unexpectedly without any change in the load vs. time curve. This phenomenon could not be explained without utilizing another characterizing technique or a deeper investigation like the frequency content of the acoustic data. Nonetheless, analysing the waveform of those signals in their time-frequency domain can also provide information about whether these signals are representatives of machine or surrounding noises. This is not the only information that can be obtained from the AE results. In JLS 1 specimen, the cumulative energy increases steeply around 1.1 s duration and it remains almost constant until the specimen reaches its final rupture region. Nonetheless, the cumulative counts kept increasing gradually beyond the 1.1 s duration and the slope took a steep increase as the material reaches failure. This means that a large number of counts have been recorded with very low energy from 1.1 s duration to the final failure. In a non-linear material, such as plastics or composite materials, the large counts with low energy represent the matrix cracking. The ratio between the cumulative energy and the cumulative count varies at different damage modes in a material. When this ratio exhibits a very low value, it can be said that the energy is low with large counts. On that basis, this can be attributed to the matrix cracking. Brittle fractures such as interlaminar crack growth and fiber breakage release large energy acoustic events, which are responsible for the initial and final ruptures. Between the stages of the initial failure and the final failure in the JLS 1 specimen, the material has experienced only matrix cracking. This indicates that shearing has occurred in the adhesive region rather than peeling which attributes to the low-energy acoustic events. This indicates that the JLS 1 specimen has carried more load before failure than the JLS 2 and JLS 3 specimens due to the shearing mode of failure. This could be one of the reasons why the JLS 1 specimen has a higher peak load when compared to JLS 2 and JLS 3.

In the cases of JLS 2 and JLS 3, they have similar load vs. time, cumulative counts and energy patterns. Both these specimens experienced the initial rupture at a very early stage, 0.1 s for JLS 2 and 0.35 s for JLS 3. The cumulative counts and energy have also increased steeply at these time periods. Beyond that point, the cumulative counts and the cumulative energy increase almost linearly with one another. This represents the delamination in the adhesive bonded region. The adhesive peeling is occurring at this stage, which is the reason for the simultaneous increase in the counts and energy. While reaching the final stage of failure, JLS 2 and JLS 3 have probably experienced fiber breakage which is indicated by the sudden increase in counts and energy near the final rupture region.

While looking at Figure 6, all the three specimens have almost similar patterns in terms of load vs. time, cumulative counts and energy. However, the simultaneous increase in the counts and energy of the SLS specimens must not be compared or to be confused with the JLS specimens. Although in both the cases the cumulative counts and cumulative energy increased linearly, the slope of the increment is entirely different. The slope of the increment in SLS specimens is higher than the JLS specimens. Therefore, they do not represent the same damage mode although it looks similar. In the SLS specimens, the cumulative counts and energy are very low during the initial stages of loading until the specimens have suffered their initial rupture. In the SLS specimens, most of the load is distributed in shearing and these low-energy and low-count AE signals indicate the shearing occurring in the adhesive region. AE signals connecting to fiber breakages and transverse cracks growth release acoustic data with higher energy and lower counts, owing to their lower order symmetric nature. Since the energy and counts recorded are low during these stages, these signals are higher order asymmetric in nature that could be connected only to shearing phenomena [20]. Beyond the region of the initial rupture, in all three SLS specimens, the cumulative counts and energy start to increase simultaneously until they reach the final failure. This indicates that the cumulative counts and energy were very low until the crack opens in the adhesive. Generally, the crack openFs at the adhesive and adherend interface, which is governed by the geometry of the overlapping area. Once the crack had opened, due to the brittle nature of the adhesive, it progressed faster through the thickness of the adhesive. This interlaminar crack growth always generates higher counts and higher energy. This pattern continued until the specimens failed at the final rupture.

3.3. Fracture Surface Analysis

To confirm the brittle failure of the adhesives in the SLS specimens and to identify the peeling and fiber breakage failure in JLS specimens, the fractured surfaces were observed under optical microscope. The fractographic results of specimen JLS is presented in Figure 7.

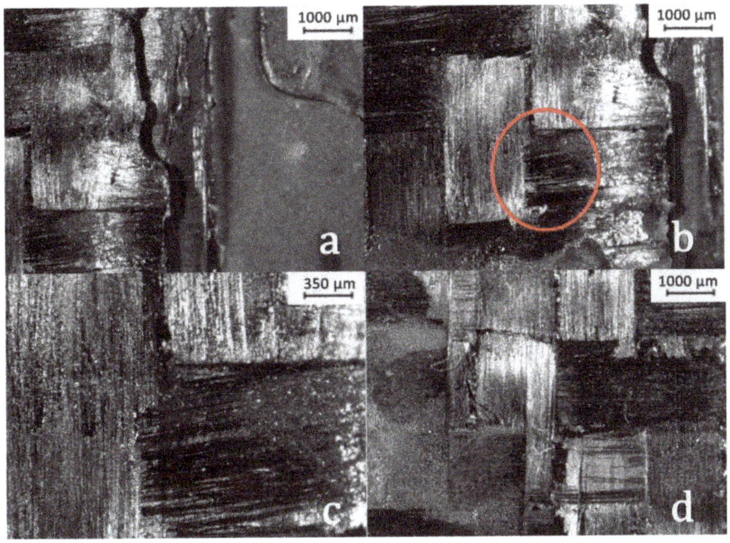

Figure 7. Fractographic analysis of JLS specimens. (**a**) Peeling at the knee of adhesive; (**b**) Debonding of laminates; (**c**) Ruptured Fiber; (**d**) Fiber/Matric Debonding.

In Figure 7a, the adhesive layer peeling off at the knee of the curvature in the joggled overlapping region can be observed. The debonding of the laminates along the applied adhesive layers can be viewed in Figure 7b. The adhesive not only peeled off from the lamina but also inflicted fiber/matrix debonding in the same lamina. Figure 7c is the zoomed view of the circled region in Figure 7b. From Figure 7c, the ruptured fiber at the end of the lamina can be observed. Figure 7d shows the presence of more fiber/matrix debonding and fiber breakage near the knee of the adhesive joint. This consolidates the explanations provided in Section 3.2 regarding the failure modes by acoustic emission parameters.

Similarly, the optical images of SLS specimens are presented in Figure 8. Figure 8a shows the complete lack of matrix/fiber debonding in the area where the adhesive is applied. The fibers are left unharmed in the fractured region of the adhesive. The adhesive has delaminated from the outermost ply, but it did not affect the fiber/matrix integrity. Figure 8b shows the brittle fracture and the direction of the crack grew along the width of the adhesive under loading. Figure 8c shows the fiber breakage at the end of the specimen. Figure 8d is the zoomed image of the ruptured fibers. So, the fiber breakage has occurred in the SLS specimens, nonetheless, only at the edge of the lamina.

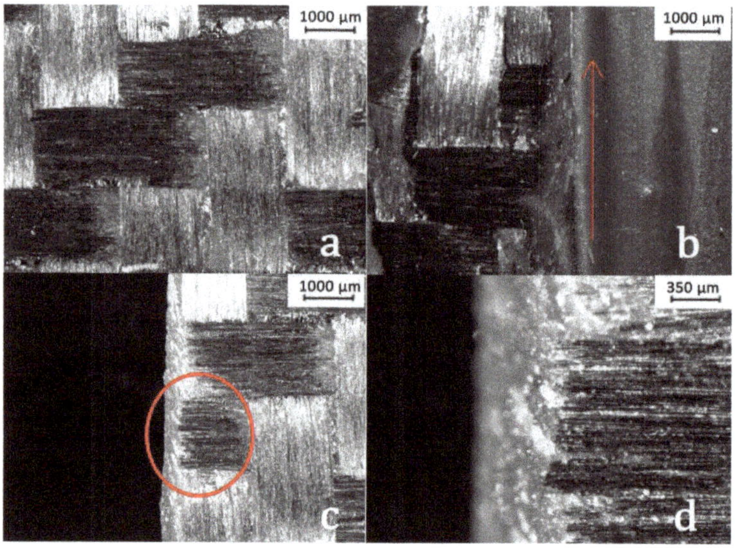

Figure 8. Fractographic analysis of SLS specimens. (**a**) Lack of Fiber/Matrix debonding; (**b**) Direction of cracking growth along the length of adhesive; (**c**) Fiber breakage; (**d**) Close in view of fiber breakage.

Therefore, by comparing Figures 7 and 8, it can be concluded that the crack growth initiated at the knee of the joggled lap in the JLS specimens, which in turn results in the peeling. The peeling also has induced the debonding between the fiber and matrix in the area where the adhesive is applied. However, in SLS specimens, the crack opened in the adhesive region and travelled through the width of the specimen resulting in a brittle failure.

The fractographic results are sufficient to prove the damage modes explained in the acoustic emission result sections. Thus, the AE proves to be a powerful tool, not only in predicting the failure in an adhesive bonded FRP but also in identifying the different damage modes in the failure. By setting up a proper experimental campaign and understanding the different AE descriptors, it is efficient to analyse the failure modes in an entire structure comprising adhesive-bonded FRP.

4. Conclusions

The bonding characteristics of adhesive-bonded FRP specimens were tested using the acoustic emission technique. The differences in the failure mechanisms were observed by comparing the cumulative counts and cumulative energy under loading. The slope of the cumulative counts and energy were different in JLS and SLS specimens indicating that both the specimen groups failed under different damage modes. Moreover, the cumulative energy and cumulative counts can provide information on the critical points of failure. In the joggled-lap specimens, the initial failure was due to the peeling of the adhesive layer at the knee of the curvature region and it induced the fiber/matrix debonding. The final failure was due to the fiber breakage in the lamina. In the single-lap specimens, the initial failure was due to the crack opening of the adhesive and the crack propagated through the width of the adhesive resulting in the final failure. The explanations provided by the AE results were evidentially proved by the fractographic analysis of the specimens. Acoustic emission proves to be a powerful tool in analysing the damage characteristics of fiber-reinforced plastics.

Author Contributions: Conceptualization, P.K.V. and C.B.; methodology, P.K.V., C.B. and V.M.; validation, C.B. and P.K.V.; formal analysis, P.K.V. and C.B.; investigation, P.K.V. and C.B.; data curation, C.B., V.M. and P.K.V.; writing—original draft preparation, P.K.V.; writing—review and editing, P.K.V. and C.B.; supervision, C.C. and C.P.; project administration, C.C. and C.P. All authors have read and agreed to the published version of the manuscript.

Funding: This research received no external funding.

Conflicts of Interest: The authors declare no conflict of interest.

References

1. Sugita, Y.; Winkelmann, C.; La Saponara, V. Environmental and chemical degradation of carbon/epoxy lap joints for aerospace applications, and effects on their mechanical performance. *Compos. Sci. Technol.* **2010**, *70*, 829–839. [CrossRef]
2. Campilho, R.D.S.G.; de Moura, M.F.S.F.; Domingues, J.J.M.S. Using a cohesive damage model to predict the tensile behaviour of CFRP single-strap repairs. *Int. J. Solids Struct.* **2008**, *45*, 1497–1512. [CrossRef]
3. Yu, Q.; Gao, R.; Gu, X.; Zhao, X.; Chen, T. Bond behavior of CFRP-steel double-lap joints exposed to marine atmosphere and fatigue loading. *Eng. Struct.* **2018**, *15*, 76–85. [CrossRef]
4. Kucher, N.K.; Zemtsov, M.P.; Zarazovskii, M.N. Deformation behavior and strength of unidirectional carbon fiber laminates. *Mech. Compos. Mater.* **2006**, *42*, 407–418. [CrossRef]
5. Tong, L.; Steven, G.P. *Analysis and Design of Structural Bonded Joints*; Kluwer Academic Publishers: Boston, MA, USA, 1999.
6. Barile, C.; Casavola, C.; Gambino, B.; Mellone, A.; Spagnolo, M. Mode-I fracture behavior of CFRPs: Numerical model of the experimental results. *Materials* **2019**, *12*, 513. [CrossRef] [PubMed]
7. Benatar, A.; Gillespie, J., Jr.; Kedward, K. Joining of composites. In *Advanced Composites Manufacturing*; Gutowski, T.G., Ed.; John Wiley: New York, NY, USA, 1997; pp. 487–512.
8. Jones, R.M. *Mechanics of Composite Materials*, 2nd ed.; Taylor and Francis Publishers: New York, NY, USA, 1999.
9. Barile, C.; Casavola, C.; Pappalettera, G.; Vimalathithan, P.K. Characterization of adhesive bonded CFRP laminates using full-field digital image stereo-correlation and finite element analysis. *Compos. Sci. Technol.* **2019**, *169*, 16–25. [CrossRef]
10. Li, G.; Pang, S.S.; Woldesenbet, E.; Stubblefield, M.A.; Mensah, P.F.; Iberkwe, S.I. Investigation of prepreg bonded composite single lap joint. *Compos. Part B Eng.* **2001**, *32*, 651–658. [CrossRef]
11. Taib, A.A.; Boukhili, R.; Achiou, S.; Gordon, S.; Boukehili, H. Bonded joints with composite adherends. Part I. Effect of specimen configuration, adhesive thickness, spew filler and adherend stiffness on fracture. *Int. J. Adhes. Adhes.* **2006**, *26*, 226–236. [CrossRef]
12. Yang, C.; Tomblin, J.S. *Investigation of Adhesive Behavior in Aircraft Applications*; DOT/FAA/AR-01/57; US Department of Transportation Federal Aviation Administration, Office of Aviation Research: Washington, DC, USA, 2001.
13. Grosse, C. Introduction. In *Acoustic Emission Testing*; Grosse, C., Ohtsu, M., Eds.; Springer: Berlin/Heidelberg, Germany, 2008.
14. Balázs, G.L.; Grosse, C.; Koch, R.; Reinhardt, H.W. Damage accumulation on deformed steel bar to concrete interaction detected by acoustic emission technique. *Mag. Concr. Res.* **1996**, *48*, 311–320. [CrossRef]
15. Grosse, C.; Linzer, L. Signal-based AE analysis. In *Acoustic Emission Testing*; Grosse, C., Ohtsu, M., Eds.; Springer: Berlin/Heidelberg, Germany, 2008.
16. Barile, C.; Casavola, C.; Pappalettera, G. Acoustic emission waveform analysis in CFRP under Mode I test. *Eng. Fract. Mech.* **2018**, *210*, 408–413. [CrossRef]
17. Barile, C.; Casavola, C.; Pappalettera, G.; Vimalathithan, P.K. Damage characterization in composite materials using acoustic emission signal-based and parameter-based data. *Compos. Part B-Eng.* **2019**. [CrossRef]
18. Barile, C.; Casavola, C.; Pappalettera, G.; Vimalathithan, P.K. Investigation of structural integrity of composite materials using wavelet packet transform. *Procedia Eng.* **2019**, *17*, 582–588. [CrossRef]
19. Hamstad, M.A. Frequencies and Amplitudes of AE Signals in a Plate as a Function of Source Rise Time. In Proceedings of the 29th European Conference on Acoustic Emission Testing; Available online: https://www.ndt.net/events/EWGAE%202010/proceedings/papers/20_Hamstad.pdf (accessed on 4 March 2020).
20. Oz, F.E.; Ersoy, N.; Lomov, S.V. Do high frequency acoustic emission events always represent fibre failure in CFRP laminates? *Compos. Part A Appl. Sci.* **2017**, *103*, 230–235. [CrossRef]
21. Liu, P.F.; Chu, J.K.; Liu, Y.L.; Zheng, J.Y. A study on the failure mechanisms of carbon fiber/epoxy composite laminates using acoustic emission. *Mater. Design* **2012**, *37*, 228–235. [CrossRef]

22. De Groot, P.J.; Wijnen, P.A.M.; Janssen, R.B.F. Real-time frequency determination of acoustic emission for different fracture mechanisms in carbon/epoxy composites. *Compos. Sci. Technol.* **1995**, *55*, 405–412. [CrossRef]
23. Njuhovic, E.; Bräu, M.; Wolff-Fabris, F.; Starzynski, K.; Altstädt, V. Identification of failure mechanisms of metallised glass fibre reinforced composites under tensile loading using acoustic emission analysis. *Compos. Part B-Eng.* **2015**, *81*, 1–13. [CrossRef]
24. Alia, A.; Fantozzi, G.; Godin, N.; Osmani, H.; Reynaud, P. Mechanical behaviour of jute fibre-reinforced polyester composite: Characterization of damage mechanisms using acoustic emission and microstructural observations. *J. Compos. Mater.* **2019**. [CrossRef]
25. ASTM D5868-01(2014). *Standard Test Method for Lap Shear Adhesion for Fiber Reinforced Plastic (FRP) Bonding*; ASTM International: West Conshohocken, PA, USA, 2014; Available online: www.astm.org (accessed on 4 March 2020).
26. Mangalgiri, P.D.; Johnson, W.S.; Everett, R.A., Jr. Effect of adherend thickness and mixed mode loading on debond growth in adhesively bonded composite joints. *J. Adhes.* **1987**, *23*, 263–288. [CrossRef]
27. Hsiao, K.T.; Alms, J.; Advani, S.G. Use of epoxy/multiwalled carbon nanotubes as adhesives to join graphite fibre reinforced polymer composites. *Nanotechnology* **2003**, *14*, 791–793. [CrossRef]
28. Avila, A.F.; Bueno, P.O. An experimental and numerical study on adhesive joints for composites. *Compos. Struct.* **2004**, *64*, 531–537. [CrossRef]
29. Liu, X.; Shao, X.; Li, Q.; Sun, G. Experimental study on residual properties of carbon fibre reinforced plastic (CFRP) and aluminum single-lap adhesive joints at different strain rates after transverse pre-impact. *Compos. Part A Appl. Sci. Manuf.* **2019**, *124*, 105372. [CrossRef]
30. Finkel, P.; Mitchell, J.R.; Carlos, M.F. Experimental study of 'Auto Sensor Test-Self Test Mode' for acoustic emission system performance verification. *AIP. Conf. Proc.* **2000**, *509*, 1995–2002.
31. Harris, D.O.; Bell, R.L. The measurement and significance of energy in acoustic-emission testing. *Exp. Mech.* **1977**, *17*, 347–353. [CrossRef]
32. Kishore, A.N.; Prasad, N.S. An experimental study of Flat-Joggle-Flat bonded joints in composite laminates. *Int. J. Adhes. Adhes.* **2012**, *35*, 55–58. [CrossRef]
33. Taib, A.A.; Boukhili, R.; Achiou, S.; Boukhili, H. Bonded joints with composite adherends. Part II. Finite element analysis of joggle lap joints. *Int. J. Adhes. Adhes.* **2006**, *26*, 237–248. [CrossRef]

applied sciences

MDPI

Review

Mechanical Strength Evaluation of Elastic Materials by Multiphysical Nondestructive Methods: A Review

Huiting Huan [1,2], Lixian Liu [1,2,*], Andreas Mandelis [2], Cuiling Peng [1], Xiaolong Chen [1] and Jinsong Zhan [1,*]

1 School of Mechano-Electronic Engineering, Xidian University, Xi'an 710071, China; huanht90@sina.com (H.H.); ckpcl96@gmail.com (C.P.); xlchen@mail.xidian.edu.cn (X.C.)

2 Center for Advanced Diffusion-Wave and Photoacoustic Technologies (CADIPT), Department of Mechanical and Industrial Engineering, University of Toronto, Toronto, ON M5S 3G8, Canada; mandelis@mie.utoronto.ca

* Correspondence: lixianliu@xidian.edu.cn (L.L.); zhanjinsong@xidian.edu.cn (J.Z.)

Received: 24 December 2019; Accepted: 19 February 2020; Published: 27 February 2020

check for updates

Abstract: The main purpose of industrial nondestructive testing (NDT) is to diagnose the stability, reliability and failure probability of materials, components and structures. Industrial component mechanical strength is one of the most important properties NDT is used to characterize. Subtle but perceptible changes in stress-strain behavior can be reliable indicators of defect formation. A detailed review on the state-of-the-art NDT methods using optical-radiation, photoacoustic, and photothermal techniques for mechanical strength evaluation and defect pre-diagnosis is presented in this article. Mechanical strength is analyzed in terms of the deformation/strain field, the stress-strain relation, and the residual stress in an elastic material subjected to tensile or compressive loading, or impact. By introducing typical NDT experiments, the history and features of each methodology are revisited and typical applications are discussed. This review also aims to be used as a reference toward further research and development of NDT technologies characterizing mechanical strength of materials and components.

Keywords: stress-strain relation; nondestructive testing; optical-radiation; photoacoustic; photothermal

1. Introduction

Nondestructive testing/evaluation (NDT/NDE) is a collection of scientific techniques which are able to evaluate the properties and condition of materials without inflicting permanent damage on them. They were originally designated for inspecting parts, products, and structures of high value. The number of published research reports on NDT rapidly increased in the beginning of the 20th century because of industrialization and the two World Wars. There have been wide applications for NDT technologies in numerous fields such as industrial manufacturing [1,2], aerospace and aeronautics [3], civil engineering [4,5], and material science and technology [6,7]. Practical NDT techniques used for materials characterization can reduce maintenance costs and enforce lifetime management [8]. From the viewpoint of NDT on components made from solid elastic materials such as alloys, polymers and composites, quality is assessed by inspecting the possible existence of defects and changes in mechanical strength [9,10]. Conventionally, the mechanical strength of a material is measured with a tensile machine and the stress-strain relation is determined by dynamic tensile testing from load-free to failure [11–13]. The stress-strain data are recorded on samples of cylindrical, center-necked shape and contact stress/strain measurement modules such as adhesive transducers are usually part of the machine. Such tensile tests are destructive because samples undergo irreversible deformation during the tensile test. To resolve the problem associated with mechanical strength testing, several nondestructive methodologies have been developed for the purpose of stress-strain evaluation. Some of them focus on noncontact approaches which characterize the change of stress-strain state by other physical fields during tensile testing. Other methodologies do not involve tensile testing but are devoted to measuring surface or internal residual stress distribution in an elastic sample both qualitatively and quantitatively. Nevertheless, the ultimate goal always is pre-diagnosing the mechanical strength of a component before actual cracks and failure occur.

This review article discusses typical NDT methodologies that measure stress or strain by means of both contact and noncontact configurations. Noncontact methods can be used under extreme testing conditions or for in-situ inspection. Features of these noncontact NDT modalities are discussed with regard to their physical basis and state-of-the-art applications. No ranking of the various techniques is made or conclusions drawn with respect to their relative advantages because they all have distinctive features and limitations. It is expected that this review article will help interested readers to learn about the underlying physical principles, development histories and modern applications of a number of major NDT methods used for industrial quality control to-date.

2. Noncontact Mechanical Property Testing Methods

2.1. Optical/Radiation Techniques

The stress-strain relation of elastic materials is important because it is not only a way to characterize the hardness and brittleness of raw materials, but also reveals the mechanical status of a sample under test [14,15]. Normally, the strain value is measured with a strain gauge affixed on the sample surface. Under extreme conditions such as high temperature, strain sensors along with special designs of insulating and high-Curie-temperature materials, are used for stress-strain measurements in contact configurations [16–19]. Such a scheme cannot satisfy the requirements of many in-field applications due to difficulties with transducer bonding. To perform totally noncontact measurements, digital-image correlation (DIC), a.k.a. digital laser speckle (DLS), a purely optical imaging technique, was devised. This method was first proposed by Lyons et al. [20] with the basic principle shown in Figure 1. To avoid blackbody radiation from a high temperature surface, a visible-range expanded-beam laser is usually adopted in DIC measurement to generate diffuse-reflection-induced speckle patterns. A camera (with optical filter) is placed in front of the sample to record speckle images during tensile tests. By cross-correlating the digital images before and after tensile loading, the displacement field can be determined using the following equation [21]:

$$\begin{pmatrix} u_x \\ u_y \end{pmatrix} = \frac{1}{V_i}\begin{pmatrix} \Delta m P_x \\ \Delta n P_y \end{pmatrix} \tag{1}$$

where u_x and u_y represent the two orthogonal displacement fields in the camera, V_i is the image size and Δm, Δn are the respective cross-correlation values. P stands for the pixel pitch which is determined by the resolution of camera. The ultimate strain field of interest can be further derived from the gradients of u_x and u_y. Using the proposed method, Anwander et al. achieved strain measurements in tensile tests at 1200 Celsius degree with aluminum [21]. Four years later, Vǒlkl et al. further increased the temperature to 3000 degree with an Ohmic heater and conducted strain field measurements using a similar DIC schematic [22]. Zhu et al. developed a time-dependent DIC system and dynamically recorded the change of strain field under a constant loading rate with respect to non-metallic composites [23]. By using laser extensometry, the local strain rate value can be characterized by time-dependent phase shift of correlated signal thus real-time measurement is possible.

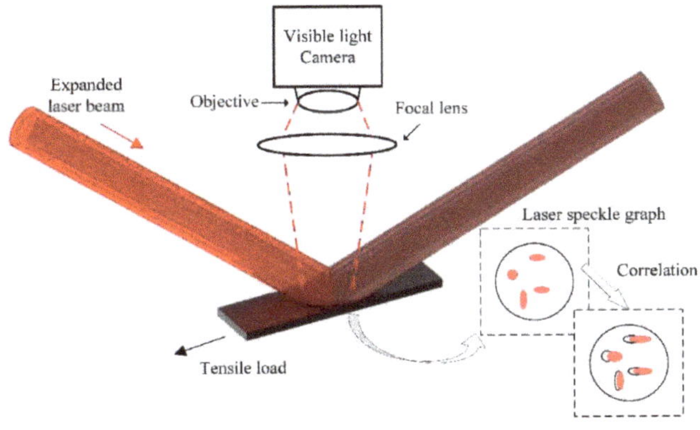

Figure 1. The schematic of digital-image correlation (DIC) strain evaluation.

Another interesting topic is exploring the stress-strain behavior of alloys under high temperature which can reveal the strength of materials under extreme conditions, and also investigating how they react to thermal stress at such high temperatures. Such a thermal reactive character can be used to determine thermal compatibility with other materials in a heterogeneous structure [24]. Pan et al. introduced an optical bandpass imaging system shown in Figure 2 to evaluate surface thermal strain without a surface irradiating laser source [25]. The bandpass filter allows only the violet-to-blue light to pass and avoids interferences from blackbody radiation at high temperatures. The use of an incoherent light source makes the system less costly but still sensitive [26]. The DIC of incoherent speckle imaging technique can also be used for real-time strain recording, as proposed by Yang et al., to assist optimization of supersonic aircraft coating-material thermal shock processes [27].

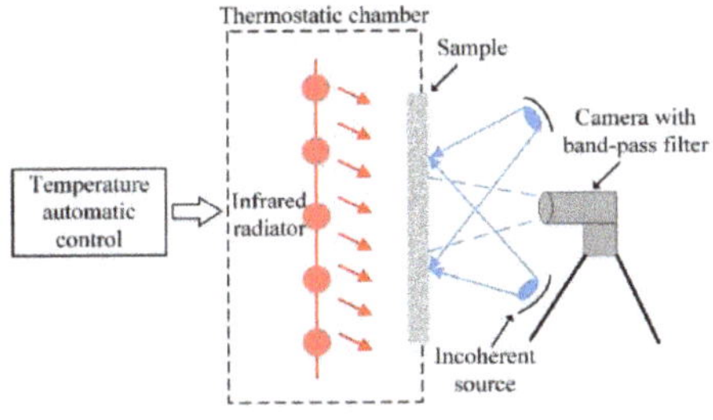

Figure 2. Strain measurement configuration with blue light source, redrawn from [24].

Apart from industrial applications, noncontact optical measurements can also be used to perform strength evaluation of concrete in civil engineering using an alternative configuration of DLS. A narrow collimated visible laser beam impinges perpendicularly on a sample surface, while the camera monitors and captures scattered laser light from the incident spot at a non-normal angle in a scanning patterns over the surface so as to determine roughness mappings. This method is usually named laser triangulation. A series of tests with laser triangulation scanning were carried out [28–30] to systematically quantify mechanical bonding strength between different concrete layers. The correlation between adhesive performance of concrete and roughness, grain size, texture etc. was investigated. Optical non-contact measurements represent an accurate and effective modality for evaluating mechanical strength of inelastic building materials.

DIC is a successful approach to remotely detect small displacements and reconstruct strain fields. It relies heavily on the resolution, processing speed and aberration performance of the camera. However, there are some limitations with respect to its applications. The processing algorithm requires input of speckle images, therefore, the signal-to-noise ratio from highly reflective surfaces can easily deteriorate. Furthermore, the strain field inside the sample cannot be visualized with purely optical means. The existence of local stress/strain should be justified with other techniques to yield a comprehensive stress-strain evaluation. Ionizing radiative electromagnetic waves have deep penetration and thus can be used for internal stress-strain evaluation.

Among the established methods, X-ray diffraction is regarded as a classic and powerful tool for diagnosing subsurface or internal stress without contact. Having very small wavelength, X-ray radiation penetrates metallic materials and diffracts from irregular-lattice area caused by internal stress/strain, a common case in the production of composites consisting of non-metal and metal layers with different thermal expansion/contraction properties [31,32].

X-rays have also been shown to be capable of evaluating the stress-strain relation from mechanical tests by single-angle, two-angle and $\sin^2 \Psi$ scattering modes [33–35]. The schematic of X-ray diffraction and the associated relevant coordinates is shown in Figure 3. The basic principle for evaluating strain due to external load can be represented as a function of latitudinal lattice spacing $d_{\Psi\Phi}$ as [32,33]:

$$\varepsilon_{\psi\phi} = \frac{d_{\psi\phi} - d_0}{d_0} \tag{2}$$

where, d_0 is the lattice spacing before deformation. The strain component can be determined by analyzing the X-ray radiation signal in different directions, i.e., at angles Ψ and Ψ. Based on this basic principle, a numerical analysis was developed to obtain tomographic images of stressed materials in recent reports [36–39]. X-ray computed tomography was further extended to evaluate mechanical performance and defects. Youssef et al. [40] and Patterson et al. [36] investigated the stress-strain property of polymers by X-ray tomography. They used a finite element method to model and in-situ test to validate elastic and hyper-elastic deformation properties of polymers under compression; Fieres et al. focused on the new technology of 3D printing and used X-ray to test the failure possibility of printed parts [37]; Xing et al. inspected jointed rocks with X-ray and scanning electron microscope to find evidence of cracks under mechanical compression [39]. Although X-ray diffraction provides sensitive, reliable and quantitative stress-strain measurements and usually acts as a reference, such radiographic NDT has the obvious disadvantage of complicated instrumentation and harmful ionizing radiation which requires high level of operation and testing standards. The extremely high sensitivity of X-ray diffraction to stress makes it more applicable to small- and micro-scale structures in laboratories. Other types of NDT such as ultrasonic scanning and thermography should be explored to find the possibility for stress-strain evaluation and treat macro-scale components.

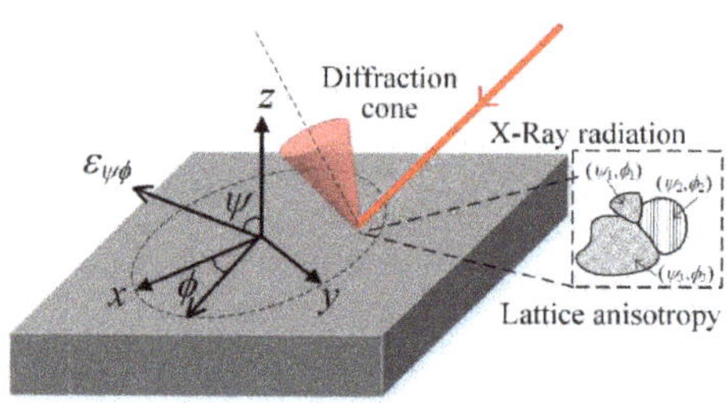

Figure 3. Scheme of residual stress detection by X-ray diffraction.

2.2. Photoacoustic (PA) Techniques

Regarding optical NDT techniques, the term "photo" is used here to refer to the phenomenon of photonic-to-acoustic (specifically ultrasonic) energy conversion. The photo-thermo-elastic or simply photoacoustic (PA) effect was discovered by Alexander Graham Bell in 1880 [41], followed by a number of theoretical and experimental research reports on the physical principle of ultrasound generation in solids by transient optical power absorption [42–44]. From the acoustic point of view, the existence of stress in a sample can affect the propagation of elastic waves because of elastic inhomogeneity or anisotropy. Such a phenomenon was first discussed theoretically by seismologist Biot [45] and experimentally validated by Hughes and Kelly [46] and Bergman and Shahbender [47]. They concluded that the existence of static pre-stress can change the acoustic wave velocity inside a medium, a phenomenon described as the acoustoelastic effect [48–50]. In the case of irreversible residual stress, finite and irrecoverable deformation and strain should be considered with the result that the stress-strain relation becomes nonlinear [51]. According to Murnaghan's finite deformation of elastic materials, the stress-strain relation should be determined with the help of the free-energy function W_s which is defined as [46,52]:

$$W_s = \frac{1}{2}(\lambda + 2\mu)I_1^2 - 2\mu I_2 + \frac{1}{3}(l + 2m)I_1^3 - 2mI_1I_2 + nI_3 \tag{3}$$

Here, λ and μ are the Lamé constants with respect to infinitesimal deformation, l, m and n are Murnaghan's constants determined by the type of sample material, I_1, I_2 and I_3 are the strain invariants of first-, second- and third-order. Conservation of energy requires Hooke's Law to be expressed as:

$$\rho\delta W_s = \sigma_{ij}\frac{\partial\delta u_i}{\partial x_j} \tag{4}$$

where δW and δu_i denote the finite increment in free-energy function and displacement field, respectively. ρ is the density after deformation. Combining Equations (3) and (4) results in an acoustoelastic equation which connects static loading with the elastic wave velocity under hydrostatic pressure P:

$$\begin{aligned} \rho_0 v_c^2 &= \lambda + 2\mu - \tfrac{P}{3\lambda+2\mu}(6l + 4m + 7\lambda + 10\mu) \\ \rho_0 v_s^2 &= \mu - \tfrac{P}{3\lambda+2\mu}(3m - 0.5n + 3\lambda + 6\mu) \end{aligned} \tag{5}$$

The subscripts c and s imply compressive (longitudinal) and shear wave respectively. Subscript 0 denotes the undeformed state. The level of stress can thus be determined by measuring the wave velocity based on Equation (5) [53].

A typical ultrasonic stress measurement setup is illustrated by Figure 4. The sample is fixed on a tensile machine and subjected to uniaxial loading. Ultrasonic transducers are used to generate and receive an acoustic pulse that propagates along the sample and measure the wave velocity from the two ends. Chaki and Bourse extended this time-of-flight velocity measurement to metallic cable strands and determined their stress-strain state [54]. Gennisson et al. focused on stressed soft solids and explored shear modulus measurements [55]. Besides bulk waves, surface acoustic waves were also used to evaluate the existence and level of surface residual stress. A nonlinear acoustoelastic theory with respect to Rayleigh wave propagation was introduced by Iwashimizu and Kobori [56]. They concluded that the propagation of a Rayleigh surface wave in finitely deformed solids satisfies the wave equation similar to the linear-elastic material, only the elastic modulus tensor loses some of the symmetric conditions as the regular stress-strain relation. More recently, Duquennoy et al. used an interdigitated transducer and a laser interferometer to test superficial residual stress experimentally [57,58]. The transducer was designated for surface wave generation using a contacting approach.

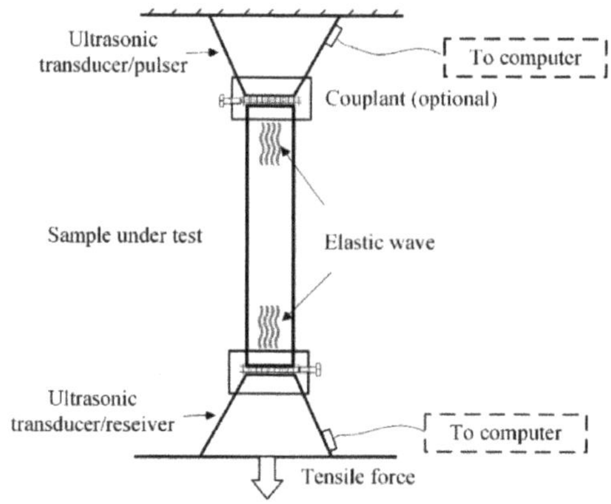

Figure 4. Stress measurement scheme by pulsed ultrasound.

The advantage of acoustic methodologies is their multimode propagation which is applicable for both internal and superficial stress-strain evaluation. However, the generation and detection of ultrasonic waves invariably relies on contact transducers. Based on the PA effect investigated in the 1980s and 1990s, it is possible to realize noncontact acoustic stress-strain characterization. Kasai and Sawada first discussed the possibility of stress distribution measurements using a PA microscope [59]. Because of the existence of photon-phonon energy conversion, the heat equation was modified as:

$$\frac{Q}{T_0} + \rho C \frac{\Delta T}{T_0} = -\left(\alpha - \frac{1}{E^2}\frac{\partial E}{\partial T}\sigma\right)\Delta\sigma \tag{6}$$

where, E is Young's modulus, α is linear expansion coefficient, σ and $\Delta\sigma$ are static stress and stress change after deformation, Q is the heat flux induced by photon (laser) irradiation and T is the temperature, the subscript 0 refers to stress-free state. Equation (6) shows the coupling of stress and temperature change. Using piezoelectric devices, the amplitude of the detected PA signal is proportional to the $\Delta\sigma$ for small signal approximation. However, Equation (6) does not seem to be comprehensive enough and thus no experimental procedure was proposed in Ref. [59].

Muratikov et al. published a series of theoretical and experimental reports [60–62] on PA residual stress evaluation. They started from the basic form of the first law of thermodynamics and Murnaghan's nonlinear theory and wrote the modified wave equation as:

$$\frac{\partial P_{ij}}{\partial x_j} = \rho_0 \frac{\partial^2 u_i}{\partial t^2} \tag{7}$$

where, u_i is the displacement field and ρ_0 is the density before deformation. Equation (7) can be regarded as an alternative form to Equation (4) by introducing the Piola-Kirchhoff stress tensor P_{ij} [63]. Repeated subscripts stand for summation and the equation is in Euler coordinates (x_1, x_2, x_3). P_{ij} is a function of strain energy due to pre-stressing (Equation (3)) and photon-phonon conversion, the latter being expressed as:

$$W_0 = \left(\frac{3}{2}\lambda + \mu\right)\alpha\left[(1 + \beta_0 I_1^0)\delta_{ij} + \beta_1 \varepsilon_{ij}^0\right]\left(\frac{\partial u_i}{\partial x_j} + \frac{\partial u_j}{\partial x_i} + \frac{\partial u_i}{\partial x_k}\frac{\partial u_j}{\partial x_k}\right)\Delta T \tag{8}$$

where, W_0 is the free-energy function of laser radiation, β_0 and β_1 are thermoelastic coupling coefficients of static strain and repeated subscript summation is used. Further derivation shows that the PA signal is proportional to the vertical displacement component. In their experiments, Muratikov et al. used several levels of force to produce indentations on a silicon nitride sample and the PA signal was plotted as a function of laser scanning distance. Figure 5 shows the result using a certain level of indentation with an obvious change of PA signal in the region of indentation-induced flaws. In their reports, Muratikov et al. did not quantify the correlation between PA signal and the actual local residual stress [60]. Moreover, the system used does not qualify as totally noncontact because of the use of contacting piezoelectric crystals as detectors.

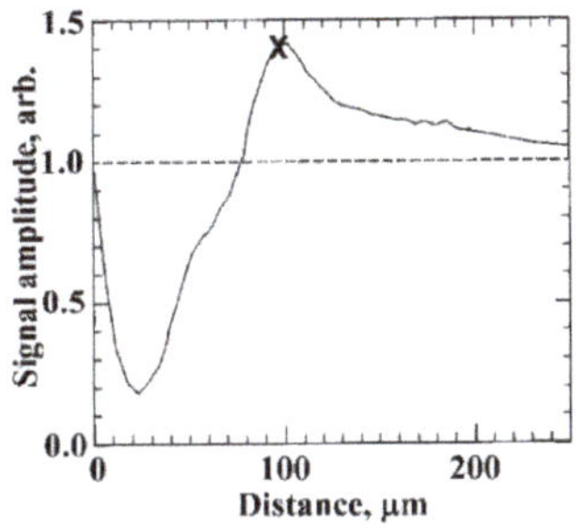

Figure 5. The scanning photoacoustic signal amplitude change. The visible defect is between 0 and 97 μm ("X" marking); reproduced from [60] with permission from The American Institute of Physics © 2000.

Huan et al. developed a noncontact PA stress-strain measurement method using a narrow-bandwidth immersion ultrasonic transducer [64]. The use of a tensile machine and adhesive strain gauge enabled simultaneous measurements of actual strain and PA signal. As shown in Figure 6, in the MHz range, the detected PA signal phase and amplitude show linear-to-nonlinear evolution of local strain from elastic to plastic deformation. The results show that a change of elastic properties by means of tensile stress exists even for linear deformation, although a corresponding explanation of the stress effect was not given in the published report. Another limitation of the proposed setup was the requirement for water coupling.

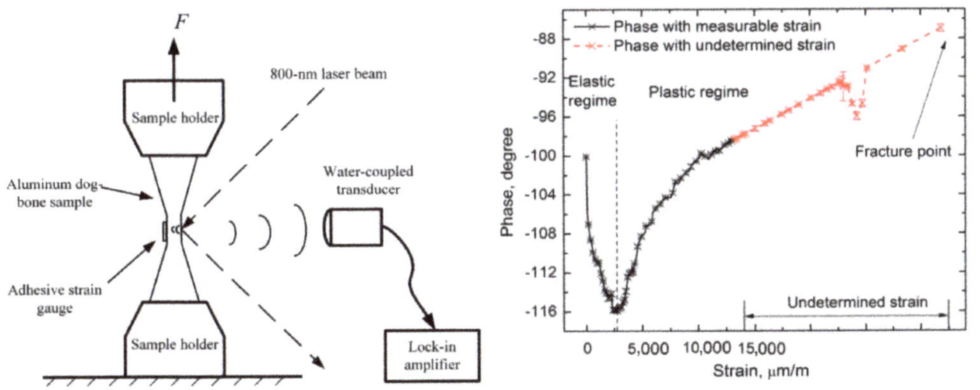

Figure 6. PA stress-strain test schematic with water coupling (left) and the stress-strain curve on the basis of PA phase (right); reproduced from [64] with permission from Springer Nature © 2017.

If the ultrasonic wave is detected with an optical method, the PA stress-strain measurement can be truly noncontact. As proposed by Sun and Zhou [65], a high sensitivity laser interferometric system was developed to measure stress-induced delamination in carbon-fiber reinforced polymer. Based on the principle of time-of-flight approach [65], Karabutov et al. measured superficial and subsurface residual stresses in a metal with laser-induced ultrasound [66]. Not only lasers, but also incoherent sources such as a helium lamp, can be used in stress-strain analysis. McDonald et al. developed a PA spectroscopic methodology to detect the stress level in some transparent materials such as a thin polymer film [67]. A Fourier Transform Infrared (FTIR) spectrometer [68,69] was adopted in the test. The presence of stress can slightly shift absorption peaks of the film, which lays a basis for quantitative stress measurement by spectral analysis.

Photoacoustic techniques have inherent advantages for mechanical strength evaluation. Ultrasonic wave mode propagation velocities exhibit a clear theoretical dependence on the stress-strain level [70], which makes PA techniques sensitive and fast. Optical generation and detection schemes can produce multi-physical and totally noncontact NDT methods. The main disadvantage of PA techniques is their low energy conversion due to thermoelasticity and thus sophisticated apparatuses must be adopted. In addition, they are not sensitive to stress accumulation and are hard to perform local tests with, because ultrasonic waves travel fast in elastic media.

2.3. Photothermal (PT) Techniques

The discussion of PA signal dependence on the stress-strain state is based on the coupling between elastic and thermophysical properties such as density and bulk modulus, the change of which has perceptible impact on the acoustic signal. However, according to the thermodynamic theory proposed by Landau and Lifshitz [71] and Love [72], a change of stress-strain energy can affect another two groups of parameters apart from elasticity, namely, thermophysical properties [69] and the thermoelastic coupling coefficient [73,74]. Therefore, it is possible to determine the mechanical performance of a material remotely by examining its thermal property changes by means of, e.g., thermal infrared detectors.

The earliest implementation of this relationship is found in a patented instrumentation system named Stress Pattern Analysis by measurement of Thermal Emission, or SPATE [75]. A sample under test is fixed on a repetitive loading machine which provides a constant strain rate. The temperature associated with the strain energy change is measured with an infrared detector. The measurement is based on the relation between temperature and stress level described in Equation (6) and as a result a quantitative measurement of stress can materialize [76]. SPATE is a passive infrared thermographic (IT) system that does not require pumping external energy: the thermal radiation is the result of stress work in the sample. In 1988, Wong et al. reported that IT method can also be used for characterizing static loading in addition to cyclic loading state such as in a SPATE system [77]. They exerted a static bending load on an aluminum plate and captured thermographic images with an infrared camera. By analyzing the images before and after static deformation, they validated the possibility of detecting residual stress in metallic materials [77]. Similar tests were carried out for thermographical residual stress analysis in titanium alloys in NASA by Gyekenyesi and Baaklini [78]. Quinn et al. tested a series of steel and aluminum alloys with a commercial tensile machine and an IT setup. The thermoelastic signal even showed a qualitative stress distribution around holes on plates [79]. For hyper-elastic materials such as polymers, the use of passive IT can also diagnose residual stress and fatigues in them [80]. More recently, the development of NDT in reinforced composites drew the attention of a number of researchers. When the laminated structure was subjected to high levels of tensile, bending and torsional loading, the formation of internal delamination and micro cracks occurred along with energy release. By observing thermal infrared emission with a camera, IT techniques enables inspection of a whole piece of material and analysis of the inelastic stress-strain performance [81,82].

Generally, the heat equation in the presence of static stress can be derived from the first law of thermodynamics, which has the following form [70,83–85]:

$$\rho C\frac{\partial T}{\partial t}-\left(k_{ii}J_{ii}{}^{2}\frac{\partial^{2}T}{\partial {x_i}^{2}}\delta_{ij}+(1-\delta_{ij})(k_{ij}+k_{ji})J_{ii}J_{jj}\frac{\partial^{2}T}{\partial x_i\partial x_j}\right)=g \tag{9}$$

where, J_{ij} is a Jacobian indicating the coordinate transformation due to loading (repeated subscripts denote summation), g is the source term and k_{ij} is the thermal conductivity tensor. It can be shown that the thermal conductivity tensor has a direct connection with the static stress-strain state [45,60,71,83]. As a result, instead of the SPATE mode which is suitable for evaluating cyclic loading, the static stress-strain performance can be determined based on thermal anisotropy analysis [85,86] with a proper heat source, usually a laser, which brings photothermal (PT) techniques into this consideration. Compared with PA techniques, PT detection decouples the stress-strain dependence of many physical properties by focusing only on thermal fields. The PT testing modality is local, quantitative, and absolutely noncontact.

Photothermal radiometry (PTR) is an active infrared emission testing modality. High power lasers are typically used to generate transient or harmonic local heating in solid and liquid samples, followed by free diffusion or convection. Monitoring the behavior of the diffusing field, the thermophysical properties can be evaluated. Since the 1980s, Long et al. [87,88], Milner et al. [89,90], Busse et al. [91] and Mandelis et al. [92–94] have reported on a series of PT-NDT techniques and applications, mainly for measuring thermophysical properties. The first report on mechanical performance monitoring published by Yarai et al. [95] used a pyroelectric device to capture the thermal field and evaluate the residual stress. The system was simple but could not achieve fully noncontact detection. Pron et al. [96,97] constructed a totally noncontact system using an Ar-ion laser and an infrared camera to generate and record the PT field on a stressed sample as shown in Figure 7. The beam splitter in that figure ensured most of the laser power was directed to the sample while an optical detector provided a reference signal for amplitude and phase data acquisition. The relative change in thermal conductivity was calculated based on the diffusion field amplitude and phase images. For a carbon steel sample, a few percent change in its thermal conductivity was found in the direction paralleled to the uniaxial loading within the elastic regime [96]. Based on a similar setup, Paoloni et al. examined the PT signal of a plastically deformed metal sample [98]. The sample was stretched to fracture so that a high level of residual stress was distributed non-uniformly around the failure surface. The diffusivity changed as much as 65% between the fracture region and the surrounding intact part. Mzali et al. used a halogen lamp instead of a laser to test the change in thermal properties of a plastically deformed metal subjected to tensile loading [99]. Since the test was continuous, a thermocouple was attached to the sample surface to measure the temperature. The contact measurement makes the system inflexible and rigid. An elastic-to-plastic full-range tensile photo-thermo-mechanical radiometry (PTMR) test was carried out by Huan et al. using both a Mercury Cadmium Telluride (MCT) detector (single point measurement, shown in Figure 8 [100] and an infrared camera for imaging [84]. In those systems, a fiber-coupled high power diode laser was used to provide more operational flexibility. The pump laser was configured at oblique incidence and the tested data were processed with a normalization algorithm. The results are shown in Figure 9. The stress-strain relation of full-range tensile loading was plotted as a function of the thermal parameter κ which is defined as the sample thickness over square-root of thermal diffusivity. The curves show a good analogy with the conventional stress-strain relation and the relative change of thermal conductivity is at the same level as [97]. The aforementioned PTMR development shows the possibility of using a PTR setup as a non-contact strain gauge.

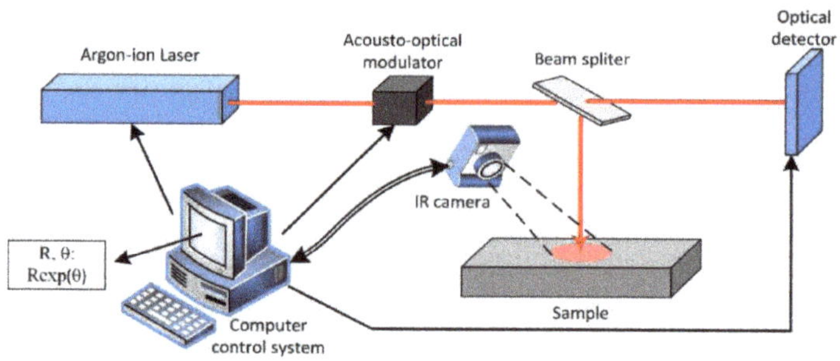

Figure 7. Schematic of a lock-in photothermal technique for stress-strain evaluation, redrawn based on [97].

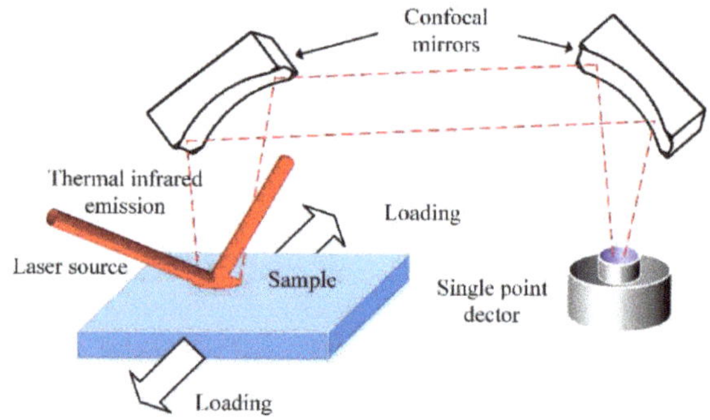

Figure 8. Schematic of lock-in photo-thermo-mechanical radiometry (PTMR) testing with a single-point detector and a confocal radiation collection system, redrawn based on [100].

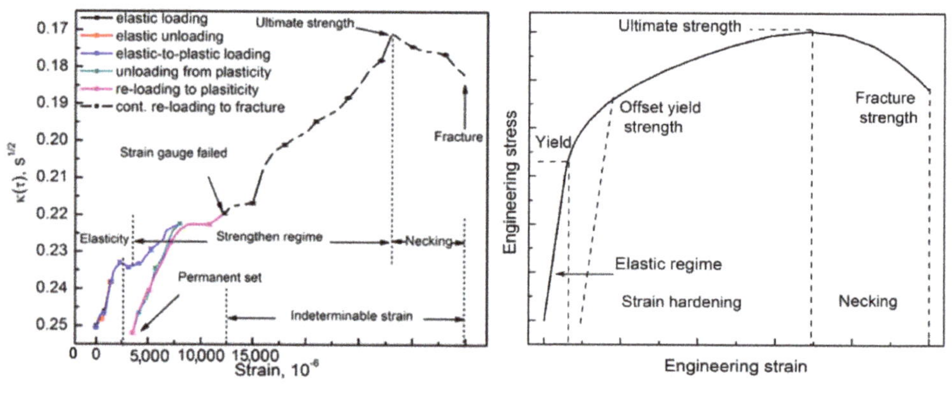

Figure 9. Composite thermal parameter κ as a function of strain that shows good analogy with a typical engineering stress-strain curve, reproduced from [100] with permission from Elsevier © 2016.

In addition to regular metallic materials, polymers and composites are also treated with PT techniques to assess their mechanical performance, especially after plastic deformation occurs. Wang and Wright studied the change of polymer thermal diffusivity in the plastically deformed range [101]. They derived the principal diffusivity tensor as a function of the Cauchy-Green deformation tensor. A much more conspicuous change in diffusivity was observed in polymers compared with metals as expected. However, such change is reversible because of the hyperelasticity of polymers. Huan et al. further used a single point PTR configuration to test nano-coated aluminum composites [102]. A linear dependence of thermal diffusivities on the load level was found within the elastic regime and only a minimal 0.6% diffusivity change occurred the in nano-coated samples. Based on these reports, the PTMR research was extended to a number of complex materials and structures with different elasticity. Mechanical properties of intact, defective and multi-structured samples were evaluated by studying the stability of thermo-mechanical parameters under loading.

Photothermal techniques have advantages over other methods in characterizing stress-strain relations. Compared with PA, PTMR detection usually evaluates mechanical performance through diffusivity changes. Coupling between thermoelastic properties can be completely separated and the mechanical stress-strain relation can be represented by a single parameter. In addition, dynamic PT techniques generate thermal-wave fields inside a sample where the effective thermal diffusion length can be controlled by the modulation frequency or the delay time of the photothermal source which can be made to impinge on specific locations and thus can be more sensitive to regionally distributed stresses. The use of continuous wave lasers and thermal infrared detectors and cameras makes the PTMR methodology totally non-contact. Nevertheless, PT techniques analyze the thermophysical properties by sensing thermal-wave field changes and thus are time consuming and optimally performing when confocal optical systems are used which can be complicated and expensive.

3. Conclusions

Nondestructive stress-strain characterization is of fundamental importance in modern industry for the early diagnosis of mechanical failure and prediction of safety issues. Many new nondestructive testing methodologies for mechanical performance assessment of industrial manufactured parts have been developed since the first application of X-ray tomography machines for industrial component residual stress monitoring in the 1940s. This article has reviewed three categories of typical NDT/NDE approaches which have their own unique features toward characterizing the stress-strain state. The highlights and drawbacks of each method were discussed by reviewing the development history and some results from state-of-the-art research. A brief summary is listed in Table 1 for each testing method.

Table 1. Summary of popular stress-strain NDT/NDE methods.

Category	Methodology	Physical Field	System Complexity	Contact/Non-Contact	Measurement Range
Optical/radiative	DIC	Electromagnetic	Low	Non-contact	Surface only
	X-ray	Electromagnetic	High	Non-contact	Internal
Acoustic	Pure acoustic	Elastic	Low	Contact	Surface/internal
	PA	Thermoelastic	Medium	Semi-/total-noncontact	Surface/internal
Thermal	SPATE	Electromagnetic	Medium	Non-contact	Average
	PTMR	Electromagnetic	Low	Non-contact	Surface/internal

In future, the following points should be considered to further boost related technologies and in-field applications:

Stress-strain imaging and distribution reconstruction. Until now most reports on mechanical measurements are based on simple loaded state configurations such as hydrostatic or uniaxial loading. Unfortunately, they are insufficient for dealing with real-world stress-strain conditions. New mathematical modeling and numerical approaches should be sought to develop in-field and quantitative stress distribution analyses.

Involvement of artificial intelligence (AI) and big data (BD) technologies in NDT. Multiphysical interfaces allow inputs of data from different physical fields. This approach incorporates advantages of various processes and can bring to bear testing of different properties such as thermal, optical, elastic etc. More data may be generated from a single group of tests and provide more powerful assistance in industrial applications. AI and BD technologies such as deep-learning and machine-vision enable automatic and massive data analysis with reliable and quantitative outputs. They are also useful for diagnosing components of irregular shape and dimension. Pioneering works that treated cylinders, spheres, corners and wedges etc. can be found in [103–107].

Author Contributions: Methodology, H.H. and A.M.; resources, L.L., C.P., and X.C.; writing—original draft preparation, H.H.; writing—review and editing, J.Z.; supervision, A.M.; funding acquisition, L.L. All authors have read and agreed to the published version of the manuscript.

Funding: This research was funded by National Natural Science Foundation of China (NNSFC), grant number 61801358, 61805187 and 61727804; the NSERC Discovery grants and the Canada Research Chairs Program; the China Postdoctoral Science Foundation, grant number 2019M65346; the Fundamental Research Funds for the Central Universities, grant number JB20190412 and XJS190505; the National Key R&D Program of China, grant number 2017YFF0106705. The APC was funded by 61801358 and 2019M65346.

Acknowledgments: The authors want to acknowledge the School of Optoelectronic Science and Engineering, University of Electronic Science and Technology of China for a supporting program.

Conflicts of Interest: The authors declare no conflict of interest.

References

1. Sposito, G.; Ward, C.; Cawley, P.; Nagy, P.B.; Scruby, C. A review of non-destructive techniques for the detection of creep damage in power plant steels. *NDT&E Int.* **2010**, *43*, 555–567.
2. Breysse, D. Nondestructive evaluation of concrete strength: An historical review and a new perspective by combining NDT methods. *Constr. Build. Mater.* **2012**, *33*, 139–163. [CrossRef]
3. Kamsu-Foguem, B. Knowledge-based support in Non-Destructive Testing for health monitoring of aircraft structures. *Adv. Eng. Inform.* **2012**, *26*, 859–869. [CrossRef]
4. Hola, J.; Schabowicz, K. State-of-the-art non-destructive methods for diagnostic testing of building structures—Anticipated development trends. *Arch. Civ. Mech. Eng.* **2010**, *10*, 5–18. [CrossRef]
5. Hola, J.; Beiń, J.; Sadowski, L.; Schabowicz, K. Non-destructive and semi-destructive diagnostics of concrete structures in assessment of their durability. *Civ. Eng.* **2015**, *63*, 87–96.
6. Ibrahim, M.E. Nondestructive evaluation of thick-section composites and sandwich structures: A review. *Compos. Part A-Appl. Sci. Manuf.* **2014**, *64*, 36–48. [CrossRef]
7. Duchene, P.; Chaki, S.; Ayadi, A.; Krawczak, P. A review of non-destructive techniques used for mechanical damage assessment in polymer composites. *J. Mater. Sci.* **2018**, *53*, 7915–7938. [CrossRef]
8. Burte, H.M. A science base for NDE and its coupling to technology. In *International Advances in Nondestructive Testing*; McGonnagle, W.J., Ed.; CRC Press: Boca Raton, FL, USA, 1979; Volume 6, pp. 19–38.
9. Adams, R.D.; Cawley, P. A review of defect types and nondestructive testing techniques for composites and bonded joints. *NDT&E Int.* **1988**, *21*, 208–222.
10. Malhotra, V.M.; Carino, N.J. *Handbook on Nondestructive Testing of Concrete*; CRC Press: Boca Raton, FL, USA, 2003; pp. 1–13.
11. Harding, J.; Wood, E.O.; Campbell, J.D. Tensile testing of materials at impact rates of strain. *J. Mech. Eng. Sci.* **1960**, *2*, 88–96. [CrossRef]
12. Ramberg, W.; Osgood, W.R. *Description of Stress-Strain Curves by Three Parameters*; National Advisory Committee for Aeronautics Technical Note; NACA-TN-902; National Advisory Committee for Aeronautics: Washington, DC, USA, 1943.
13. Hill, H.N. *Determination of Stress-Strain Relations from "Offset" Yield Strength Values*; National Advisory Committee for Aeronautics Technical Note; NACA-TN-927; National Advisory Committee for Aeronautics: Washington, DC, USA, 1944.
14. McCullough, K.Y.G.; Fleck, N.A.; Ashby, M.F. Uniaxial stress-strain behaviour of aluminium alloy foams. *Acta Mater.* **1999**, *47*, 2323–2330. [CrossRef]
15. Ling, Y. Uniaxial true stress-strain after necking. *AMP J. Technol.* **1996**, *5*, 37–48.
16. Turner, R.C.; Fuierer, P.A.; Newnham, R.E.; Shrout, T.R. Materials for high temperature acoustic and vibration sensors: A review. *Appl. Acoust.* **1994**, *41*, 299–324. [CrossRef]
17. Cegla, F.B.; Cawley, P.; Allin, J.; Davies, J. High-temperature (>500 °C) wall thickness monitoring using dry-coupled ultrasonic waveguide transducers. *IEEE Trans. Ultrason. Ferroelectr. Freq. Control* **2011**, *58*, 156–167. [CrossRef] [PubMed]

18. Kobayashi, M.; Jen, C.K.; Bussiere, J.F.; Wu, K.T. High-temperature integrated and flexible ultrasonic transducers for nondestructive testing. *NDT&E Int.* **2009**, *42*, 157–161.
19. Liu, Z.; Wu, H.; Paterson, A.; Luo, Z.; Ren, W.; Ye, Z.G. High Curie-temperature (TC) piezo-/ferroelectric single crystals with bismuth-based complex perovskites: Growth, structures and properties. *Acta Mater.* **2017**, *136*, 32–38. [CrossRef]
20. Lyons, J.S.; Liu, J.; Sutton, M.A. High-temperature deformation measurements using digital-image correlation. *Exp. Mech.* **1996**, *36*, 64–70. [CrossRef]
21. Anwander, M.; Zagar, B.G.; Weiss, B.; Weiss, H. Noncontacting strain measurements at high temperature by the digital laser speckle technique. *Exp. Mech.* **2000**, *40*, 98–105. [CrossRef]
22. Võlkl, R.; Fischer, B. Mechanical testing of ultra-high temperature alloys. *Exp. Mech.* **2004**, *44*, 121–127. [CrossRef]
23. Zhu, D.; Mobasher, B.; Rajan, S.D. Non-contacting strain measurement for cement-based composites in dynamic tensile testing. *Cem. Concr. Compos.* **2012**, *34*, 147–155. [CrossRef]
24. Wolverton, M.; Bhattacharyya, A.; Kannarpady, G.K. Efficient, flexible, noncontact deformation measurements using video multi-extensometry. *Exp. Tech.* **2009**, *33*, 24–33. [CrossRef]
25. Pan, B.; Wu, D.; Wang, Z.; Xia, Y. High-temperature digital image correlation method for full-field deformation measurement at 1200 °C. *Meas. Sci. Technol.* **2011**, *22*, 015701. [CrossRef]
26. Pan, B.; Qian, K.; Xie, H.; Asundi, A. Two-dimensional digital image correlation for in-plane displacement and strain measurement: A review. *Meas. Sci. Technol.* **2009**, *20*, 062001. [CrossRef]
27. Yang, X.; Liu, Z.; Xie, H. A real time deformation evaluation method for surface and interface of thermal barrier coatings during 1100 °C thermal shock. *Meas. Sci. Technol.* **2012**, *23*, 105604. [CrossRef]
28. Hoła, J.; Sadowski, Ł.; Reiner, J.; Stach, S. Usefulness of 3D surface roughness parameters for nondestructive evaluation of pull-off adhesion of concrete layers. *Constr. Build. Mater.* **2015**, *84*, 111–120. [CrossRef]
29. Czarnecki, S.; Hoła, J. Evaluation of the height 3D roughness parameters of concrete substrate and the adhesion to epoxy resin. *Int. J. Adhes. Adhes.* **2016**, *67*, 3–13.
30. Sadowski, Ł. Non-destructive identification of pull-off adhesion between concrete layers. *Automat. Constr.* **2015**, *57*, 146–155. [CrossRef]
31. Barrett, C.S.; Predecki, P. Stress measurement in graphite/epoxy uniaxial composites by X-ray. *Polym. Compos.* **1980**, *1*, 2–6. [CrossRef]
32. Ledbetter, H.M.; Austin, M.W. Internal strain (stress) in an SiC-Al particle-reinforced composite: An X-ray diffraction study. *Mater. Sci. Eng.* **1987**, *89*, 53–61. [CrossRef]
33. Hemley, R.J.; Mao, H.; Shen, G.; Badro, J.; Gillet, P.; Hanfland, M.; Häusermann, D. X-ray imaging of stress and strain of diamond, iron, and tungsten at megabar pressures. *Science* **1997**, *276*, 1242–1245. [CrossRef]
34. Prevey, P.S. X-ray diffraction residual stress techniques. *ASM Handb.* **1986**, *10*, 380–392.
35. Hughes, D.J.; Mahendrasingam, A.; Martin, C.; Oatway, M.B.; Heeley, E.L.; Bingham, S.J.; Fuller, W. An instrument for the collection of simultaneous small and wide angle X-ray scattering and stress–strain data during deformation of polymers at high strain rates using synchrotron radiation sources. *Rev. Sci. Instrum.* **1999**, *70*, 4051–4054. [CrossRef]
36. Patterson, B.M.; Cordes, N.L.; Henderson, K.; Williams, J.J.; Stannard, T.; Singh, S.S.; Ovejero, A.R.; Xiao, X.; Robinson, M.; Chawla, N. In situ X-ray synchrotron tomographic imaging during the compression of hyper-elastic polymeric materials. *J. Mater. Sci.* **2016**, *51*, 171–187. [CrossRef]
37. Fieres, J.; Schumann, P.; Reinhart, C. Predicting failure in additively manufactured parts using X-ray computed tomography and simulation. *Procedia Eng.* **2018**, *213*, 69–78. [CrossRef]
38. Subramanian, J.; Seetharaman, S.; Gupta, M. Processing and properties of aluminum and magnesium based composites containing amorphous reinforcement: A review. *Metals* **2015**, *5*, 743–762. [CrossRef]
39. Xing, J.; Zhao, C.; Yu, S.; Matsuda, H.; Ma, C. Experimental study on rock-like specimens with single flaw under hydro-mechanical coupling. *Appl. Sci.* **2019**, *9*, 3234. [CrossRef]
40. Youssef, S.; Marie, E.; Gaertner, R. Finite element modelling of the actual structure of cellular materials determined by X-ray tomography. *Acta Mater.* **2005**, *53*, 710–730. [CrossRef]
41. Bell, A.G. On the production and reproduction of sound by light. *Am. J. Sci.* **1880**, *20*, 305–324. [CrossRef]
42. Scruby, C.B.; Dewhurst, R.J.; Hutchins, D.A.; Palmer, B. Quantitative studies of thermally generated elastic waves in laser-irradiated metals. *J. Appl. Phys.* **1981**, *51*, 6210–6216. [CrossRef]

43. Rose, L.R.F. Point-source representation for laser-generated ultrasound. *J. Acoust. Soc. Am.* **1984**, *75*, 723–732. [CrossRef]
44. Hutchins, D.A. Mechanisms of pulsed photoacoustic generation. *Can. J. Phys.* **1986**, *64*, 1247–1264. [CrossRef]
45. Biot, M.A. The influence of initial stress on elastic waves. *J. Appl. Phys.* **1940**, *11*, 522–530. [CrossRef]
46. Hughes, D.S.; Kelly, J.L. Second-order elastic deformation of solids. *Phys. Rev.* **1953**, *92*, 1145–1149. [CrossRef]
47. Bergman, R.H.; Shahbender, R.A. Effect of statically applied stresses on the velocity of propagation of ultrasonic waves. *J. Appl. Phys.* **1958**, *29*, 1736–1738. [CrossRef]
48. Tokuoka, T.; Saito, M. Elastic wave propagations and acoustical birefringence in stressed crystals. *J. Acoust. Soc. Am.* **1968**, *45*, 1241–1246. [CrossRef]
49. Husson, D.; Kino, G.S. A perturbation theory for acoustoelastic effects. *J. Appl. Phys.* **1982**, *53*, 7250–7258. [CrossRef]
50. Shams, M.; Destrade, M.; Ogden, R.W. Initial stresses in elastic solids: Constitutive laws and acoustoelasticity. *Wave Motion* **2011**, *48*, 552–567. [CrossRef]
51. Davies, G.F. Quansi-harmonic finite strain equations of state of solids. *J. Phys. Chem. Solids* **1973**, *34*, 1417–1429. [CrossRef]
52. Murnaghan, F.D. Finite deformations of an elastic solid. *Am. J. Math.* **1937**, *59*, 235–260. [CrossRef]
53. Tylczyński, Z.; Mróz, B. The influence of uniaxial stress on ultrasonic wave propagation in ferroelastic $(NH_4)_4LiH_3(SO_4)_4$. *Solid State Commun.* **1997**, *101*, 653–656. [CrossRef]
54. Chaki, S.; Bourse, G. Guided ultrasonic waves for non-destructive monitoring of the stress levels in prestressed steel strands. *Ultrasonics* **2009**, *49*, 162–171. [CrossRef]
55. Gennisson, J.L.; Renier, M.; Catheline, S.; Barriere, C.; Bercoff, J.; Tanter, M.; Fink, M. Acoustoelasticity in soft solids: Assessment of the nonlinear shear modulus with the acoustic radiation force. *J. Acoust. Soc. Am.* **2007**, *122*, 3211–3219. [CrossRef] [PubMed]
56. Iwashimizu, Y.; Kobori, O. The Rayleigh wave in a finitely deformed isotropic elastic material. *J. Acoust. Soc. Am.* **1978**, *64*, 910–916. [CrossRef]
57. Duquennoy, M.; Ouaftouh, M.; Deboucq, J.; Lefebvre, J.E.; Jenot, F.; Ourak, M. Influence of a superficial field of residual stress on the propagation of surface waves—Applied to the estimation of the depth of the superficial stressed zone. *Appl. Phys. Lett.* **2012**, *101*, 234104. [CrossRef]
58. Duquennoy, M.; Ouaftouh, M.; Ourak, M.; Jenot, F. Theoretical determination of Rayleigh wave acoustoelastic coefficients: Comparison with experimental values. *Ultrasonics* **2002**, *39*, 575–583. [CrossRef]
59. Kasai, M.; Sawada, T. Non-destructive evaluation of the distribution of stress by means of the photoacoustic microscope. In *Photoacoustic and Photothermal Phenomena II*; Murphy, J.C., Maclachlan Spicer, J.W., Aamodt, L.C., Royce, B.S.H., Eds.; Springer: Berlin, Germany, 1990; pp. 33–36.
60. Muratikov, K.L.; Glazov, A.L.; Rose, D.N.; Dumar, J.E. Photoacoustic effect in stressed elastic solids. *J. Appl. Phys.* **2000**, *88*, 1948–2955. [CrossRef]
61. Muratikov, K.L. Theory of stress influence on the photoacoustic thermoelastic signal near the vertical crack tips. *Rev. Sci. Instrum.* **2003**, *74*, 722–724. [CrossRef]
62. Muratikov, K.L. Theory of the generation of mechanical vibrations by laser radiation in solids containing internal stresses on the basis of the thermoelastic effect. *Tech. Phys.* **1999**, *44*, 792–796. [CrossRef]
63. Keller, J.B. Finite elastic deformation governed by linear equations. *J. Appl. Mech.* **1986**, *53*, 819–820. [CrossRef]
64. Huan, H.; Mandelis, A.; Lashkari, B.; Liu, L. Frequency-domain laser ultrasound (FDLU) non-destructive evaluation of stress-strain behavior in an aluminum alloy. *Int. J. Thermophys.* **2017**, *38*, 62. [CrossRef]
65. Sun, G.; Zhou, Z. Non-contact detection of delamination in layered anisotropic composite materials with ultrasonic waves generated and detected by lasers. *Optik* **2016**, *127*, 6424–6433. [CrossRef]
66. Karabutov, A.; Devichensky, A.; Ivochkin, A.; Lyamshev, M.; Pelivanov, I.; Rohdgi, U.; Solomatin, V.; Subudhi, M. Laser ultrasonic diagnostics of residual stress. *Ultrasonics* **2008**, *48*, 631–635. [CrossRef] [PubMed]
67. McDonald, W.F.; Goettler, H.; Urban, M.W. A novel approach to photoacoustic FT-IR spectroscopy: Rheo-photoacoustic measurements. *Appl. Spectrosc.* **1989**, *43*, 1387–1393. [CrossRef]
68. Liu, L.; Mandelis, A.; Huan, H.; Melnikov, A. Step-scan T cell-based differential Fourier transform infrared photoacoustic spectroscopy (DFTIR-PAS) for detection of ambient air contaminants. *Appl. Phys. B* **2016**, *122*, 268. [CrossRef]

69. Liu, L.; Mandelis, A.; Huan, H.; Michaelian, K.H. Step-scan differential Fourier transform infrared photoacoustic spectroscopy (DFTIR-PAS): A spectral deconvolution method for weak absorber detection in the presence of strongly overlapping background absorptions. *Opt. Lett.* **2017**, *42*, 1424–1427. [CrossRef] [PubMed]
70. Qian, M. New thermoelastic technique for detection of residual stress distribution in solids. *Acta Acust.* **1995**, *14*, 97–106, (Chinese Version).
71. Landau, L.D.; Lifshitz, E.M. *Theory of Elasticity*; Pergamon: Oxford, UK, 1959; pp. 119–121.
72. Love, A.E.H. *A Treatise on the Mathematical Theory of Elasticity*, 2nd ed.; Cambridge University Press: Cambridge, UK, 1906; pp. 90–107.
73. Wong, A.K.; Jones, R.; Sparrow, J.G. Thermoelastic constant or thermoelastic parameter? *J. Phys. Chem. Solids* **1987**, *48*, 749–753. [CrossRef]
74. Wong, A.K.; Sparrow, J.G.; Dunn, S.A. On the revised theory of the thermoelastic effect. *J. Phys. Chem. Solids* **1988**, *49*, 395–400. [CrossRef]
75. Belgen, M.H. Structural stress measurements with an infrared radiometer. *ISA Trans.* **1967**, *6*, 49–53.
76. Stanley, P.; Chan, W.K. Quantitative stress analysis by means of the thermoelastic effect. *J. Strain Anal. Eng.* **1985**, *20*, 129–137. [CrossRef]
77. Wong, A.K.; Dunn, S.A.; Sparrow, J.G. Residual stress measurement by means of the thermoelastic effect. *Nature* **1988**, *332*, 613–615. [CrossRef]
78. Gyekenyesi, A.L.; Baaklini, G.Y. Thermoelastic stress analysis: The mean stress effect in metallic alloys. In Proceedings of the Nondestructive Evaluation Techniques for Aging Infrastructures and Manufacturing, Newport Beach, CA, USA, 3–5 March 1999. NASA/TM-1999-209376.
79. Quinn, S.; Dulieu-Barton, J.M.; Langlands, J.M. Progress in thermoelastic residual stress measurement. *Strain* **2004**, *40*, 127–133. [CrossRef]
80. Menczel, J.D.; Prime, R.B. *Thermal Analysis of Polymers Fundamentals and Applications*; John Wiley & Sons: Hoboken, NJ, USA, 2009; pp. 384–495.
81. Zhang, H.; Sfarra, S.; Sarasini, F.; Santulli, C.; Fernandes, H.; Avdelidis, N.P.; Ibarra-Castanedo, C.; Maldague, X.P.V. Thermographic non-destructive evaluation for natural fiber-reinforced composite laminates. *Appl. Sci.* **2018**, *8*, 240. [CrossRef]
82. Liu, J.; Gong, J.; Liu, L.; Qin, L.; Wang, Y. Investigation on stress distribution of multilayered composite structure (MCS) using infrared thermographic technique. *Infrared Phys. Technol.* **2013**, *61*, 134–143. [CrossRef]
83. Huan, H.; Mandelis, A.; Liu, L. Characterization of the mechanical stress–strain performance of aerospace alloy materials using frequency-domain photoacoustic ultrasound and photothermal methods: An FEM approach. *Int. J. Thermophys.* **2018**, *39*, 55. [CrossRef]
84. Huan, H.; Mandelis, A.; Liu, L.; Melnikov, A. Local-stress-induced thermal conductivity anisotropy analysis using non-destructive photo-thermo-mechanical lock-in thermography (PTM-LIT) imaging. *NDT&E Int.* **2017**, *91*, 79–87.
85. Salazar, A.; Sanchez-Lavega, A.; Ocariz, A.; Guitonny, J.; Pandey, G.C.; Fournier, D.; Boccara, A.C. Thermal diffusivity of anisotropic materials by photothermal methods. *J. Appl. Phys.* **1996**, *79*, 3984–3993. [CrossRef]
86. Iravani, M.V.; Nikoonahad, M. Photothermal waves in anisotropic media. *J. Appl. Phys.* **1987**, *62*, 4065–4071. [CrossRef]
87. Long, F.H.; Anderson, R.R.; Deutsch, T.F. Pulsed photothermal radiometry for depth profiling of layered media. *Appl. Phys. Lett.* **1987**, *51*, 2076–2078. [CrossRef]
88. Prahl, S.A.; Vitkin, I.A.; Bruggemann, U.; Wilson, B.C.; Aderson, R.R. Determination of optical properties of turbid media using pulsed photothermal radiometry. *Phys. Med. Biol.* **1992**, *37*, 1203–1217. [CrossRef]
89. Milner, T.E.; Smithies, D.J.; Goodman, D.M.; Lau, A.; Nelson, J.S. Depth determination of chromophores in human skin by pulsed photothermal radiometry. *Appl. Opt.* **1996**, *35*, 3379–3385. [CrossRef]
90. Milner, T.E.; Katzir, A.; Jacques, S.L. Pulsed photothermal radiometry of port-wine stains. *Proc. SPIE* **1993**, *1882*, 34–42.
91. Busse, G.; Wu, D.; Karpen, W. Thermal wave imaging with phase sensitive modulated thermography. *J. Appl. Phys.* **1992**, *71*, 3962–3965. [CrossRef]
92. Mandelis, A.; Abrams, S.H.; Nicolaides, L.; Garcia, J.A. Method and Apparatus for Detection of Defects in Teeth. U.S. Patent US6584341, 24 June 2003.

93. Wang, C.; Mandelis, A.; Liu, Y. Thermal-wave nondestructive evaluation of cylindrical composite structures using frequency-domain photothermal radiometry. *J. Appl. Phys.* **2005**, *97*, 014911. [CrossRef]
94. Balderas-López, J.A.; Mandelis, A.; Garcia, J.A. Thermal-wave resonator cavity design and measurements of the thermal diffusivity of liquids. *Rev. Sci. Instrum.* **2000**, *71*, 2933–2937. [CrossRef]
95. Yarai, A.; Yokoyama, Y.; Nakanishi, T. New non-destructive photothermal measurement of anisotropically distributed residual stress inside samples. In Proceedings of the IEEE Ultrasonics Symposium, Cannes, France, 31 October–3 November 1994; pp. 683–686.
96. Pron, H.; Henry, J.F.; Offermann, S.; Bissieux, C.; Beaudoin, J.L. Analysis of stress influence on thermal diffusivity by photothermal infrared thermography, Estimation of local thermophysical properties by means of front-face photothermal infrared thermography: Application to mechanical stress analysis. *High Temp.-High Press.* **2000**, *32*, 473–477. [CrossRef]
97. Pron, H.; Bissieux, C. 3-D thermal modelling applied to stress-induced anisotropy of thermal conductivity. *Int. J. Therm. Sci.* **2004**, *43*, 1161–1169. [CrossRef]
98. Paoloni, S.; Tata, M.E.; Scudieri, F.; Mercuri, F.; Marinelli, M.; Zammit, U. IR thermography characterization of residual stress in plastically deformed metallic components. *Appl. Phys. A* **2010**, *98*, 461–465. [CrossRef]
99. Mzali, F.; Albouchi, F.; Nasrallah, S.B.; Petit, D. Optimal experiment design and thermo-physical characterization of a plastically deformed solid. *Inverse Probl. Sci. Eng.* **2009**, *17*, 335–345. [CrossRef]
100. Huan, H.; Mandelis, A.; Liu, L.; Melnikov, A. Non-destructive and non-contacting stress-strain characterization of aerospace metallic alloys using photo-thermo-mechanical radiometry. *NDT&E Int.* **2016**, *84*, 47–53.
101. Wang, Y.; Wright, N.T. A relationship between thermal diffusivity and finite deformation in polymers. *Int. J. Thermophys.* **2005**, *26*, 1849–1859. [CrossRef]
102. Huan, H.; Mandelis, A.; Liu, L.; Melnikov, A. Evaluation of mechanical performance of NiCo nanocoated aerospace aluminum alloy using quantitative photo-thermo-mechanical radiometry as a non-contact strain gauge. *NDT&E Int.* **2017**, *87*, 44–49.
103. Tai, R.; Zhang, J.; Wang, C.; Mandelis, A. Thermal-Wave Fields in Solid Wedges Using the Green Function Method: Theory and Experiment. *J. Appl. Phys.* **2013**, *113*, 133501. [CrossRef]
104. Liu, L.; Wang, C.; Yuan, X.; Mandelis, A. Curvature-insensitive methodology for thermal-wave depth-profilometry in curvilinear solids. *J. Phys. D-Appl. Phys.* **2010**, *43*, 285403. [CrossRef]
105. Celorrio, R.; Mendioroz, A.; Apiñaniz, E.; Salazar, A.; Wang, C.; Mandelis, A. Reconstruction of radial thermal conductivity depth profile in case hardened steel rods. *J. Appl. Phys.* **2009**, *105*, 083517. [CrossRef]
106. Zhang, J.; Xie, G.; Wang, C.; Mandelis, A. Laser induced thermal-wave fields in multi-layered spherical solids based on Green function method. *J. Appl. Phys.* **2012**, *112*, 033521. [CrossRef]
107. Wang, M.; Mandelis, A.; Melnikov, A.; Wang, C. Quantitative lock-in thermography imaging of thermal-wave spatial profiles and thermophysical property measurements in solids with inner corner geometries using thermal-wave field theory. *J. Appl. Phys.* **2018**, *124*, 205106. [CrossRef]

applied sciences

MDPI

Article

Inducing Damage Diagnosis Capabilities in Carbon Fiber Reinforced Polymer Composites by Magnetoelastic Sensor Integration via 3D Printing

Dimitrios G. Dimogianopoulos [1], Panagiotis J. Charitidis [2] and Dionysios E. Mouzakis [3,*]

1 Department of Industrial Design and Production Engineering, University of West Attica, 12241 Athens, Greece; dimogian@uniwa.gr

2 Environmental Engineering School, Democritus University of Thrace, 67100 Xanthi, Greece; nanocomp@mail.com

3 Sector of Mathematics and Engineering Applications, Mechanics Laboratory, Hellenic Army Academy, PO Vari P.O. 16673 Attica, Greece

* Correspondence: demouzakis@sse.gr; Tel.: +30-210-890-4000

Received: 15 January 2020; Accepted: 31 January 2020; Published: 4 February 2020

check for updates

Abstract: This study investigates the possibility of inducing damage diagnosis capabilities in carbon fiber reinforced polymer composite slabs using custom-built integrated sensors and conventional, affordable equipment. The concept utilizes magnetoelastic strips integrated via 3D printing procedures in composite slabs. Under external mechanical loading, the strip magnetization changes due to the magnetoelastic phenomenon. Accordingly, electrical signals may be passively induced in conventional reception coil circuits placed at a distance from the slab. Since these signals quantify the vibrating slab's response, which is affected by the slab's structural integrity, damage may be detected when specific signal characteristics change. Two main issues are examined, namely the ability of receiving meaningful (with respect to noise) electrical signals from the built-in strips despite their contact-less passive reception, and the potential of diagnosing damage using such signals. Hence, slabs of various sizes and levels of structural damage (notches) have been vibrated at different frequencies and amplitudes. Treating the experimental data from integrated strips by applying the proposed processing framework allows for calculating eigenfrequencies sensitive to occurring damage (and its severity), as verified by finite element models of the vibrating slabs. Accordingly, damage may be detected and evaluated via the currently proposed experimental testing and analysis framework.

Keywords: damage detection; damage assessment; smart sensor; magnetoelastic strip; 3D printing; additive manufacturing

1. Introduction

In the past decade considerable advances have been made in the technology of rapid manufacturing. Not only have the techniques per se evolved, but the range of available materials for the so-called 3D printing technologies (or additive manufacturing) has also widened [1]. Numerous works in the literature already refer to the next level of meta-structures which can be designed and manufactured under the provocative term "4D printing", referring to materials and structures with time-dependent response or geometry [2]. Another interesting feature of the wide range of additive manufacturing technologies is that sensors may be easily incorporated within the layered structure during its building phase, resulting in smart systems, i.e., objects with built-in sensing capabilities [3]. Several review articles (see [4,5] for typical examples) classify such smart systems according to their manufacturing process, operating principle, and/or functionality. Different types of sensors, e.g., optical for interactive devices [6], or embedded wires for capacitive sensing applications [7,8] have been suggested as potential smart systems. Mechanical strain sensing via various incorporated sensors in 3D printed systems has also been proposed [9–11]. Magnetite/thermoplastic composite in 3D printing as direct replacements for commercially available flow sensors [12], or custom-built devices (whiskers) with built-in magnetostrictive materials for specific flow measuring applications [13] have been successfully manufactured. Magnetic force sensors [14], or devices for sensing human tactile force and imperceptible skin deformation due to heart pulses [15] for biomechanical applications are reported as feasible options for 3D printed smart objects. Specifically, some applications related to stretchable sensors for biomedical purposes are also reported in [16].

On the other hand, several custom-built sensors designed to monitor structures for damage diagnosis purposes, have been proposed over the recent years, for instance [17–22]. Such sensors are mostly designed to be patched onto the structure, so that the sensing element such as, fiber bragg gratings [18,19], magnetoelastic strips [21], and PZT impedance [22] or MFC (macro-fiber composite) transducers [23] to name but a few, could measure specific dynamic properties of the structure under vibration.

Magnetoelastic sensors have been under intense investigation for many applications in the past decade. Since they react to changes in imposed mechanical strain, by changing their magnetic constant, they can be used as wireless and passive sensors at the same time. Magnetoelastic sensors are usually made by rapid melt quenching, in the form of amorphous ferromagnetic ribbons. Most members of this family are iron-rich alloys, such as the well-known $Fe_{40}Ni_{38}Mo_4B_{18}$ marketed under the commercial name Metglas®2826 MB. The authors have successfully used this material as a damage sensor for polymers and composite systems [21] and have gathered considerable experience in sensor application, data gathering, and processing over the last years. A main advantage of such sensors against all other types of relevant devices is that they require no wiring, electrical nor optical, and can be interrogated wirelessly and in a passive state.

In general, the sensor is carefully configured in order to interfere as little as possible with the monitored structure's dynamics. Therefore, the sensor weight and volume are kept as low as possible, meaning that in some cases even thermal-spraying techniques to achieve deposition of sensing element (copper electrodes) onto the structure [24] may be used. A notable exception to this rule is found in [25], where the sensor placed on the structure is deliberately chosen of considerable weight so as to change at will the underlying dynamics.

In general, 3D printing procedures have been used for the sensor's construction [17] or the integration of the necessary circuitry/devices in order to facilitate operation of laser displacement sensors [20] or MFC sensors/transducers [23]. In other words, the 3D printing process has very often served as a means of facilitating the sensor's construction process, or it's mounting onto a given structure and its effective operation, but it has rarely been used for manufacturing a structure with a sensing element integrated within.

The aim of the current study is to use the additive manufacturing principle for integrating sensing elements in a structure during its 3D printing process, and to diagnose structural damage based on data from these built-in sensing elements and conventional, affordable equipment. The considered structures are carbon fiber reinforced polymer (CFRP) composite slabs of different dimensions, which incorporate magnetoelastic strips acting as sensing elements under external slab loading. Then, due to the magnetoelastic phenomenon, such loading lead to changes in the strip's magnetization, which causes the induction of electrical signals in conventional reception coil circuits placed at a distance from the slab. Since these signals quantify the vibrating slab's response (itself depending on the slab's structural integrity) damage is detectable when changes of specific signal characteristics (selected eigenfrequencies) occur. Even if, at times, this detection principle may prove not very practical to use, it is a traditional approach for structural health monitoring purposes, as it can be sensitive to even small damages [25].

This detection principle has also been used in a recent study [26], which features a similar operational and testing concept to that proposed in the current work, but for the important detail of the magnetoelastic sensing material being attached onto (rather than integrated within) the structure. Due to this critical difference, two main issues must be dealt with, namely the possibility of recording meaningful (with respect to noise) electrical signals from the built-in strips despite their contact-less reception, and the potential of diagnosing damage using such signals. Therefore, slabs of various dimensions and damage levels have been vibrated at different frequencies and amplitudes. Via a dedicated test data processing framework along with finite element method (FEM) analysis, it is shown that magnetoelastic strips integrated within the material structure may successfully act as passive embedded sensors, transmitting meaningful signals during mechanical loading and providing qualitative and quantitative damage diagnosis results.

The rest of the paper is organized as follows: Section 2 presents the 3D printing process of CFRP slabs with the integration of sensing elements, and the testing protocol used for the experiments. In Section 3, the experimental data along with the relevant framework for data analysis and validation via FEM are shown. Based on this framework, damage detection and assessment results are obtained and commented upon. Finally, Section 4 presents some concluding remarks.

2. Materials and Methods

2.1. Slab Preparation by 3D Printing

A reprap project series clone model of a 3D printer (CTC i3, Zhuhai Electronic Ltd, Zhuhai City, Guangdong, China) operated by an Arduino® controller, was used to prepare the slabs. The printer operated in Fused Deposition Modeling (FDM) mode, had a single nozzle (diameter of 0.4 mm) and used a 1.75 mm polymer filament. The filament was a short carbon fiber reinforced PET-G polymer composite by 20 wt% (NEEMA3D™ Carbon: plus) of NEEMA3D™, Petroupolis, Greece. The filament had a Young's modulus of E = 3800 MPa and a yield strength of 52.5 MPa according to the manufacturer's specifications.

Rectangular slabs of 60 × 12 × 3 mm^3 (hereafter referred to as type-A slabs) were prepared by means of the FDM printer, which was paused once the desired layer height was reached, so as to allow for attaching the 2826 MB Metglas® magnetoelastic (20 μm) strip of 25 × 5 mm^2 (kindly donated by Prof. D. Kouzoudis, University of Patras). The magnetoelastic strips were attached to the desired depth by means of an epoxy adhesive. Subsequently, the magnetoelastic strip was covered by successive material layers until obtaining the final slab thickness (Figure 1). In this way, one set of type-A slabs with "hidden" built-in magnetoelastic strip sensors, hereafter referred to as M-sensors, was prepared. Again, another series of type-A slabs without magnetoelastic strips (type-A blanc) was also prepared for testing purposes.

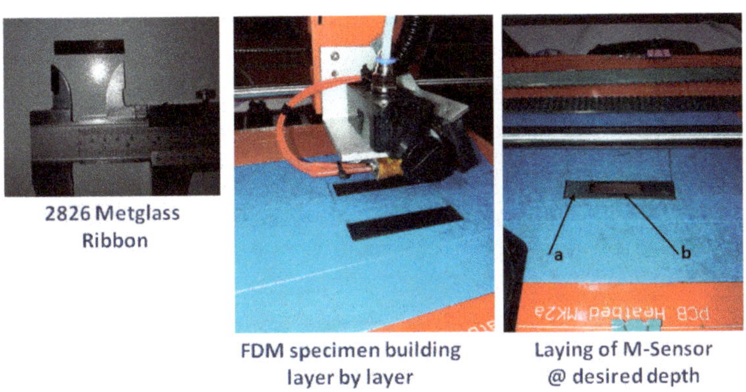

Figure 1. Smart slab preparation showing the strip of magnetoelastic 2826 Metglas® ribbon material and two views of the carbon fiber PET-G layers (**a**) during the process of ribbon (**b**) integration.

Two sets of larger slabs of 150 × 25 × 3 mm^3 were also prepared for the testing and may be seen in Figure 2. As seen in Figure 2, the magnetoelastic strip was placed inside the specimens during the 3D printing procedure, at heights of h = 1.15 mm for specimen's type A and h = 1.75 mm for specimen Types B and C, with respect to the specimen's total thickness, measuring from the bottom of the slabs.

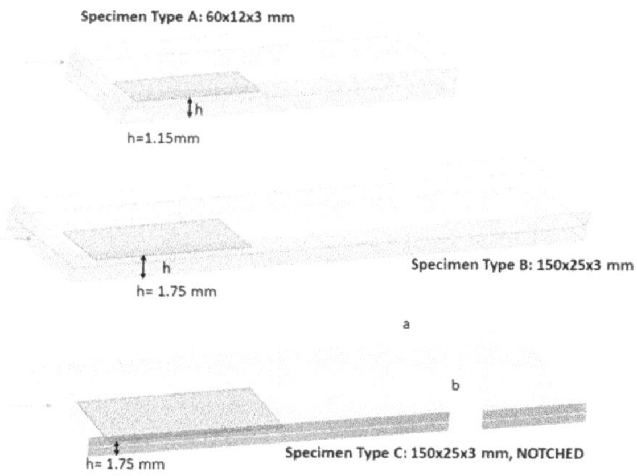

Figure 2. The three main slab types tested with built-in M-sensor. Type-A, -A blanc: slabs for DMA testing, Type-B: slabs for high-amplitude vibration, Type-C1, -C2: notched slabs for damage diagnosis with one (-C1), or two (-C2) notches at points a, b. Arrows on the left sides of slabs denote neutral axis in bending levels.

The first set involved M-sensor slabs (hereafter referred to as type-B slabs) and the series contained artificially damaged M-sensor slabs with one or two transverse notches (hereafter referred to as type-C1 or -C2 slabs, respectively), made with a diamond saw and featuring a depth of 1.5 mm.

Not that preparing two different slab sizes (type-A and type-B/-C1/-C2) was necessary because of technical limitations imposed by the testing procedures, as explained in the next subsection. Details on the objectives to be achieved and the testing protocol are given in the following subsection.

2.2. Experimental Procedures

Successful M-sensor operation depends on achieving two objectives:

- O1: That meaningful (with respect to noise) electrical signals passively emitted from M-sensors could be generated/recorded, and
- O2: That the analysis of recorded signals could provide damage detection and assessment (diagnosis) results, especially at high amplitude/low frequency vibrating operation.

The reason for focusing on high amplitude/low frequency operation in O2 is because both objectives were achieved in a high frequency/low amplitude vibrating context [21], even though the magnetoelastic strip was attached on the surface of the vibrating slab. Nonetheless, objective O1 was checked under all frequency—amplitude combinations, simply to ensure that the slabs with built-in strips do operate as sensors under such conditions. Two testing procedures have been used for checking whether O1 and O2 were achievable:

- P1: The procedure simulated high frequency and low amplitude operational conditions, typical of systems with fast dynamics and involved type-A slabs both in M-sensor and blanc (type-A blanc) forms. Testing was performed on a Dynamic Mechanical Analyzer (DMA Q800) of TA—Instruments (Figure 3), which led to choosing the specific dimensions of type-A slabs. Slabs were subjected to three-point dynamic bending under a linear frequency scan of 1–100 Hz @ RT and at v = 25 μm constant static deflection. Electrical signals created by induction to the remote interrogation coil (placed at 20 mm from the slab) were recorded by PioneerHill-Spectraplus© Software for Data Acquisition at 4096 Hz via a PC-Soundcard 3.5 mm jack. Comparing signals from type-A and -A blanc slabs allows for validating whether the hidden magnetoelastic material could provide sensing capabilities to the (otherwise passive) slab.

- P2: The procedure involved series of type-B and -C1, -C2 slabs, fixed as cantilevers (Figure 4). The free end suffered a load of quasi-sinusoidal form with frequency f = 4 Hz, mean value equal to 12.5 mm and amplitude d = 12.5 mm. In other words, the free end featured deflections between 0 and 25 mm, applied via a custom-built mechanical exciter, which allowed for using a (larger than type-A) size of type-B/-C1/-C2 slabs. Hence, low frequency and high amplitude operational conditions typical of systems (structures) with slow dynamics were simulated. Electrical signals created by induction to a low-cost interrogation coil (Vishay IWAS) placed at 20 mm from the vibrating slab's end, were recorded via a conventional digital oscilloscope at 200 KHz. C comparing signals from pristine (type-B) or damaged (-C1, -C2) slabs should conclude on the existence of frequency shifts and ensure that damage diagnosis results could be obtained using the M-sensor concept.

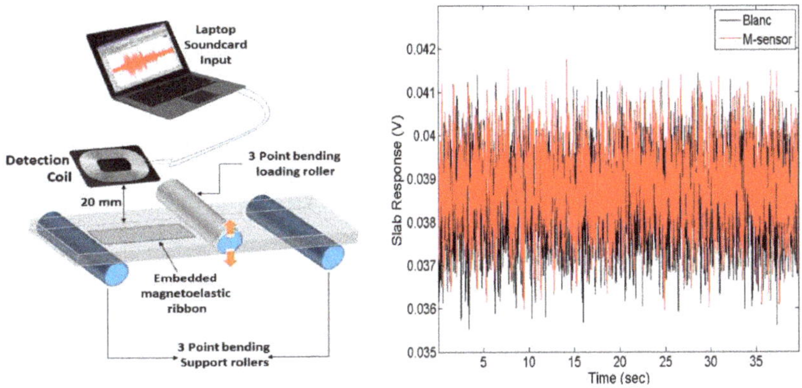

Figure 3. (**a**): DMA oscillation of smart type-A and -A blanc slabs. (**b**): Response of type-A, and A blanc slabs (in V) versus time.

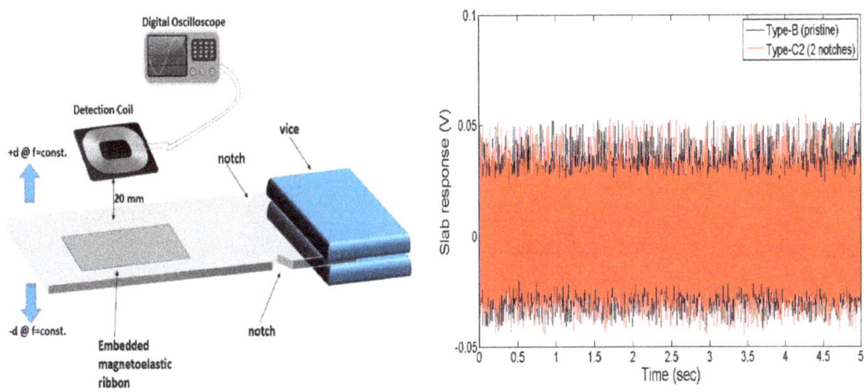

Figure 4. (**a**): Single cantilever arrangement for testing type-B, -C1 and -C2 slabs. (**b**): Response of type-B and -C2 (the most damaged) slabs (in V) versus time.

Note that the coil distance of 20 mm from the slab was defined by a trial-and-error process, and corresponds to the maximum distance for recording a meaningful signal with the given set up. Obviously, placing the coil very close (or onto) the surface should provide better signal recording capabilities, but it would undermine the practical aspects of the concept and could induce sensitivity to the "lift-off" effects noted, for instance, in applications based on Eddy currents. All recorded signals were analyzed with respect to their power spectra, estimated via the Welch method (usually coded in various software packages such as MATLAB®) with results presented in Section 3.

2.3. Finite Element Method Analysis

For achieving objective O2, the eigenfrequencies of type-B, -C1 and -C2 slabs involved in testing should be experimentally computed, and compared to their corresponding theoretical values. The latter were computed via FEM analysis, thus enabling to model and study complete type-B, -C1 and -C2 slabs, meaning that the magnetoelastic sensor (type-B), or even notches (-C1 and -C2 slabs) were incorporated. FEM COMSOL Multiphysics 5.4 software was used to model the process and compute the eigenfrequencies and mode shapes of the cantilever slabs with point loading (see Section 3.2). Free tetrahedral and boundary triangular elements were selected for meshing. Mesh size was selected as finer, because extremely finer meshes did not have a significant effect on the results. The required number of elements for meshing was found equal to 166.301 in total, while minimum element quality was set to 0.05. In all cases, the computations employed second-order polynomials.

3. Experimental Results and Analysis

3.1. Testing for Meaningful Signals with Respect to Noise (O1)

As described in Section 2.2, the DMA testing provided a conventional, yet rigorous means of testing whether the objective O1 is achievable. The procedure P1 involved a relatively high vibration frequency peak of 100 Hz and low vibration amplitude of 25 μm defined in the DMA operational settings. Two type-A blanc and two -A slabs were initially tested in dynamic three-point bending see Figure 3a. Typical responses of one type-A and one -A blanc in form of electrical signals (in V) versus time are presented in Figure 3b. Obviously, based on the time plot alone, no quantitative distinction (for instance, differences on signal variance or mean) between signals from type-A and -A blanc slabs may be noted. This behavior is not unreasonable, because the external mechanical loading cannot generate significant electrical power by means of the magnetoelastic phenomenon. Consequently, when similar passive techniques (i.e., no excitation on the reception coil) are utilized, one should at best expect qualitative rather than quantitative changes in the signals recorded. In other words, the spectral characteristics of such signals should be analyzed.

Figure 5 presents the power spectra corresponding to response signals from two type-A blanc and two A slabs showing clear differences. The insets show significant spikes at around 740 and 1220 Hz for signals from the two type-A slabs (red and green lines, respectively). At these frequencies, at least, there is a visible contribution from the M-sensor, superseding the effects of other phenomena (noise and so on). The latter show up as an output response signal on the coil even when a type-A blanc slab is tested.

Furthermore, in view of Figure 5, note that the slab excitation takes place at frequencies up to 100 Hz, yet the slab response shows mainly frequency content (harmonics) at higher frequencies (740 and 1220 Hz). The explanation may be two-fold. First, Figure 5 suggests that, at low frequencies (up to 200 Hz), a type-A slab actually produces a signal which has insignificant frequency content, barely distinguishable from parasitic noise. At higher frequencies and especially at 740, 1220 Hz, or more, the magnetoelastic strip seems to be generating a signal with more significant frequency content. This lack of sensitivity might be alleviated if a more efficient (and, thus, expensive) coil was used, but this would compromise one of the main goals of this work, namely performing damage diagnosis via conventional, affordable solutions. Second, there is the possibility of the vibrating slab behaving in a nonlinear manner. External loading of the slab at a given frequency produces a response signal with a shift of that frequency (and its harmonics), see Ch. 5 in [27]. Even though it is not trivial to state which explanation is valid, the fact is that, at a specific frequency band, the M-sensor can provide a signal qualitatively different than that recorded by vibrating type–A blanc slabs, thus achieving objective O1.

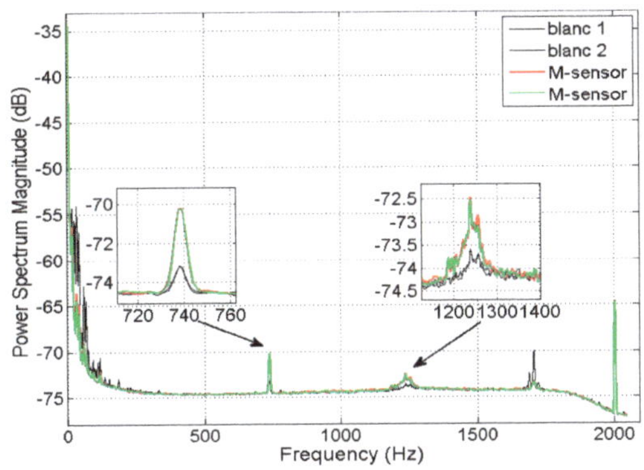

Figure 5. Power spectrum of response of two type-A (in green and red) and two -A blanc (in dark grey and black) slabs tested following procedure P1. Evidence that the hidden strip acts like a sensor are given in insets.

3.2. *Testing for Damage Detection Capabilities (O2)*

Section 2.2 described the second objective (O2) which should be achieved by following the testing procedure P2, namely that of obtaining signals useful for detecting and assessing (diagnosing) damage in the corresponding slabs, especially for vibrations of low frequency and high amplitude.

Procedure P2 was applied to three type-B slabs (pristine condition), two type-C1 slabs (thus, involving one notch), and two more -C2 slabs (with two notches-see Figure 2). Testing involved vibrating the free slab end at 4 Hz, with mean value of 12.5 mm and amplitude of 25 mm, with the response signals (as generated by the reception coil) recorded at 200 KHz via a conventional oscilloscope. Even though the sampling frequency is larger than those used in traditional structural monitoring systems, it is by no means unobtainable, since no particular exotic equipment solutions are needed.

Figure 6 shows the power spectra of the signal responses from the type-B slabs, with insets presenting frequency regions with the largest spikes from test data of at least two out of the three type-B slabs tested. From tests with type-A (see Section 3.1), it was deduced that an M-sensor slab provides meaningful signal (with respect to its frequency content) at high frequencies. At the same time, a convenient fact is that small damage may be quite clearly indicated at higher modes of vibration [25]. Hence, one should check whether the spikes noted in Figure 6 actually correspond to eigenfrequencies, so that they could be used for damage diagnosis purposes. Therefore, the FEM analysis described in Section 2.3 was carried out.

Figure 7 presents indicative results of type-B, -C1 and -C2 slabs for two frequency ranges (around 47,000 and 90,000 Hz). These plots resulted from the finite elements modal vibration analysis and show the eigenfrequencies and eigenmodes of the combined structure (slab and ribbon) at two significant frequencies noted in signal power spectral analysis (Figure 6). As seen, type-B slabs show quite different behavior to that of type-C1, -C2 slabs for both frequency bands. The results from eigenfrequencies determination for a type-B (among others) slab are presented in Table 1 and compared with those resulting from Figure 6, whereas their percent error is also shown. Interestingly, the six largest spikes in Figure 6 are at frequency values extremely close to the eigenfrequencies of type-B slabs. The percent error shown in Table 1 is almost insignificant, except for the lowest eigenfrequency at 4982.4 Hz, where it reaches almost 5%. This is in agreement with findings of type-A slab testing (Section 3.1), where the M-sensor's output frequency content was significant mainly at higher frequencies. So, finite element modal analysis has confirmed the values of significant eigenfrequencies of the specimens as detected in the experimentally obtained signals.

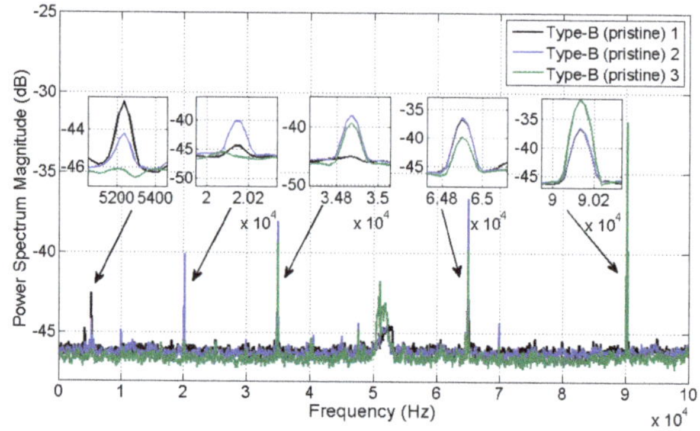

Figure 6. Power spectrum of response of type-B slab tested following procedure P2. Evidence that the hidden strip acts like a sensor providing specific eigenfrequencies are given in the insets.

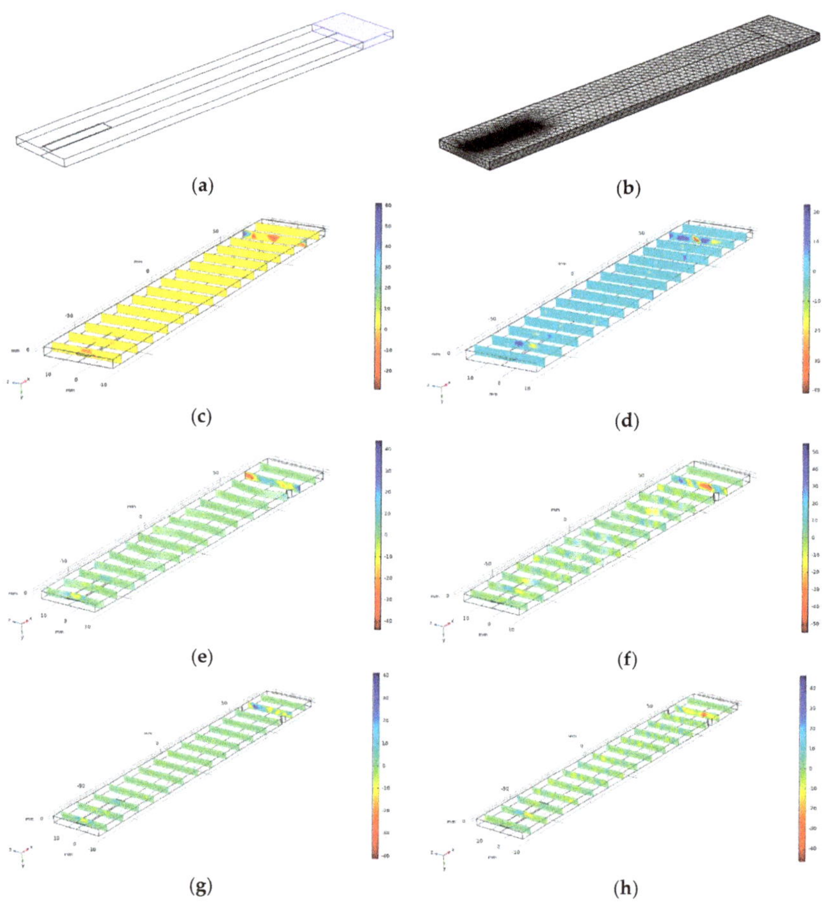

Figure 7. FEM setup and results: (**a**) Fixed constraints; (**b**) mesh., Displacements (mm) (**c**), (**d**) type-B at 47,402 and 89,994 Hz; (**e**,**f**) type-C1 at 47,253 and 90,038 Hz; (**g**,**h**) type-C2 at 47,017 and 90,064 Hz.

Table 1. Eigenfrequencies of type-B, -C1 and -C2 slabs as determined by the solution of the finite elements model compared with the ones picked up in signal power spectra.

FEM Eigenfrequencies in Hz for Indicated Slab: (I)			Power Spectra Eigenfrequencies in Hz for Indicated Slab: (II)			Absolute Value of % Error: abs{(I)-(II)}/(I)		
-B	-C1	-C2	-B	-C1	-C2	-B	-C1	-C2
4982.4	4986.9	5158.5	5235	5230	5233	5	5	1.4
20,389	20,373	19,985	20,142	20,140	20,140	1.2	1.1	0.8
34,804	34,798	35,257	34,868	34,865	34,868	0.2	0.2	1.1
47,402	47,253	47,017	47,553	47,555	47,555	0.3	0.6	1.1
65,038	65,007	65,061	64,898	64,900	64,895	0.2	0.2	0.3
89,994	90,038	90,064	90,135	90,133	90,133	0.2	0.2	0.1

This result could be anticipated, even without the FEM analysis. In fact, every realistic implementation of a quasi-sinusoidal excitation would trigger harmonics in the response signal, with their magnitude in the signal power spectrum following a decreasing trend with growing frequency values. On the rare occasion of spikes actually becoming larger with growing frequencies, one may start thinking of such harmonics being close to eigenfrequencies of the tested structure.

Figure 8 shows the power spectra of the signal responses from two type-C1 and another two type-C2 slabs. Power spectra of signals from type-B slabs are also presented for comparison. Considering the 6 largest spikes (as for the case of test data from type-B slabs), it results that type-C1 and -C2 slabs show spikes at approximately the same frequencies as type-B slabs (see Table 1) except for the spikes at 50,110 Hz and 40,090 Hz.

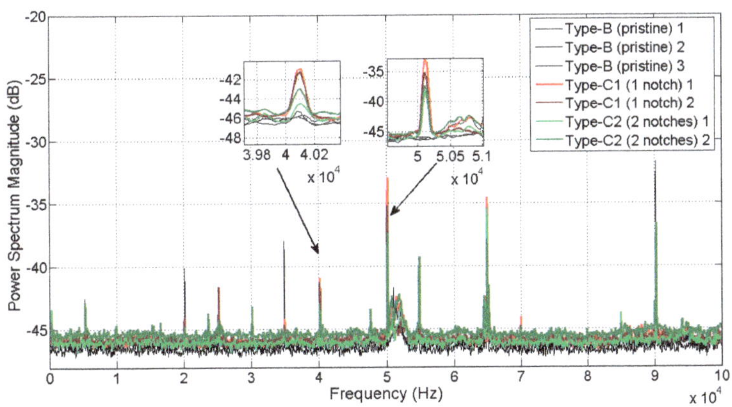

Figure 8. Power spectrum of response of type-C1 and -C2 slabs tested following P2 (type-B shown for comparison): Hidden strip acts as a sensor and damage causes spikes at 40 and 50 KHz (see insets).

These values were also cross-validated via FEM analysis: Eigenfrequencies at the region of 50 KHz were computed at 49,545 Hz and 49,946 Hz for -C1 and -C2 slabs, respectively. Again, as observed at the region of 40 KHz, eigenfrequencies equal to 39,936 Hz and 40,088 Hz were computed for -C1 and -C2 slabs, respectively. In all cases, the percent errors of the experimentally obtained (via power spectra) eigenfrequency values with respect to those computed via FEM analysis are extremely small. In both cases, the spike occurrences indicate alterations in structural condition, because at these frequencies no significant content was present in signals from pristine slabs (see also Figure 6). Note that in many other (mainly lower) frequency areas, one may find circumstantial evidence of spikes associated to signals from some but not all damaged slabs, in some cases somewhat undermined by noise. Hence, such frequency areas have been systematically disregarded, in favor of frequency bands where a selected number of the most prominent spikes may be clearly denoted.

There is also evidence that the damage level (one or two notches) may lead to spikes of different amplitude, since at both frequency ranges of interest (around 40 and 50 kHz), -C1 slabs feature power spectra with larger spikes than those from -C2 slabs. The analysis in [28] (specifically in Ch. 10), proposes an estimation of the damping coefficient from such plots, by associating spikes of narrower angle (i.e., more abrupt) to lower damping for a given frequency.

The fact that, at both frequencies, the spikes from -C1 slabs (in red and dark red) have a slightly narrower angle than those from -C2 slabs (in green and dark green) may indicate that damping slightly increases with notches, as should theoretically be the case [25].

Hence, damage is definitely detectable by considering a number of the most prominent spikes of the tested slab (mainly at higher frequencies) and comparing with the corresponding group of spikes from pristine slabs. Currently, frequencies around 40 and 50 KHz seem suitable for detecting damage, with occurrence of new spikes pointing to structural alteration (notches) of the tested slabs.

3.3. *Testing for Damage Assessment (O2)*

Even though detecting damage (notch) occurrence based on signals from testing with procedure P2 is definitely feasible, assessing the different damage levels is less trivial. Recall that damage detection proved possible at two frequency regions (40 and 50 KHz- see Section 3.2). At these regions, spikes from signals associated to -C1 and -C2 slabs differ only slightly. For an answer as to whether damage level (one or two notches) may be assessed via P2, a final series of tests has been carried out. Four -C1 and another four -C2 slabs were tested following P2 and the associated power spectra were plotted in Figure 9. Since four slabs of each damage level are available, ranges of spectrum magnitudes at 40 and 50 KHz for both damage levels may be formed. These ranges indicate the variation of spectral magnitude for signals associated to slabs with one or two notches.

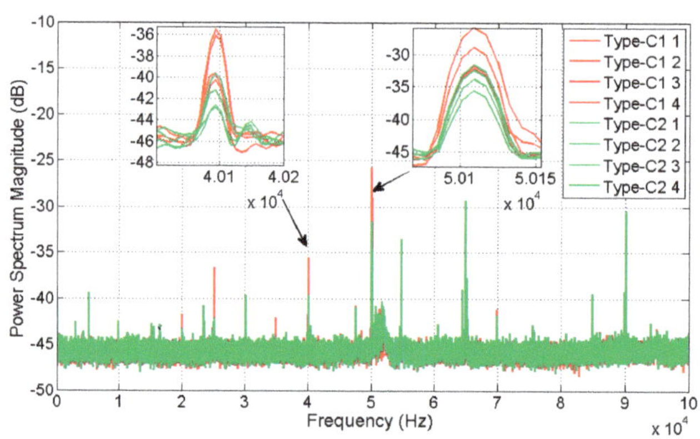

Figure 9. Power spectrum of response of type-C1 and -C2 slab tested following procedure P2. Evidence that the damage level is only slightly short of being detectable at 40 and 50 KHz is given in insets.

At both frequencies, an obvious trend for signals from more damaged (-C2) slabs having spectra of lower magnitude than their counterparts from less damaged (-C1) slabs is obvious. Spectra from -C1 slabs range between −35.5 to −40.5 dB at 40 KHz and −26 to −32.5 dB at 50 KHz. Spectra from -C2 slabs range between −40 to −43 dB at 40 KHz and −32 to −36.5 dB at 50 KHz. In both frequency areas, there is one -C1 slab whose magnitude overlaps that corresponding to a -C2 slab. Hence, since most of -C1 slabs have spectra of larger magnitude than those from -C2, procedure P2 provides indicative (rather than definitive) answers on the damage level of the tested slab.

These conclusions are based on the (rather empirical) estimation of the magnitude of damping values at each frequency found in [28]. Various approaches (for instance using dedicated equipment to analyze mode shape) may help to characterize differences induced by the damage level to the slab vibration modes. In the current study, quantitative damping data have been obtained by the recorded signals by investing effort in signal modeling procedures, as follows:

1. Response signals recorded by testing each -C1 or -C2 slab using procedure P2 were treated as stochastic time-series sequences. This was quite realistic, since the recorded signals had considerable noise due to electromagnetic parasitic phenomena related to the contact-less recording principle. Furthermore, given that damage may be assessed by means of two frequency regions (around 40,000 and 50,000 Hz), it was beneficial to perform high-pass filtering of the raw signals, allowing for a lower number of frequency components to be modeled. One simple means of doing so, was by differencing the raw signals (that is, forming the differences of each signal value at instant t minus that at $t-1$ for all t) as many times as necessary for effectively undermining lower frequency (below 40 KHz) components. Naturally, a traditional high pass filter may also be applied. Here, the raw signals recorded from testing with -C1 and -C2 slabs (with their spectral characteristics shown in Figure 9) were differenced six times, for obtaining signals with insignificant frequency content below 40 KHz.
2. Each signal was modeled by means of stochastic output–only AutoRegressive representations with constant coefficients (CC-AR), identified via standard algorithms. These may be coded in various programming languages, or even be found in software packages such as MATLAB®.
3. For each identified AR representation, poles at 40 and 50 KHz were monitored with respect to their damping. Characteristic changes in damping values (at these frequencies) between different data sets lead to damage assessment conclusions about the slabs generating these data.

The CC-AR representation is a standard yet effective means of modeling dependencies among values of a stochastic sequence. The stochastic output-only CC-AR representation relating the current (at time instant t) value of the modeled output signal to older ones has the following generic form:

$$\begin{aligned} y[t] &= \sum_{i=1}^{L} \theta_i \cdot p_i[t] + e[t] \\ e[t] &\sim NID\left(0, \sigma_e^2\right) \end{aligned} \quad (1)$$

with t designating the normalized discrete time and $y[t]$, $e[t]$ the output and one-step-ahead prediction error [or residual, assumed to be a zero-mean uncorrelated sequence with variance σ_e^2 signals, respectively. The term $NID(\mu,\sigma_e^2)$ stands for Normally Independently Distributed (with the indicated mean μ and variance σ_e^2). The terms $p_i[t]$ are referred to as regressors and involve lagged output values ranging from $y[t-1]$ up to $y[t-ny]$, with ny designating the maximum value of lags admitted for the signal $y[t]$ in (1) (model order). The i-th regressor coefficient is noted θ_i and has constant value. For instance, a CC-AR representation designated as $y[t] = 0.3 \cdot y[t-1] + 0.7 \cdot y[t-2] - 0.1 \cdot y[t-3] + e[t]$ has an order ny equal to 3 and is usually referred to as a CC-AR(3) model. Its regressor coefficients are $\theta_1 = 0.3$, $\theta_2 = 0.7$ and $\theta_3 = -0.1$.

The objective is to identify the CC-AR representation, that is, to select the regressors $p_i[t]$ and to estimate the associated coefficients θ_i in order to most effectively represent the dynamics of the considered signal. The identification of CC-AR representation along with the estimation of coefficients θ_i are trivial procedures extensively presented in, among others, [21]. The interested reader may refer to this work for details on determining the model order and estimating the coefficients. Briefly, a signal is effectively modeled by a CC-AR(ny) representation if one determines the lowest ny value for which the associated CC-AR(ny) model accurately describes the frequency content of the signal (by assigning poles at those frequencies) and the residual sequence $e[t]$ is uncorrelated. The latter means that the values of $e[t]$ for various t feature insignificant mutual dependencies, as checked by specific hypothesis tests [21].

At this point, note that identifying models for fault diagnosis applications is a somewhat different task than performing identification for purposes such as control, signal pre-whitening and so on. For fault diagnosis, the identified model should include poles with frequencies as close as possible to the peak locations (frequencies) of interest in the power spectrum plot. At the same time, the identified CC-AR(ny) model should have as few regressors as possible: Extra regressors (over-modeling) may lead to fewer correlated residuals $e[t]$ but tend to assign poles even at frequency spikes corresponding to noise or other parasitic phenomena. Over-modeling should, thus, be avoided, as far as possible.

In the current case, these peaks of interest were located at 40 and 50 KHz. Ideally, the user should determine the value range for ny, for which all associated identified CC-AR(ny) models include poles at the desired frequency peaks. In other words, one should plot CC-AR(ny) poles at each frequency as a function of the corresponding value for ny (stabilization plot), and select a value range for ny leading to obtaining poles at the desired frequency regions. At the same time, these CC-AR(ny) models should produce residuals as uncorrelated as possible. Ideally, the model structure selected should have the smallest ny of those in the range, thus avoiding any over-modeling issues.

In Figure 10a typical stabilization plot from a type-C1 test data and another from a type-C2 slabs are shown. The considered values for ny ranged from 4 up to 180. With data from either -C1 or -C2 slabs, for $ny \geq 80$ the associated CC-AR(ny) models consistently assign poles at frequencies around 40 and 50 KHz. This means that for a given ny satisfying the condition $ny \geq 80$, there is always a set of poles assigned to frequencies close to 40 KHz and another set for those close to 50 KHz. Hence, for the selected value of ny, the relative difference of pole damping values at each frequency may provide quantitative information on the damage level, exactly as when the power spectra of test signals are (qualitatively) examined.

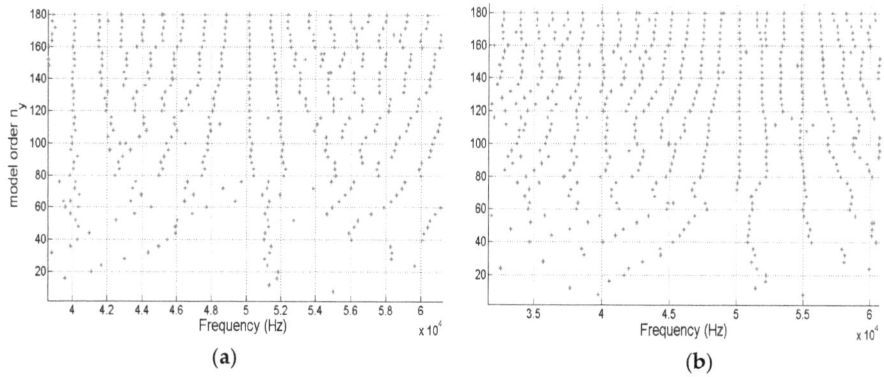

Figure 10. Stabilization plots with '*' indicating poles at each frequency: (**a**) type-C1 and (**b**) type-C2 slabs. For values $ny \geq 80$, poles are consistently assigned at the monitored frequencies of interest (around 40 and 50 KHz).

Four such models have been identified on data from type -C1 slabs. Another four CC-AR(80) models have been identified on data from -C2 slabs. As shown in Figure 11a, all eight models feature poles located at the frequency band of interest (around 40 and 50 KHz). The autocorrelation coefficients of CC-AR(80) residuals computed with test data from one -C1 and one -C2 slabs (with an indicative case shown in Figure 11b) appear uncorrelated, since they are located in-between the dashed lines designating statistical limits at the 5% risk level (see [21]). Then, clearly a CC-AR(80) representation is an adequate compromise for modeling test data from -C1 and -C2 slabs.

The pole-zero maps of these eight CC-AR(80) models in Figure 11a show that at 40 and 50 KHz there is one group of four poles associated to models identified on data from -C1 slabs (red crosses) which slightly overlaps another group of four poles associated to -C2 slabs (green crosses). It is, thus, obvious that it is not possible to completely distinguish between the two groups, due to the small overlap.

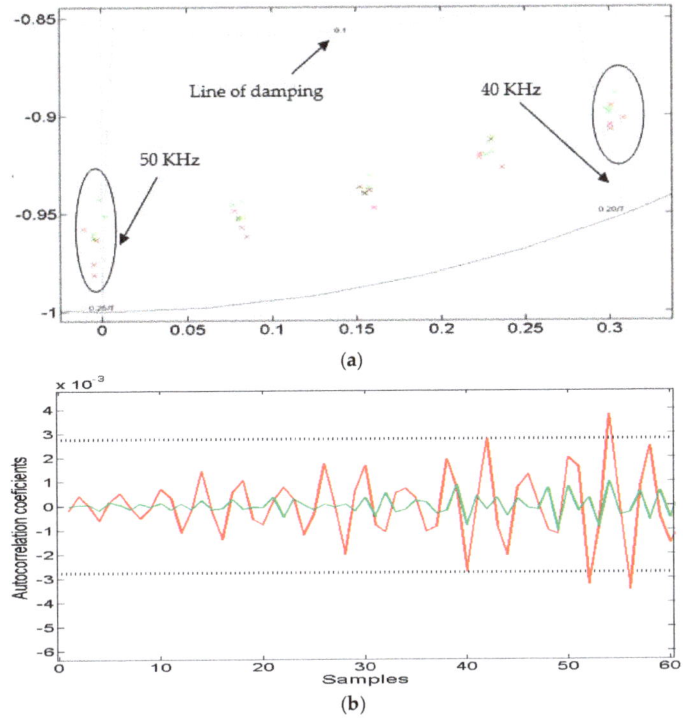

Figure 11. (**a**): Pole-zero maps in the complex Im-Re plane of CC-AR(80) models with red crosses associated to poles corresponding to -C1 slabs and green crosses associated to poles corresponding to -C2 slabs. (**b**): Autocorrelation plots indicating uncorrelated CC-AR(80) residuals (values mostly in-between the dashed limits) for one signal from -C1 and another from -C2 slabs.

However, it is also obvious that the poles associated to -C2 slabs (green crosses) have larger damping than that from poles associated to -C1 slabs (red crosses). This stems from the fact that the green crosses are closer to the dashed damping line (designating region of poles with damping coefficient equal to 0.1) than the red ones. In fact, damping of poles at 40 KHz ranges from 0.0389 to 0.0458 for -C1 slabs and from 0.0436 to 0.0508 for -C2 slabs. Damping of poles at 50 KHz ranges from 0.012 to 0.0272 for -C1 slabs and from 0.024 to 0.037 for -C2 slabs. These conclusions are identical to those resulting from the analysis based on the power spectral magnitude plots, presented above.

Moreover, tests with ny values up to 130 were also attempted only to find that, even for such high values of ny, the values in the damping of poles corresponding to -C1 and -C2 data respected the same relative order. Poles from -C1 consistently featured higher damping than those from -C2 slabs. Hence, if the relative order in damping values of poles resulting from slabs with different damage is monitored, then procedure P2 is able to provide quantitative indications about the level of damage in the tested slabs. Accordingly, the objective O2 was deemed as achievable, if only partly with respect to damage assessment. It is expected that fine-tuning P2 (or dropping the affordability condition by investing in high-tech equipment) should lead to smaller amplitude ranges for the power spectra of signals, thus achieving easier distinction of damage levels.

4. Conclusions

This study proposed a methodology for manufacturing smart slabs with embedded magnetoelastic sensors aiming at providing damage diagnosis capabilities with conventional, affordable equipment. Successful application of the proposed damage diagnosis concept requires that the smart slab must be under continuous monitoring, with the latter starting for a slab in perfect operational condition. Comparisons of experimental data over time will then detect and evaluate any occurring damage. The concept considered magnetoelastic strips integrated via 3D printing procedures in CFRP slabs. During the two testing procedures proposed, the smart slabs emitted (in a passive manner) electrical signals under vibration loading. These signals were received in a contact-less manner by conventional low-cost reception coils. The analysis of the power spectra of these signals validated the sensing concept, since they allow for computing the slab eigenfrequencies at high frequency regions. Such frequencies may thus be monitorable by combining conventional oscilloscopes, low-cost coils along with extensive analysis of the signal involved. Furthermore, it was shown that damage occurrence in slabs caused spikes in specific (high) frequency areas of the signals, and that following a specific signal modeling procedure, quantitative indications for assessing damage severity could also be found. Future work is underway to extend the concept from the current basic state (of a cantilever setup suffering from notches) to damage diagnosis cases of actual structural components. These include specimens of larger dimensions, implemented in representative to real-life structures, suffering from various damage profiles such as cracks, delaminations, and so on.

Author Contributions: Conceptualization D.E.M.; methodology D.E.M. and D.G.D.; software D.G.D. and P.J.C.; validation D.G.D. and P.J.C.; formal analysis D.G.D.; investigation D.E.M. and D.G.D.; resources D.E.M. and D.G.D.; data curation D.E.M. and D.G.D.; writing—original draft preparation, D.G.D.; writing—review and editing D.G.D. and D.E.M.; visualization D.G.D. and D.E.M.; supervision, project administration, D.E.M. and D.G.D. All authors have read and agreed to the published version of the manuscript.

Funding: This research received no external funding.

Acknowledgments: The authors would like to thank Professors Evangelia Kontou for kindly providing access to experimental facilities and Evangelos Hristoforou for the useful discussions and suggestions on this work. Both Professors are from the National Technical University of Athens (NTUA).

Conflicts of Interest: The authors declare no conflict of interest.

Abbreviations

The following abbreviations are used in this manuscript:

AR	autoregressive representation (model)
CC-AR	autoregressive representation with constant coefficients (model)
CFRP	carbon fiber reinforced polymer
DMA	dynamic mechanical analyzer
FBG	fiber bragg gratings
FEM	finite element method (model)
FDM	fused deposition modeling
Hz	Hertz
M-sensor	specimen with integrated magnetoelastic strip acting as integrated sensor
MFC	macro-fiber composite
MPa	mega Pascal
NID	Normally Independently Distributed
O1	objective 1 (Section 2.2)
O2	objective 2 (Section 2.2)
P1	experimental procedure 1(Section 2.2)
P2	experimental procedure 2(Section 2.2)
PC	personal computer
PET-G	polyethylene terephthalate glycol
PZT	piezoelectric transducer
type-A	slab of $60 \times 12 \times 3$ mm^3 with integrated Metglas® ribbon
type-B	slab of $150 \times 25 \times 3$ mm^3 with integrated Metglas® ribbon
type-C1	slab of $150 \times 25 \times 3$ mm^3 with integrated Metglas® ribbon and one notch
type-C2	slab of $150 \times 25 \times 3$ mm^3 with integrated Metglas® ribbon and two notches

References

1. Mouzakis, D.E. Advanced technologies in manufacturing 3D-layered structures for defense and aerospace. In *Lamination: Theory and Application*; Intechopen: London, UK, 2018; Volume 5, pp. 571–596.
2. Khoo, Z.X.; Teoh, J.E.M.; Liu, Y.; Chua, C.K.; Yang, S.; An, J.; Leong, K.F.; Yeong, W.Y. 3D printing of smart materials: A review on recent progresses in 4D printing. *Virt. Phys. Prototyp.* **2015**, *10*, 103–122. [CrossRef]
3. Ota, H.; Emaminejad, S.; Gao, Y.; Zhao, A.; Wu, E.; Challa, S.; Chen, K.; Fahad, H.M.; Jha, A.K.; Kiriya, D.; et al. Application of 3D printing for smart objects with embedded electronic sensors and systems. *Adv. Mater. Technol.* **2016**, *1*, 1600013. [CrossRef]
4. Ni, Y.; Ji, R.; Long, K.; Bu, T.; Chen, K.; Zhuang, S. A review of 3D-printed sensors. *Appl. Spectroc. Rev.* **2017**, *52*, 623–652. [CrossRef]
5. Xu, Y.; Wu, X.; Guo, X.; Kong, B.; Zhang, M.; Qian, X.; Mi, S.; Sun, W. The Boom in 3D-Printed Sensor Technology. *Sensors* **2017**, *17*, 1166. [CrossRef]
6. Willis, K.; Brockmeyer, E.; Hudson, S.; Poupyrevet, I. Printed optics: 3D printing of embedded optical elements for interactive devices. In Proceedings of the 25th annual ACM symposium on User interface software and technology (UIST '12), Cambridge, MA, USA, 7–10 October 2012.
7. Shemelya, C.; Cedillos, F.; Aguilera, E.; Maestas, E.; Ramos, J.; Espalin, D.; Muse, D.; Wicker, R.; MacDonald, E. 3D printed capacitive sensors. In Proceedings of the Sensors, 2013 IEEE, Baltimore, MD, USA, 3–6 October 2013; pp. 1–4.
8. Shemelya, C.; Cedillos, F.; Aguilera, E.; Espalin, D.; Muse, D.; Wicker, R.; MacDonald, E. Encapsulated copper wire and copper mesh capacitive sensing for 3-D printing applications. *IEEE Sens. J.* **2015**, *15*, 1280–1286. [CrossRef]
9. Muth, J.T.; Vogt, D.M.; Truby, R.L. Embedded 3D Printing of Strain Sensors within Highly Stretchable Elastomers. *Adv. Mater.* **2014**, *26*, 6307–6312. [CrossRef]
10. Agarwala, S.; Goh, G.L.; Yap, Y.L.; Goh, G.D.; Yu, H.; Yeong, W.Y.; Tran, T. Development of bendable strain sensor with embedded microchannels using 3D printing. *Sens. Actuators A Phys.* **2017**, *263*, 593–599. [CrossRef]
11. Christ, J.F.; Aliheidari, N.; Ameli, A.; Pötschke, P. 3D printed highly elastic strain sensors of multiwalled carbon nanotube/thermoplastic polyurethane nanocomposites. *Mater. Des.* **2017**, *131*, 394–401. [CrossRef]
12. Huber, C.; Abert, C.; Bruckner, F.; Groenefeld, M.; Muthsam, O.; Schuschnigg, S.; Sirak, K.; Thanhoffer, R.; Teliban, I.; Vogler, C.; et al. 3D print of polymer bonded rare-earth magnets, and 3D magnetic field scanning with an end-user 3D printer. *J. Appl. Phys. Lett.* **2016**, *109*, 162401. [CrossRef]
13. Na, S.M.; Park, J.J.; Jones, N.J.; Werely, N.; Flatau, A.B. Magnetostrictive whisker sensor application of carbon fiber-alfenol composites. *Smart Mater. Struct.* **2018**, *27*, 105010. [CrossRef]
14. Chatzipirpiridis, G.; Erne, P.; Ergeneman, O.; Pane, S.; Nelson, B.J. A magnetic force sensor on a catheter tip for minimally invasive surgery. In Proceedings of the 37th Annual International Conference of the IEEE Engineering in Medicine and Biology Society (EMBC), Milan, Italy, 25–29 August 2015.
15. Lee, H.B.; Kim, Y.W.; Yoon, J.; Lee, N.K.; Park, S.-H. 3D customized and flexible tactile sensor using a piezoelectric nanofiber mat and sandwich-molded elastomer sheets. *Smart Mater. Struct.* **2017**, *26*, 045032. [CrossRef]
16. Amjadi, M.; Kyung, K.U.; Park, I.; Sitti, M. Stretchable, skin-mountable and wearable strain sensors and their potential applications: A review. *Adv. Funct. Mater.* **2016**, *26*, 1678–1698. [CrossRef]
17. Kong, Q.; Fan, S.; Bai, X.; Mo, Y.L.; Song, G. A novel embeddable spherical smart aggregate for structural health monitoring: Part I. Fabrication and electrical characterization. *Smart Mater. Struct.* **2017**, *26*, 095050. [CrossRef]
18. Bocherens, E.; Bourasseau, S.; Dewynter-Marty, V.; Py, S.; Dupont, M.; Ferdinand, P.; Berenger, H. Damage detection in a radome sandwich material with embedded fiber optic sensors. *Smart Mater. Struct.* **2000**, *9*, 310. [CrossRef]
19. Ding, G.; Cao, H.; Xie, C. Multipoint cure monitoring of temperature and strain of carbon fibre-reinforced plastic shafts using fibre Bragg grating sensors. *Nondestruct. Test. Eval.* **2019**, *34*, 117–134. [CrossRef]
20. Chang, S.W.; Lin, T.K.; Kuo, S.Y.; Huang, T.-H. Integration of high-resolution laser displacement sensors and 3D printing for structural health monitoring. *Sensors (Basel)* **2018**, *18*, 19. [CrossRef]

21. Mouzakis, D.E.; Dimogianopoulos, D.G. Magnetoelastic metglas® sensors: Application of wireless detection principle and stochastic nonlinear modelling for damage diagnosis in smart systems. In *Glass Materials Research Progress*; Wolf, J.C., Lange, L., Eds.; Nova Science Publishers: New York, NY, USA, 2008; pp. 225–257.
22. Yang, Y.; Liu, H.; Annamdas, V.G.M.; Soh, C.K. Monitoring damage propagation using PZT impedance transducers. *Smart Mater. Struct.* **2009**, *18*, 045003. [CrossRef]
23. De Medeiros, R.; Sartorato, M.; Vandepitte, D.; Tita, V. A comparative assessment of different frequency based damage detection in unidirectional composite plates using MFC sensors. *J. Sound Vib.* **2016**, *383*, 171–190. [CrossRef]
24. Ogawa, M.; Huang, C.; Nakamura, T. Damage detection of CFRP laminates via self-sensing fibres and thermal-sprayed electrodes. *Nondestruct. Test. Eval.* **2013**, *28*, 1–16. [CrossRef]
25. Khammassi, M.; Wali, R.; Al-Mutory, A.; Yousaf, A.; Sassi, S.; Gharib, M. Simplified modal-based method to quantify delamination in carbon fibre-reinforced plastic beam. *Nondestruct. Test. Eval.* **2019**, *34*, 283–298. [CrossRef]
26. Samourgkanidis, G.; Kouzoudis, D.A. Pattern matching identification method of notchs on cantilever beams through their bending modes measured by magnetoelastic sensors. *Theor. Appl. Fract. Mech.* **2019**, *103*, 102266. [CrossRef]
27. Landau, L.D.; Lifshitz, E.M. *Mechanics*, 3rd ed.; Elsevier, Butterworth-Heinenann: Oxford, UK, 1981.
28. Rao, S.S. *Mechanical Vibrations*, 5th ed.; Prentice Hall: Upper Saddle River, NJ, USA, 2011.

applied sciences

MDPI

Article

Stress Estimation Using the Acoustoelastic Effect of Surface Waves in Weak Anisotropic Materials

Jihyun Jun [1], **Young-Dae Shim** [1] **and Kyung-Young Jhang** [2,*]

[1] Department of Mechanical Convergence Engineering, Graduate School, Hanyang University, Seoul 04763, Korea; jjhhelen@naver.com (J.J.); shim.youngdae@gmail.com (Y.-D.S.)
[2] School of Mechanical Engineering, Hanyang University, Seoul 04763, Korea
* Correspondence: kyjhang@hanyang.ac.kr; Tel.: +82-2-2220-0434

Received: 20 November 2019; Accepted: 18 December 2019; Published: 24 December 2019

check for updates

Abstract: This paper proposes a novel stress measurement method using the acoustoelastic effect of surface wave to estimate the stress of a homogeneous material plate with orthogonal anisotropy, in which the surface wave velocities are measured in three different directions before and after loading stress. The effectiveness of the proposed method was verified by numerical simulations and experiments. For the simulations, the surface wave velocities in three directions were obtained from a conventional perturbation model for weak anisotropic materials. The simulation results showed that the stress estimation error was less than 3% for an anisotropic rate up to 2% under stress conditions up to 90 MPa. Two specimens were prepared for the experiments, one was almost isotropic and another that had a relatively larger anisotropy rate of 2.6%. Then, the stresses loaded by a tensile test machine were estimated. The results showed good agreement with the given stresses for both specimens. These results confirm that the proposed method can be applied to estimate the surface stress state in anisotropic material plates. The proposed method is simple, practical, and is expected to be useful for monitoring changes of surface stress before and after machining such as the punching or bending of plate.

Keywords: acoustoelastic effect; stress estimation; surface waves; weak anisotropy material

1. Introduction

Monitoring of applied or residual stress is important for quality control of industrial parts manufacturing and health management of structures. The conventional methods of measuring stress include mechanical methods such as hole-drilling and nondestructive techniques such as X-ray diffraction (XRD) and acoustoelastic method. However, mechanical methods are complicated and not suitable for online monitoring. X-ray diffraction has the disadvantage of being harmful to the human body and requires separate pretreatment of the specimen. Unlike these approaches, the acoustoelastic method is harmless and has advantages that it can be applied to real products without pre-treatment and it is suitable for online monitoring.

The acoustoelastic method utilizes the propagation velocity of an elastic wave, which changes with the stress state of a medium [1]. The acoustoelastic effect appears in all kinds of ultrasonic waves including longitudinal, transverse and surface waves. For many applications, however, surface stresses are needed and in this case, surface waves are suitable since they only penetrate to a depth of approximately one wavelength. In addition, the transmitter and receiver of the surface waves are located on the same plane so that it is possible to perform inspection using only one side. Furthermore, it is not necessary to know the exact thickness of the applied object to measure the wave velocity. These features enhance the application of surface waves in the field. Accordingly, various studies on stress estimation using acoustoelastic effect of surface wave have been conducted. In particular, many studies have been carried out to analyze stresses in weld [2] or rail specimen [3,4]. In those studies, however, the direction of the dominant stress is already known, so it is only possible to determine the presence of the stress or to estimate the stress in the known direction of the dominant stress.

Meanwhile, there is also a method using longitudinal critical refraction (Lcr) waves, which have advantage of high sensitivity to stress [2,5]. However, the Lcr wave is only applicable in contact manner because it uses the critical refraction at the contact interface between the wedge and the test object. On the other hand, surface waves are easy to extend in a non-contact manner such as laser ultrasonic technology.

Early studies regarding the acoustoelastic effect of surface wave were conducted to identify the linear relationships between stress and surface wave velocity in isotropic material [6,7]. However, even if a material is initially isotropic, changes of the micro-structure occur during material production and processing, which may result in anisotropic properties. In this case, no matter how weak the anisotropy is, the use of the isotropic theory causes a large error in stress estimation and therefore cannot be applied.

Only a very limited number of studies have investigated on the acoustoelastic effect in anisotropic material. Delsanto et al. [8] and Mase and Delsanto [9] proposed a model that separates the contribution of the anisotropic effect and the contributions of the acoustoelastic effect to evaluate how the surface wave velocity changes from the isotropic state when a stress is applied in a weakly anisotropic material. Although, this method has not been experimentally verified, it is worthwhile to analyze the effect of anisotropy. However, it is very inconvenient for actual application, because it is necessary to know in advance how the actual elastic modulus of the material differs from the isotropic elastic modulus before the stress is applied. Another study by Thompson et al. [10] proposed a general solution for plane wave propagation in a symmetry plane of an orthorhombic, biaxial stressed anisotropic material. However, it does not provide a direct solution to the case of using surface waves. Additionally, it is necessary to know the direction of anisotropy in advance, which leads to an inconvenient step of measuring the shear wave velocity in many different polarization directions.

In this study, we propose a simpler and more practical method for estimating surface stress using the acoustoelastic effect of surface waves in a plate that has orthogonal anisotropy on its surface. In the proposed method, we modified the isotropic theory, in which the rate of change of surface wave velocity according to the stress is expressed using two acoustoelastic coefficients, two principal stresses and the angle between the principal stress and the wave propagation direction. The difference between the proposed method and isotropic theory is that in isotropic theory, the surface wave velocity at the unstressed state is constant regardless of the propagation direction, while in the proposed method it is dependent on the propagation direction due to anisotropy.

Proposed method can estimate not only magnitude but also direction of principal stress by measuring the rate of change of the surface wave velocity in any three directions before and after stress is applied, which does not require any information of stress direction in advance. The requirement to measure the initial surface wave velocities in three directions before stress is applied also will not be a problem at all in the comparison of the stress states before and after the machining such as punching or bending. Furthermore, this method is very simple and practical compared to the previous methods of Delsanto et al. [8] and Thompson et al. [10]. Since there is no need to pre-measure the directional elasticity of the material, or pre-check the direction of anisotropy. In addition, this approach requires measuring only the surface wave velocities in any three directions before and after the stress is applied.

For verification of the proposed method, numerical simulations were performed. For the simulation, we generated the surface wave velocity data with respect to the propagation direction using Delsanto's model for several typical cases when both anisotropy and stress exist. We then analyzed the error that occurs when the stress is estimated by the proposed method using this data. In the simulation, three different combinations of the anisotropy direction, principal stress direction and measurement direction, and two different stress states, uniaxial stress and biaxial stress, were evaluated.

To verify the proposed method experimentally, stresses applied by a tensile testing machine were estimated by measuring the surface wave velocities in three directions. The experiments were carried out for two types of aluminum plates, one nearly isotropic and one with relatively greater anisotropy. Two specimens were prepared for each type where one was used for the measurement of the acoustoelastic coefficients and the other was used for stress estimation. Stress up to 90 MPa was applied and the performance of stress estimation was verified by comparing the estimated stress with the applied stress.

2. Theory

2.1. Stress Estimation Method in Isotropic Material

The acoustoelastic effect of surface wave in isotropic materials is expressed as follows,

$$\frac{V_\theta - V_R}{V_R} = \frac{K_1 + K_2}{2}(\sigma_1 + \sigma_2) + \frac{K_1 - K_2}{2}(\sigma_1 - \sigma_2)\cos 2\theta \quad (1)$$

which represents the relationship between the surface wave velocity change and the principal stresses [9]. Here, V_R is the reference velocity of the surface wave in the unstressed state, V_θ is the surface wave velocity in the stressed state in the θ direction, and θ is the angle between the surface wave propagation direction and the principal stress σ_1. σ_1 and σ_2 are principal stresses, and K_1 and K_2 are the acoustoelastic coefficients.

The acoustoelastic coefficients K_1 and K_2 represent the linear proportional coefficients between the velocity change rate and stress when the surface wave propagates in the stress direction and in the direction perpendicular to the stress direction under the uniaxial stress condition, respectively. That is, when $\sigma_1 = \sigma$ and $\sigma_2 = 0$, the coefficient can be expressed as follows:

$$K_1 = \frac{V_0 - V_R}{V_R}\frac{1}{\sigma} \quad (2)$$

$$K_2 = \frac{V_{90} - V_R}{V_R}\frac{1}{\sigma} \quad (3)$$

where, V_0 is the surface wave velocity in the direction of stress and V_{90} is the velocity in the direction perpendicular to the stress.

To determine the three terms (σ_1, σ_2, θ) in Equation (1), three-directional surface wave velocities can be used. As an example, using the velocities in the three-directions with a 45° difference, as shown in Figure 1, the following three equations can be obtained.

$$\frac{V_\theta - V_R}{V_R} = \frac{K_1 + K_2}{2}(\sigma_1 + \sigma_2) + \frac{K_1 - K_2}{2}(\sigma_1 - \sigma_2)\cos 2\theta \quad (4)$$

$$\frac{V_{\theta-45} - V_R}{V_R} = \frac{K_1 + K_2}{2}(\sigma_1 + \sigma_2) + \frac{K_1 - K_2}{2}(\sigma_1 - \sigma_2)\cos 2(\theta - 45) \quad (5)$$

$$\frac{V_{\theta-90} - V_R}{V_R} = \frac{K_1 + K_2}{2}(\sigma_1 + \sigma_2) + \frac{K_1 - K_2}{2}(\sigma_1 - \sigma_2)\cos 2(\theta - 90) \quad (6)$$

In the above equations, V_θ, $V_{\theta-45}$, and $V_{\theta-90}$ are surface wave velocities propagating in the θ, $\theta - 45°$, and $\theta - 90°$ directions, respectively, in the stressed material, and V_R is the reference velocity in the unstressed state. From Equations (4)–(6), the principal stresses, σ_1 and σ_2, and the angle θ can be determined as follows:

$$\sigma_1 = \frac{1}{2}\left[\frac{1}{K_1 + K_2}\left(\frac{V_\theta}{V_R} + \frac{V_{\theta-90}}{V_R} - 2\right) + \frac{1}{K_1 - K_2}\left(\frac{V_\theta}{V_R} - \frac{V_{\theta-90}}{V_R}\right)\cdot\frac{1}{\cos 2\theta}\right] \quad (7)$$

$$\sigma_2 = \frac{1}{2}\left[\frac{1}{K_1+K_2}\left(\frac{V_\theta}{V_R}+\frac{V_{\theta-90}}{V_R}-2\right)-\frac{1}{K_1-K_2}\left(\frac{V_\theta}{V_R}-\frac{V_{\theta-90}}{V_R}\right)\cdot\frac{1}{\cos 2\theta}\right] \tag{8}$$

$$\theta = \frac{1}{2}\left[\tan^{-1}\left(\frac{\frac{V_\theta}{V_R}+\frac{V_{\theta-90}}{V_R}-2\frac{V_{\theta-45}}{V_R}}{\frac{V_\theta}{V_R}-\frac{V_{\theta-90}}{V_R}}\right)\right] \tag{9}$$

Using these three equations, when the acoustoelastic coefficients (K_1, K_2) and the reference surface wave velocity (V_R) in the unstressed condition are known, the principal stresses (σ_1, σ_2) and the angle (θ) can be estimated using the three-directional measurements of the surface wave velocity.

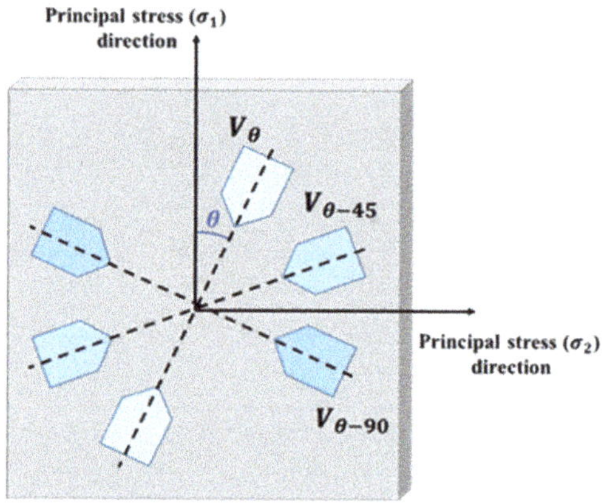

Figure 1. Schematic of the relationships of the angles between the principal stress direction and three-measuring directions in 45° increments, in which θ is the angle from a measuring direction to the principal stress (σ_1) direction, and V_θ, $V_{\theta-45}$, $V_{\theta-90}$ are surface wave velocities propagating in the θ, $\theta - 45°$, and $\theta - 90°$ directions, respectively.

2.2. Stress Estimation Method in Weakly Anisotropic Material

The method mentioned above is based on the theory for an ideal isotropic material, in which the surface wave velocity in the unstressed condition is constant regardless of the direction of propagation. However, the actual material is not always produced in a perfectly isotropic condition and has anisotropic properties. In this case the surface wave velocity is directionally dependent, and the isotropic method cannot be directly applied. As will be shown later in the numerical simulations, a very large error occurs even in the case of a very small anisotropy.

To compensate for this anisotropic effect, in this study, Equation (1) for the isotropic case was modified as follows,

$$\frac{V_\theta - V_{\theta,R}}{V_{\theta,R}} = \frac{K_1+K_2}{2}(\sigma_1+\sigma_2)+\frac{K_1-K_2}{2}(\sigma_1-\sigma_2)\cos 2\theta \tag{10}$$

where V_R in Equation (1) is replaced with $V_{\theta,R}$ which is the surface wave velocity in the θ direction in the unstressed state. This equation represents the rate of change of the surface wave velocity before and after applying stress in the θ direction.

Then, the acoustoelastic coefficients K_1 and K_2 can be rederived as follows under the uniaxial stress condition ($\sigma_1 = \sigma$ and $\sigma_2 = 0$):

$$K_1 = \frac{V_0 - V_{0,R}}{V_{0,R}}\frac{1}{\sigma} \tag{11}$$

$$K_2 = \frac{V_{90} - V_{90,R}}{V_{90,R}}\frac{1}{\sigma} \tag{12}$$

where $V_{0,R}$ and $V_{90,R}$ are the reference surface wave velocities parallel and perpendicular to the stress direction in the unstressed state, respectively. As a result, Equations (11) and (12) replace V_R in Equations (2) and (3) with the initial directional velocities, $V_{0,R}$ and $V_{90,R}$, respectively.

Next, for stress estimation using three-directional measurements of the surface wave velocity in the same way as isotropic theory, we derive the following three equations from Equation (10),

$$\sigma_1 = \frac{1}{2}\left[\frac{1}{K_1+K_2}\left(\frac{V_\theta}{V_{\theta,R}} + \frac{V_{\theta-90}}{V_{\theta-90,R}} - 2\right) + \frac{1}{K_1-K_2}\left(\frac{V_\theta}{V_{\theta,R}} - \frac{V_{\theta-90}}{V_{\theta-90,R}}\right)\cdot\frac{1}{\cos 2\theta}\right] \tag{13}$$

$$\sigma_2 = \frac{1}{2}\left[\frac{1}{K_1+K_2}\left(\frac{V_\theta}{V_{\theta,R}} + \frac{V_{\theta-90}}{V_{\theta-90,R}} - 2\right) - \frac{1}{K_1-K_2}\left(\frac{V_\theta}{V_{\theta,R}} - \frac{V_{\theta-90}}{V_{\theta-90,R}}\right)\cdot\frac{1}{\cos 2\theta}\right] \tag{14}$$

$$\theta = \frac{1}{2}\left[\tan^{-1}\left(\frac{\frac{V_\theta}{V_{\theta,R}} + \frac{V_{\theta-90}}{V_{\theta+90,R}} - 2\frac{V_{\theta-45}}{V_{\theta-45,R}}}{\frac{V_\theta}{V_{\theta,R}} - \frac{V_{\theta-90}}{V_{\theta-90,R}}}\right)\right] \tag{15}$$

where, $V_{\theta,R}$, $V_{\theta-45,R}$, and $V_{\theta-90,R}$ are the reference surface wave velocities in the θ, $\theta - 45°$ and $\theta - 90°$ directions in the unstressed state, respectively, and V_θ, $V_{\theta-45}$, and $V_{\theta-90}$ are the surface wave velocities in the stressed state. As a results, Equations (13)–(15) replace the reference velocity V_R in the unstressed condition in Equations (7)–(9) with the initial velocity in each measurement direction.

3. Numerical Simulations

Numerical simulations were performed to verify the stress estimation performance in typical stress states when the proposed method was applied to anisotropic materials. For this, the anisotropy of a material is assumed to be orthogonal, and the acoustic anisotropy rate (η) is defined as the ratio of the difference between surface wave velocity in the direction in which the wave velocity becomes maximum (V_{max}) and the surface wave velocity in the direction in which the wave velocity becomes minimum (V_{min}) to the minimum surface wave velocity (V_{min}) as follows:

$$\eta = \frac{V_{\max} - V_{\min}}{V_{\min}} \tag{16}$$

The numerical simulation requires three directional surface wave velocities in unstressed and stressed states. For this, the model suggested by Delsanto et al. [8] and Mase and Delsanto [9] was used. For the simple analysis applied in the study of Mase, only one elastic constant, C_{11}, was changed by anisotropy, and thus Equation (17) can be applied

$$\begin{aligned}\frac{V_\theta - V_R}{V_R} = \frac{1}{2\mu}A_{2222}\cos^4\phi C'_{11} + &\left(a_{\sigma 0} + \sum_{k=1}^{3}\frac{a_{\sigma k}v_k}{\mu}\right)\frac{\sigma_1+\sigma_2}{\mu} \\ &+ \left(a_{\Delta 0} + \sum_{k=1}^{3}\frac{a_{\Delta k}v_k}{\mu}\right)\cos 2\theta\frac{\sigma_1-\sigma_2}{\mu}\end{aligned} \tag{17}$$

Here, V_R is the surface wave velocity in the unstressed isotropic medium, and V_θ is the wave velocity affected by anisotropy and the acoustoelastic effect in the θ direction. μ, A_{2222}, $\alpha_{\sigma k}$, $\alpha_{\Delta k}$ (k: 0, 1, 2, 3), and ν_k (k: 1, 2, 3) are material constants of the isotropic material, and C'_{11} is the anisotropic variation of C_{11}. The left side of Equation (17) corresponds to the rate of change of the surface wave velocity, the first term on the right side reflects the effect of anisotropy on the change of velocity, and the other terms refer to the contribution of the acoustoelastic effect on the change of velocity In this model, ϕ is the angle between the acoustic anisotropy direction and the surface wave propagation direction, and θ is the angle between the principal stress direction and the surface wave propagation direction, as shown in Figure 2. The acoustic anisotropy direction is defined as the direction in which the velocity of the surface wave is maximum.

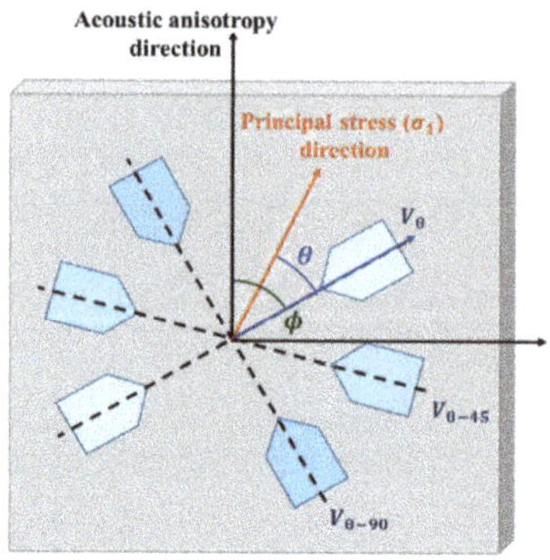

Figure 2. Principal stress (σ_1) direction, and three-measuring directions in 45° increments, in which θ is the angle from a measuring direction to the principal stress (σ_1) direction and ϕ is the angle to the anisotropy direction, and V_θ, $V_{\theta-45}$, $V_{\theta-90}$ are surface wave velocities propagating in the θ, θ – 45°, and θ – 90° directions, respectively.

The three-directional measurement technique was applied using the velocities of the surface waves propagating towards the θ direction and −45° and −90° to the θ direction. Simulations were performed for the six cases of the combinations of the different angle conditions shown in Table 1 and different stress conditions shown in Table 2. Table 1 shows the three cases for setting the anisotropy direction, principal stress direction and propagation direction. Case 1 is the case where all of the anisotropy, principal stress (σ_1), and surface wave propagation directions are the same. In Case 2, the anisotropy direction and the principal stress direction are coincident, but the surface wave propagation direction is different. Case 3 is a general case where all directions are inconsistent.

Table 2 shows the given stresses for the two groups. Group A is the case where only uniaxial stress (σ_2 = 0) is applied and σ_1 is varied among 30, 60 and 90 MPa. Group B is the case where biaxial stress is applied while fixing σ_1 at 90 MPa and varying σ_2 among three stresses (σ_2 = 30, 60, 90 MPa).

Table 1. Directions of the anisotropy, principal stress, and the surface wave propagation for numerical simulation.

	Case 1	Case 2	Case 3
∅	0°	30°	60°
θ	0°	30°	30°
	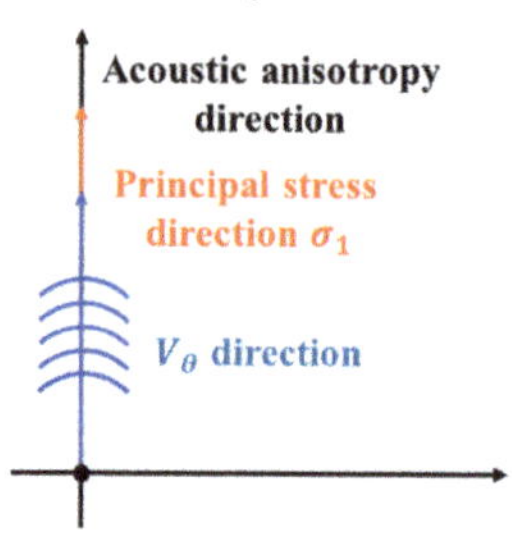	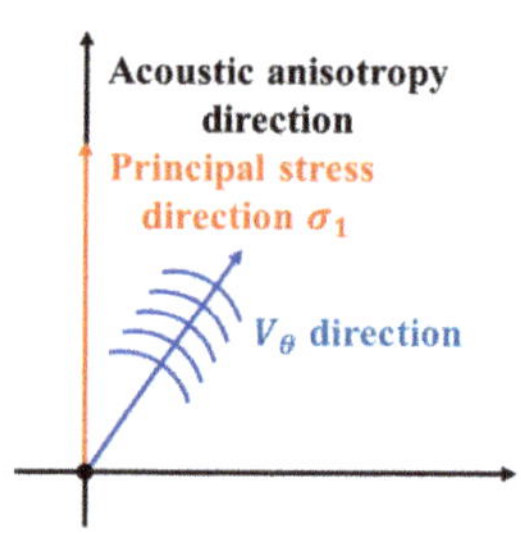	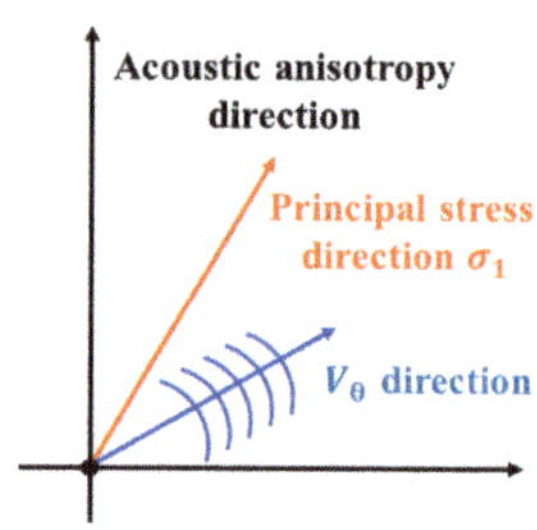

Table 2. Two groups of applied stress setting for numerical simulation.

	A	B
σ_1	30, 60, 90 MPa	90 MPa
σ_2	0 MPa	30, 60, 90 MPa

The isotropic elastic constants of aluminum in the COMSOL database, which are listed in Table 3, were used to calculate the surface wave velocity, where λ and μ are Lame constants and l, m and n are Murnaghan constants.

Table 3. Material constant used to numerical verification.

λ [GPa]	μ [GPa]	C_{11} [GPa]	l [GPa]	m [GPa]	n [GPa]
51	26	103	−250	−330	−350

The anisotropy rate η changed from 0 to 0.02 in 15 steps by varying the value of C_{11}. That is, the anisotropy was varied up to 2%. Figure 3 shows an example of the calculation results for $\eta = 0.02$, and principal stress direction is coincidence with the anisotropy direction and principal stress magnitudes are $\sigma_1 = 90$ MPa and $\sigma_2 = 0$. The velocity distribution of the unstressed state has maximum values in the directions of 0° and 180°, and a minimum value in the directions perpendicular to them. Also, it is possible to verify that the velocity distribution is slightly changed by the acoustoelastic effect when stress is applied.

To demonstrate the difficulty of applying the isotropic theory to the anisotropic material first, we applied the surface wave velocity data set obtained for the Case 1A to the isotropic theory of Equations (7)–(9), prior to applying the proposed method. Figure 4 shows the stress estimation error of σ_1, in which surface wave velocities in three directions of 0°, 45°, and 90° were used. As can be seen from the results, the stress estimation error is ridiculously large even for very weak anisotropy. Therefore, it is very difficult to estimate the stress of anisotropic materials using the isotropic theory.

Next, to verify the performance of the proposed method, the surface wave velocity data set for all six conditions shown in Tables 1 and 2 was generated. Stress estimation was performed by using surface wave velocities in three directions of 0°, 45°, and 90°. The errors between the estimated stress and given stress of σ_1 were then calculated, where the results are shown in Figure 5. In all cases tested, the errors were less than 3%. In particular, in Case 1A, where all directions of anisotropy, principal stress (σ_1) and surface wave propagation are identical, no errors are generated in the stress estimation regardless of the stress. This is because Delsanto's model in Equation (17) used in this numerical simulation is consistent with our modified Equation (10). Actually, by rearranging Equation (17) for the uniaxial stress and angle conditions of Case 1A, it becomes equal to Equation (10) (see the Appendix A).

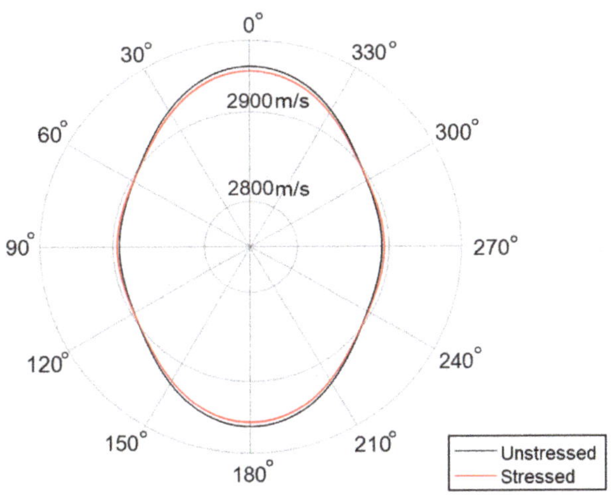

Figure 3. Surface wave velocity according to the propagation direction for unstressed and stressed states in an anisotropic material.

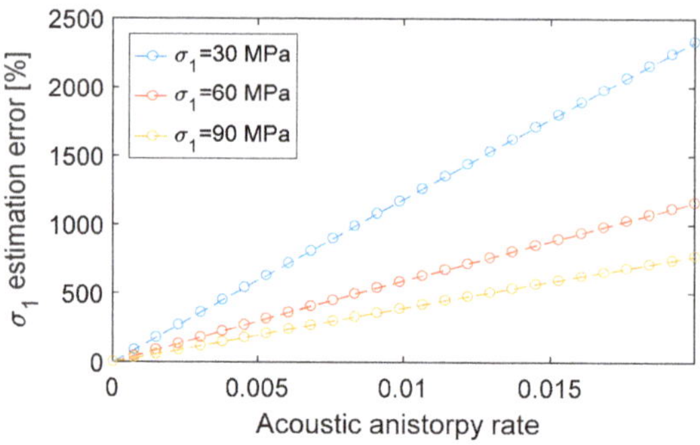

Figure 4. Stress estimation errors of the isotropic theory with respect to the anisotropy rate (Case 1A).

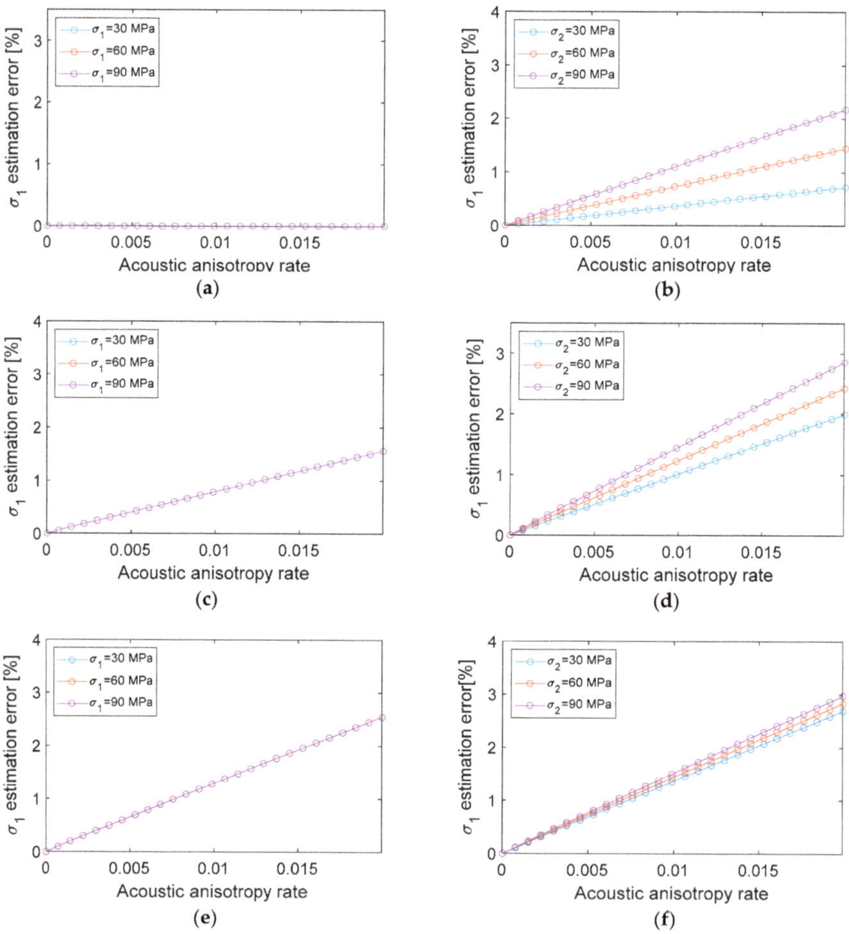

Figure 5. Stress estimation errors of the proposed method with respect to the anisotropy rate (**a**) Case 1A; (**b**) Case 1B; (**c**) Case 2A; (**d**) Case 2B; (**e**) Case 3A; (**f**) Case 3B.

Meanwhile, in the case of A with uniaxial stress, the stress estimation errors are independent of the given stress, and only increase with the anisotropy rate. In the case of B with biaxial stress, the stress estimation errors increase with both anisotropy and stress. However, it must be noted again that the estimation error in all cases are very small. Based on these results, it was confirmed that the proposed method for stress estimation can properly estimate stress in a weakly anisotropic material.

4. Specimens and Experiments

4.1. Specimens and Measurement of the Anisotropy Rate

To verify the proposed method experimentally, stress estimation experiments were performed. Two kinds of Al6061 material with different acoustic anisotropy rates were used as the test specimens. One (SA) was produced by a hot rolling process and the other (SB) was produced by an extrusion process. Two test specimens were prepared for each material as shown in Figure 6. One specimen was used to check the anisotropy and acoustoelastic coefficient, and the other was used to verify the stress estimation method. The size of the hot rolled specimen is 350 mm × 62 mm × 4 mm (length × width × thickness) and the size of the extruded specimen is 350 mm × 60 mm × 5 mm.

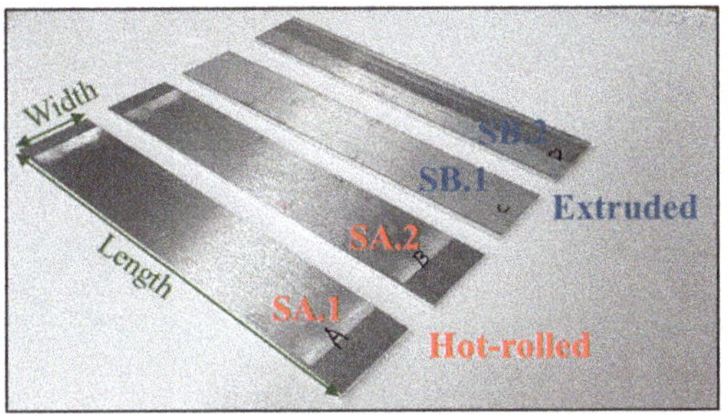

Figure 6. Aluminum plate specimens used in experiments.

To verify the surface wave velocity distribution and anisotropy rate of each specimen, the velocity was measured in 24 directions with an interval of 15°. The experimental setup is shown in Figure 7. PZT transducers with a center frequency of 2.25 MHz and wedges (ABWML-7T-90, Olympus) were used for surface wave excitation and reception. A pulser-receiver (Olympus 5077PR) was used for the excitation with a single pulse and a high-resolution digital oscilloscope (Lecroy HDO4034A) with a 10 GS/s sampling rate was used for precise measurement of small velocity changes. The surface wave propagation distance was fixed at 10 mm using a jig. The time of flight (TOF) was used to measure the change rate of the surface wave velocities or surface wave velocity ratios used in Equations (11)–(15). The TOF was measured based on the arrival time of the peak of the pulse signal. In addition, the delay time irrespective of surface wave propagation on the specimens, such as the delay in the wedge, was compensated by calibration in advance.

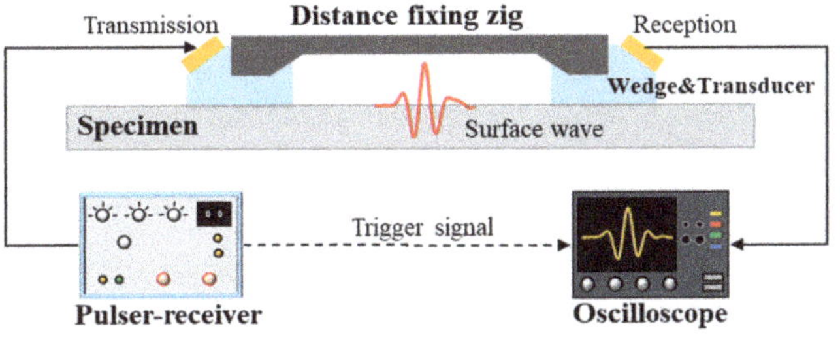

Figure 7. Surface wave velocity measurement set-up, in which the distance fixing zig holds the wedges to keep the surface wave propagation distance constant.

The measurement results of the surface wave velocity distribution are shown in Figure 8, in which the velocity was normalized with respect to the maximum velocity. Maximum velocity (V_{max}) and minimum velocity (V_{min}) in the hot-rolled specimen are 2967.1 m/s and 2957.5 m/s, respectively, and they are 2951.8 m/s and 2876.2 m/s, respectively, in extruded specimen. Figure 8a display the results of the hot-rolled aluminum specimen (SA.1) and Figure 8b shows the results of the extruded aluminum specimen (SB.1). The acoustic anisotropy rate of the hot-rolled aluminum specimen (SA.1) was 0.005 (0.5%) and that of the extruded aluminum specimen was 0.026 (2.6%). This demonstrates that specimen SA is nearly isotropic while specimen SB has a weak anisotropic feature.

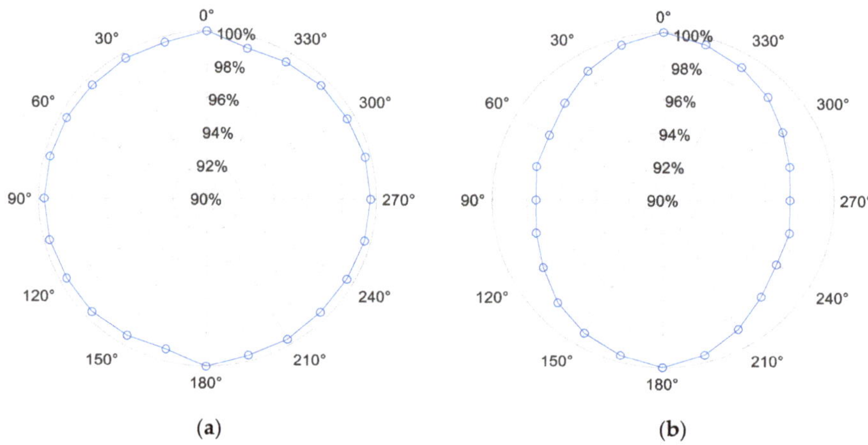

Figure 8. Surface wave velocity distributions of the two specimens: (**a**) Hot rolled specimen (SA.1); (**b**) Extruded specimen (SB.1).

4.2. Measurement of Acoustoelastic Coefficients

The acoustoelastic coefficients were obtained prior to the stress estimation experiments. The measurement set-up is shown in Figure 9. Based on the definition of the acoustoelastic coefficient, the change rates of the surface wave velocity were measured with stress in the stress direction (K_1) and in the perpendicular direction (K_2) in the uniaxial stress state. A tensile tester (Zwick-Roell) was used to apply the stress. The loading force was varied between 2500 N–22,500 N with an interval of 2500 N. The loading interval of 2500 N corresponds to a stress of about 10 MPa. The velocity measurement set-up is identical to the set-up described in Figure 7.

Figure 10 shows the change of the received pulsed surface wave signal according to the applied stress. Figure 10a corresponds to the K_1 measurement and Figure 10b reflects K_2. As shown in the zoomed-in box, the peak points of the pulse signal shift depend on the stress increment. That is, the surface wave velocity in the stress direction decreases with the stress increment, while it increases in the direction perpendicular to the stress.

Figures 11 and 12 show the change rates of the time of flight (TOF) with respect to the applied stress in each direction for the SA.1 and SB.1 specimens, respectively. The results were linearly fitted and the slopes of the fitted lines are the acoustoelastic coefficients defined in Equations (11) and (12). The obtained acoustoelastic coefficients are listed in Table 4.

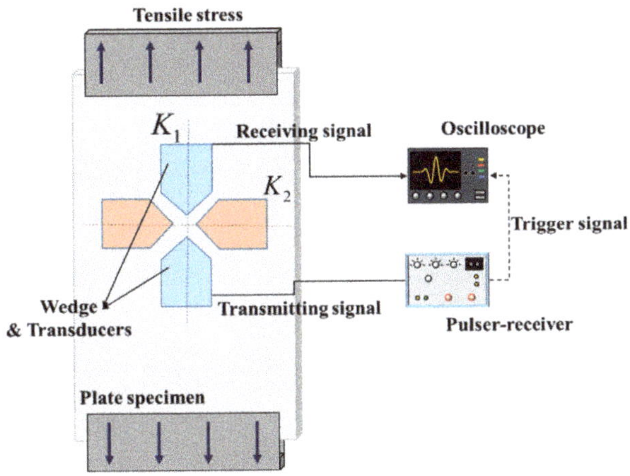

Figure 9. Set-up for acoustoelastic coefficient measurement. The change in surface wave velocity under tensile stress loading is measured. For the measurement of *K1*, transducers (blue colored) are placed in the loading direction, while for the measurement of *K2*, transducers (orange colored) are placed in the direction perpendicular to the loading direction.

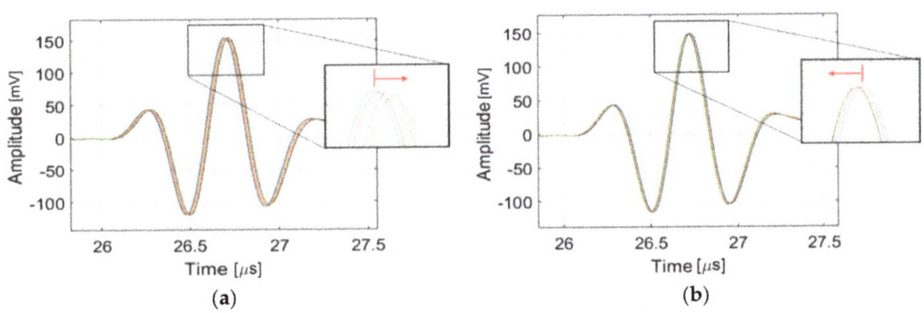

Figure 10. Received signals obtained in the acoustoelastic coefficient measurements: (**a**) Measurement of K_1—The peak moves to the right because the velocity of surface wave in the direction of tensile stress decreases with increasing stress; (**b**) Measurement of K_2—The peak moves to the left because the velocity of surface wave in the direction perpendicular to the tensile stress increases with increasing stress.

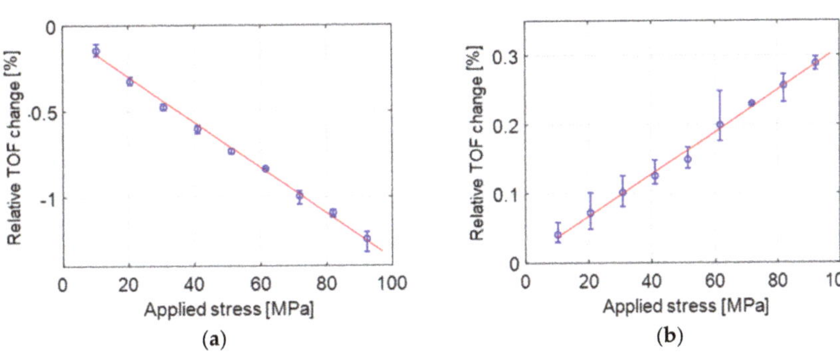

Figure 11. Change rates of the TOF with respect to the applied stress in specimen SA.1 for measurements of (**a**) K_1 and (**b**) K_2.

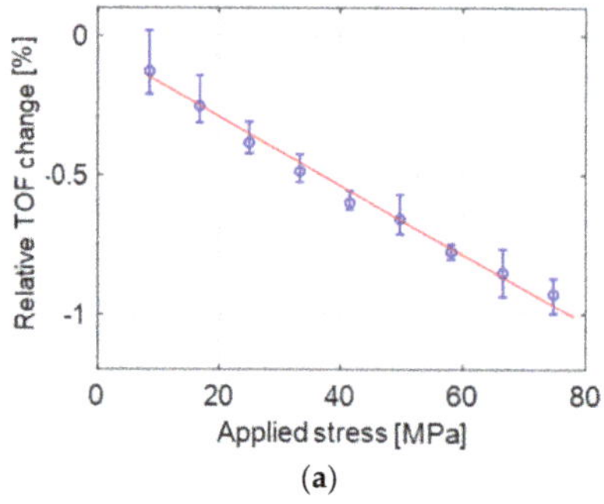

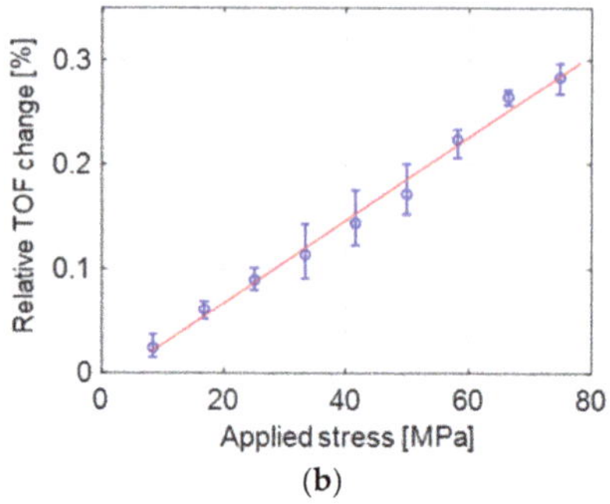

Figure 12. Change rates of the TOF with respect to the applied stress in specimen SB.1 for measurements of (**a**) K_1 and (**b**) K_2.

Table 4. Values of the acoustoelastic coefficients for specimens SA.1 and SB.1.

Specimen	K_1 [1/MPa]	K_2 [1/MPa]
A.1	-132.6×10^{-6}	30.5×10^{-6}
B.1	-124.0×10^{-6}	39.7×10^{-6}

4.3. Experiments for Stress Estimation

The stress estimation set-up is shown in Figure 13. The experimental set-up is similar to the acoustoelastic coefficient measurement except for the velocity measurement direction. The surface wave velocities were measured in three directions with difference of 45°, where the angle θ was set to 30°. This measurement condition corresponds to the Case 2B in the numerical simulation. Stress was applied by tensile loading from 7500 N to 22,500 N at an increment of 2500 N. Specimens SA.2 and SB.2 were used for the stress estimation.

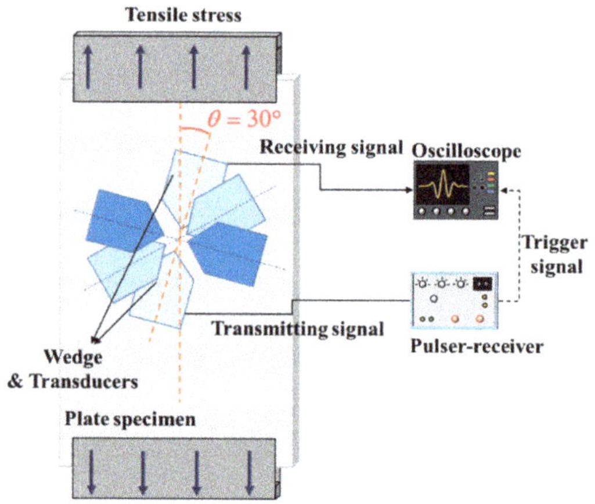

Figure 13. Experimental set-up for the proposed stress estimation.

The stress estimation results of the hot-rolled (SA.2 specimen) and extruded (SB.2 specimen) aluminum plates are plotted in Figure 14 and the detailed values are given in Tables 5 and 6, respectively. For the hot-rolled aluminum specimen, the estimated results match the given stresses and angle well. Similarly, the stress and angle estimations for the extruded specimens also exhibit a good agreement with the applied values even though this specimen has a larger anisotropy rate than the hot-rolled specimen. The maximum error was 5.14 MPa at a given stress of 90.92 MPa in the hot-rolled specimen, and the other errors were less than 5.00 MPa for both specimens. As a result, it was confirmed that stresses can be estimated in weakly anisotropic materials using the proposed method. Note that the experiments were carried out for the case where the stress in the thickness direction was uniform only to verify the proposed method. If the stress varies with depth, surface waves of different wavelengths can be used to probe the stress field at different depths.

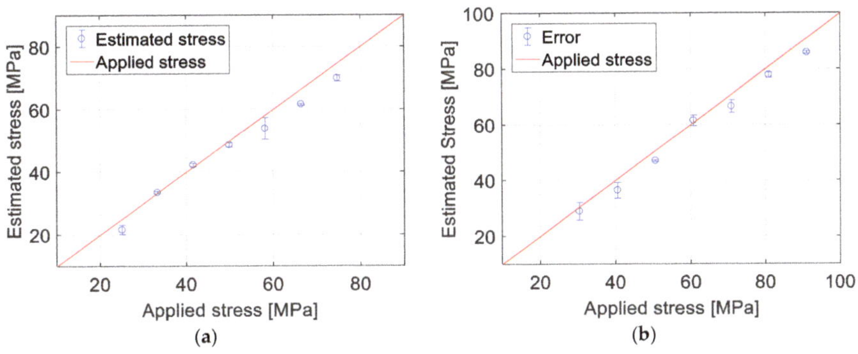

Figure 14. Stress estimation results (**a**) Hot-rolled specimen (SA.2) and (**b**) Extruded specimen (SB.2).

Table 5. Estimated results and errors for the hot-rolled specimen (SA.2).

Given stress [MPa]	30.40	40.48	50.57	60.66	70.75	80.83	90.92
Estimated stress [MPa]	28.90	36.42	47.01	61.46	66.54	77.82	85.78
Stress estimated error [MPa]	1.50	4.06	3.56	−0.80	4.21	3.01	5.14
Estimated angle θ [°]	23.39	25.41	26.18	25.35	20.17	26.23	27.73
Angle Estimated Error [°]	6.61	4.59	3.82	4.65	9.83	3.77	2.27

Table 6. Estimated results and errors for the hot-rolled specimen (SB.2).

Given stress [MPa]	24.95	33.24	41.50	49.78	58.06	66.34	74.61
Estimated stress [MPa]	21.56	33.43	42.20	48.58	53.87	61.57	69.89
Stress estimated error [MPa]	3.39	−0.19	−0.70	1.20	4.19	4.77	4.72
Estimated angle θ [°]	32.96	34.02	34.94	33.03	34.98	34.58	34.30
Angle Estimated Error [°]	−2.96	−4.02	−4.94	−3.03	−4.98	−4.58	−4.30

5. Conclusions

This study proposed a method to estimate stress in anisotropic materials, which involved modification of the equation for the isotropic theory. In this method, the ratios of the surface wave velocities in the stressed state to those in the stress-free state in three directions were utilized, enabling simultaneous estimation of the principal stresses and their directions.

Numerical simulations and experiments were conducted for verification. The results of the numerical simulation showed that the stress estimation error occurred at a very small order when the acousitc anisotropy rate was less than 2%. Experiments were performed on hot-rolled and extruded aluminum plates that had different anisotropy rates of, 0.05% and 2.6%, respectively. The stresses applied by a tensile tester were estimated by measuring the ratios of the surface wave velocities in the stressed state to those in the unstressed state in three-directions. The estimated results showed good agreement with the given stresses in the range from 30 to 90 MPa. These results demonstrate that the stress estimation method proposed in this study can effectively estimate the stress in weakly anisotropic materials. The proposed method is simple, practical and is expected to be useful for monitoring changes in surface stress before and after machining such as punching or bending of plates.

Author Contributions: Conceptualization, J.J. and Y.-D.S.; Methodology, J.J. and Y.-D.S.; Software, Y.-D.S.; Validation, J.J., Y.-D.S. and K.-Y.J.; Formal Analysis, J.J.; Investigation, J.J. and Y.-D.S.; Resources, K.-Y.J.; Data Curation, Y.-D.S.; Writing-Original Draft Preparation, J.J.; Writing-Review and Editing, K.-Y.J.; Visualization, Y.-D.S.; Supervision, K.-Y.J.; Project Administration, K.-Y.J.; Funding Acquisition, K.-Y.J. All authors have read and agreed to the published version of the manuscript.

Funding: This research was supported by the Nuclear Power Research and Development Program through the National Research Foundation of Korea (NRF) funded by the Ministry of Science, ICT & Future Planning (NRF-2013M2A2A9043241).

Conflicts of Interest: The authors declare no conflict of interest.

Nomenclature

A_{2222}	Material constants of the isotropic material
C_{11}	Elastic constant of isotropic material C′11: Variation of C11 according to anisotropy
K_1, K_2	Acoustoelastic coefficients
V_R	Velocity of the surface wave in the unstressed state for isotropic theory
V_θ	Surface wave velocities in the θ direction in the stressed state
$V_{\theta,R}$	Surface wave velocity in the θ direction in the unstressed state
V_{max}	Surface wave velocity in the direction in which the wave velocity becomes maximum
V_{min}	Surface wave velocity in the direction in which the wave velocity becomes minimum
l, m, n	Murnaghan constants
$\alpha_{\sigma k}, \alpha_{\Delta k}$ (k: 0, 1, 2, 3)	Material constants of isotropic material
η	acoustic anisotropy rate
θ	Angle between the surface wave propagation direction and the principal stress (σ_1) direction
λ, μ	Lame constants
ϕ	Angle between the anisotropy direction and the surface wave propagation direction
v_k (k: 1, 2, 3)	Material constants of isotropic material
σ_1, σ_2	Principal stresses

Appendix A

Considering the uniaxial stress ($\sigma_1 = \sigma, \sigma_2 = 0$) and angle condition of $\Phi = \theta$, Equation (17) can be rearranged as follows,

$$\frac{V_\theta - V_R}{V_R} = \frac{1}{2\mu} A_{2222} \cos^4 \theta C'_{11} + (a_{\sigma 0} + \sum_{k=1}^{3} \frac{a_{0k} v_k}{\mu}) \frac{\sigma}{\mu} - (a_{\Delta 0} + \sum_{k=1}^{3} \frac{a_{\Delta k} v_k}{\mu}) \cos 2\theta \frac{\sigma}{\mu} \quad \text{(A1)}$$

By substituting $\sigma = 0$, the velocity in the unstressed state, in the θ direction can be obtained as follows:

$$V_{\theta,R} = (\frac{1}{2\mu} A_{2222} \cos^4 \theta C'_{11} + 1) V_R \quad \text{(A2)}$$

On the other hand, the velocity in the stressed state in the θ direction is represented as follows,

$$V_\theta = \left[\frac{1}{2\mu} A_{2222} \cos^4 \theta C'_{11} + (a_{\sigma 0} + \sum_{k=1}^{3} \frac{a_{0k} v_k}{\mu}) \frac{\sigma}{\mu} + (a_{\Delta 0} + \sum_{k=1}^{3} \frac{a_{\Delta k} v_k}{\mu}) \frac{\sigma}{\mu} \cos 2\theta + 1 \right] V_R \quad \text{(A3)}$$

Considering the condition of Case 1A shown in Tables 1 and 2, for $\theta = 0°$, $45°$ and $90°$, three-directional surface wave velocities in the unstressed state can be expressed by Equations (A4)–(A6)

$$V_{0,R} = (\frac{1}{2\mu} A_{2222} \cos^4 0 C'_{11} + 1) V_R = (\frac{1}{2\mu} A_{2222} C'_{11} + 1) V_R \quad \text{(A4)}$$

$$V_{45,R} = (\frac{1}{2\mu} A_{2222} \cos^4 45 C'_{11} + 1) V_R \quad \text{(A5)}$$

$$V_{90,R} = (\frac{1}{2\mu}A_{2222}\cos^4 90C'_{11} + 1)V_R = V_R \tag{A6}$$

The three-directional velocities in the stressed state are represented by Equations (A7)–(A9).

$$V_0 = \left[\frac{1}{2\mu}A_{2222}C'_{11} + (a_{\sigma 0} + \sum_{k=1}^{3}\frac{a_{0k}v_k}{\mu})\frac{\sigma}{\mu} + (a_{\Delta 0} + \sum_{k=1}^{3}\frac{a_{\Delta k}v_k}{\mu})\frac{\sigma}{\mu} + 1\right]V_R \tag{A7}$$

$$V_{45} = \left[\frac{1}{2\mu}\frac{1}{4}A_{2222}C'_{11} + (a_{\sigma 0} + \sum_{k=1}^{3}\frac{a_{0k}v_k}{\mu})\frac{\sigma}{\mu} + 1\right]V_R \tag{A8}$$

$$V_{90} = \left[(a_{\sigma 0} + \sum_{k=1}^{3}\frac{a_{0k}v_k}{\mu})\frac{\sigma}{\mu} + 1\right]V_R \tag{A9}$$

The acoustoelastic coefficients defined in Equations (11) and (12) then become the following:

$$K_1 = \frac{(a_{\sigma 0} + \sum_{k=1}^{3}\frac{a_{0k}v_k}{\mu})\frac{1}{\mu} + (a_{\Delta 0} + \sum_{k=1}^{3}\frac{a_{\Delta k}v_k}{\mu})\frac{1}{\mu}}{(\frac{1}{2\mu}A_{2222}C'_{11} + 1)} \tag{A10}$$

$$K_2 = (a_{\sigma 0} + \sum_{k=1}^{3}\frac{a_{0k}v_k}{\mu})\frac{1}{\mu} \tag{A11}$$

Finally, Equations (A4)–(A11) are substituted into the stress estimation equations shown in Equations (13)–(15) to obtain $\sigma_1 = \sigma$, $\sigma_2 = 0$ and $\theta = 0$, which are identical to the given conditions.

References

1. Hughes, D.S.; Kelly, J.L. Second-order elastic deformation of solids. *Phys. Rev.* **1953**, *92*, 1145. [CrossRef]
2. Javadi, Y.; Plevris, V.; Najafabadi, M.A. Using LCR ultrasonic method to evaluate residual stress in dissimilar welded pipes. *Int. J. Innov. Manag. Technol.* **2013**, *4*, 170–174.
3. Gokhale, S.J.T.A. Determination of Applied Stresses in Rails Using the Acoustoelastic Effect of Ultrasonic Waves. Master's Thesis, Texas A&M University, College Station, TX, USA, 2007.
4. Hughes, J.; Vidler, J.; Khanna, A.; Mohabuth, M.; Kotooussov, A.; Ng, C. Measurement of residual stresses in rails using Rayleigh waves. In Proceedings of the International Conference on Structural Integrity and Failure, Perth, Australia, 3–6 December 2018.
5. Bray, D.E.; Tang, W.J.N.E. Subsurface stress evaluation in steel plates and bars using the LCR ultrasonic wave. *Nucl. Eng. Des.* **2001**, *207*, 231–240. [CrossRef]
6. Jassby, K.; Saltoun, D. Use of ultrasonic Rayleigh waves for the measurement of applied biaxial surface stresses in aluminum 2024-T 351 alloy. *Mater. Eval.* **1982**, *40*, 198–205.
7. Lu, W.; Peng, L.; Holland, S. Measurement of acoustoelastic effect of Rayleigh surface waves using laser ultrasonics. In *Review of Progress in Quantitative Nondestructive Evaluation*; Springer: Boston, MA, USA, 1998; pp. 1643–1648.
8. Delsanto, P.; Clark, A.V., Jr. Rayleigh wave propagation in deformed orthotropic materials. *J. Acoust. Soc. Am.* **1987**, *81*, 952–960. [CrossRef]
9. Mase, G.; Delsanto, P. Acoustoelasticity using surface waves in slightly anisotropic materials. In *Review of Progress in Quantitative Nondestructive Evaluation*; Springer: Boston, MA, USA, 1988; pp. 1349–1356.
10. Thompson, R.B.; Lee, S.; Smith, J.F. Angular dependence of ultrasonic wave propagation in a stressed, orthorhombic continuum: Theory and application to the measurement of stress and texture. *J. Acoust. Soc. Am.* **1986**, *80*, 921–931. [CrossRef]

applied sciences

MDPI

Article

Using Acoustic Emission Measurements for Ice-Melting Detection

Michael Stamm [1,2,*], Helge Pfeiffer [2], Johan Reynaert [1] and Martine Wevers [2]

1 Brussels Airlines, M&E, Brussels Airport, Building 32, 1930 Zaventem, Belgium; Johan.Reynaert@brusselsairlines.com
2 KU Leuven, Department of Materials Engineering, Kasteelpark Arenberg 44, 3001 Leuven, Belgium; helge.pfeiffer@kuleuven.be (H.P.); martine.wevers@kuleuven.be (M.W.)
* Correspondence: michael.stamm@brusselsairlines.com; Tel.: +32-470-522-889

Received: 29 October 2019; Accepted: 6 December 2019; Published: 9 December 2019

Featured Application: This work contributes to the development of a monitoring system for the remote detection, localization and quantification of melting ice. Acoustic emission measurements enable the remote monitoring in an environment where the region of interest is neither optically nor physically accessible. The kerosene tank of civil aircraft is one example where the aforementioned accessibility is only achieved with high costs and extensive labor. This is why inter alia aircraft operators that perform the daily maintenance of civil aircraft are interested in such an inspection method. Here the detection of ice inside the fuel tanks increases the operational efficiency as well as the safety for the ground personnel.

Abstract: Aircraft operators being faced with water accumulation in fuel tanks on a daily basis and are looking for reliable detection systems to determine the remaining amount of accumulated ice during maintenance after flight. Using such a technology, an increase in the safety and efficiency of the aircraft operation would be possible in this highly competitive market. This article presents the use of the Acoustic Emission Technique (AE) for the reliable and non-invasive monitoring of the melting of ice in fuel tanks. This technology is in principle based on the fact that a phase transition comes frequently along with stress relaxation that can be used for monitoring the process. Therefore, the melting of water can, in essence, be monitored with AE without accessing the ice directly. The analysis of the AE signals has been carried out in the time domain since it was the melting of ice needed to be monitored as a function of temperature rise time. The insights presented in this paper can possibly lead to new technologies for ice detection, especially in remote areas that are not easily accessible with other techniques.

Keywords: acoustic emission; melting ice; aviation; fuel system

---

1. Introduction

Since the beginning of aviation, innovation in technologies and operational procedures have steadily increased comfort, performance and safety of air transport. Especially the latter aspect leads to a continuously decreasing number of fatal accidents even though the number of passengers is steadily increasing. In this context, it is important that also aging aircraft types are retrofit equipped with modern technologies whenever feasible.

Besides new technological developments, the competition on the European as well as the global aviation market increases continuously. This pushes aircraft operators towards more efficient processes. In order to enhance efficiency, many key innovations focus on the reduction of scheduled and unscheduled grounding time (downtime) of the aircraft which is the time an aircraft is not available for flying. However, the procedure to remove liquid water from the fuel systems, called the draining procedure, requires time for the frozen water inside the tank to melt. The draining procedure makes use of the fact that liquid water, as well as ice, have a higher density than kerosene. Therefore, liquid water can be removed from the fuel tank by opening so-called drain valves at the lowest point of the tank and let the water run out of the tank. As this only removes liquid water from the tank, one must wait until all frozen water is molten. The time needed for the ice to melt depends on the inner temperature of the tank and on the outer surrounding temperature. Especially at cold winter times, scheduled downtimes are insufficient to melt all frozen water. A detailed description of the draining procedure and the problem occurring during the daily aircraft operation can be found in Stamm et al. [1].

The real-time detection of remaining ice inside the fuel tank will thus contribute to the optimization of the draining procedure that is performed to remove liquid water from the fuel tank enabling in this way a considerable reduction of costly downtimes. For the operation in the hangar by aircraft technicians, a mobile, small and easy-to-use AE system that can monitor the melting ice from outside the tank is required. To match these requirements already in laboratory experiments, a low-priced system with off-the-shelf components is used to perform presented experiments.

1.1. Water in Fuel Tanks

During the daily operation of aircraft, water accumulates inside the aircraft fuel tanks and freezes if the outside air reaches sub-zero temperatures. Different possibilities and root causes of water contamination in jet fuel are discussed in Baena-Zambrana et al. [2]. Besides the contaminated fuel that is used to fill up the aircraft tanks, water vapor condensation inside the aircraft fuel tank also leads to water accumulation inside the tank. Liquid water finally disturbs fuel quantity meters and enables microbiological corrosion whereas frozen water can harm the internal systems and structures by e.g., blocking valves or bursting pipes by freezing inside the pipes.

1.2. AE Monitoring of Melting Ice

AE is a well-established monitoring technique in the field of non-destructive testing (NDT) and many reviews mention applications in the fields of condition and process monitoring and defect detection. Li [3] gives a suitable review of using AE for monitoring metal cutting processes and discusses related signal processing methodologies including time series analysis, FFT, wavelet transform, etc.. Examples for AE-based condition monitoring and defect detection in the civil sector are described in Behnia et al. [4], Zaki et al. [5] as well as Vestrynge et al. [6] who present their work on monitoring concrete structures. Schiavi et al. [7] present their work which shows that AE can also applied to monitor the brittle materials during compression tests.

Additionally, Boyd and Varley [8] list several chemical processes that were monitored with AE such as the addition of sodium carbonate to calcium chloride [9], metal-acid reaction [10] or the solid-phase transitions of C_2Cl_6 [11].

Sawada et al. [12] measured the acoustic signals during the freezing and melting of super cooled distilled water. Melting and freezing are accompanied by acoustic activity that is more intense during the melting. This is the general behavior of phase transitions with a volume decrease being acoustically more active than transitions with volume expansion [12]. Wentzell et al. [13] also measured the initial melting phase of distilled water. Since this is believed to be predominated by crystal fracture due to the high temperature gradient between the sample and the surrounding temperature, no or only a negligible number of events that occur due to the phase transition itself are measured. Finally, the mentioned literature provides only the time histogram of acoustic events during the melting of ice and no additional analysis of the acquired data.

Unlike melting processes, AE measurements on crystallization processes are discussed in some publications, e.g., by Lube and Zlatkin [14] and Ersen et al. [15] who use AE to estimate the crystal purity during the growing of ruby and leuco-sapphire crystal and effect of malic and citric acid on the crystallization of gypsum respectively. Furthermore, Wang and Huang [16] performed AE measurement on crystallization processes of salicylic acid and concluded that AE can be used as a reliable technique to monitor the development of the crystallization process.

The possibility of using the AE technique to monitor the melting of ice, first using tab water, is further explored and documented in this paper. The experimental setup that allows a controlled environmental conditions will be explained in the following.

2. Experimental Setup

For the AE measurement during melting and freezing, an experimental setup was built that allows the controlled freezing and melting of a sample while keeping the environmental noise at a minimal level. In the following, the AE measurement setup that meets the industry requirements described above and the thermostatic setup are explained.

2.1. Data Acquisition and Processing

A good review of AE measurements of chemical processes is given in Grosse and Ohtsu [17] where the general approach, usage and advantages of acoustic emission measurements are discussed. In addition, a general AE setup is described which usually contains an acoustic sensor, followed by a filtering and amplification of the signals and dedicated signal analysis by the CPU. In the following all three parts will be explained separately.

The most common AE sensors are piezoelectric (PZT) sensors that use the piezoelectric effect which transforms elastic motions into voltages [17]. In this work, the commercial broadband piezoelectric sensor "*VS30-V*" from the Vallen Systeme GmbH with a flat frequency response between 25 to 80 kHz [18].

The PZT sensor is connected to an AC coupled preamplifier "*Model 5670*" by Olympus NDT Inc. This broadband preamplifier provides a very-low-noise 40 dB amplification of ultrasonic signals with a flat ($\pm$1.5 dB) response in a frequency range of 50 kHz–10 MHz [19]. The advantage of the bandwidth that does not match with the full operating frequency of the PZT is that frequencies below 50 kHz are not amplified which reduces the picked up ambient noise. Thereafter, a "*PicoScope 2205A*" USB oscilloscope digitizes and saves the waveforms with a sampling frequency of $f_s = 1.565$ MHz. The oscilloscope operates in a simple edge threshold trigger mode with a trigger level of $V_{THD} = +10$ mV which was approximately two time the environmental noise at the beginning of the experiments. With every trigger, a time windows with a length of t = 2 ms (trigger point at $t_{trigger}$ = 20% = 0.4 ms) is saved in a file that contains the time of occurrence.

The settings of the oscilloscope define some additional parameters that are commonly used in AE:

- hit definition time (HDT)
- hit lockout time (HLT)
- peak definition time (PDT)

Classical AE systems allow the explicit setting of these parameters for a optimized operation in certain conditions. With the setup used in this study, these parameters are given by the operating mode of the oscilloscope and no additional (software implemented) realization of these parameters is implemented. The PDT = 1.6 ms is given by the recorded time window for each AE event. The HDT and HLT are not explicitly set but implicitly given by the oscilloscope settings. The recorded time window and the duration of the AE previous event determine the HDT such as $D + HDT = 1.6$ ms. Due to the short duration of the expected AE events ($\overline{D} \approx 15\%$ of the recording time window) and the low quantity of signals, no HLT is required and it can be assumed as zero.

The post-processing of the recorded transients is done in real time in an additional module of the data acquisition software, which is coded in Python 3.7. Figure 1 shows a typical acoustic burst and some parameters that are also listed in Grosse and Ohtsu [17] and typically used for parameter-based analysis. The following parameters are calculated for each transient:

- Amplitude (AMPL) in [mV]

- Duration (D) in [ms]
- Rise time (RT) in [ms]
- Energy (E) in [ms×mV]

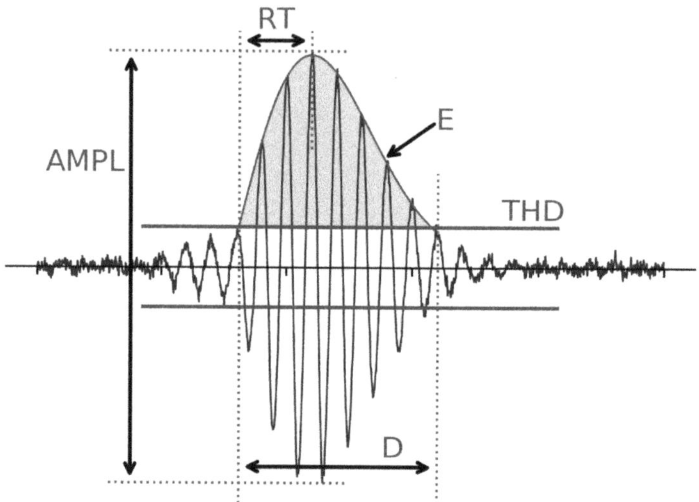

Figure 1. A sketch of an acoustic burst with some key indicators that are used for parameter-based analysis: peak-to-peak amplitude (AMPL), duration (D), rise time (RT), acoustic energy (E) and the trigger threshold (THD).

2.2. Temperature Controlling System

To measure the acoustic activity of melting ice, not only an AE measurement system but also a stable thermostatic setup is essential and this was achieved by an in-house-built device that monitors and controls the temperature in a closed volume. To reach a stable temperature in the range of −15 to +45 °C, two Peltier cooler modules (Adaptive APH-199-14-15-E) are driven by a controller of Meerstetter Engineering GmbH controller (Meerstetter TEC-1090-HV) and the accompanying software (TEC Service V 3.00).

A 10 kΩ NTC temperature sensor measures the air temperature inside the box and provides the feedback for the software to adjust the cooling power of the Peltier elements. To guarantee an equally distributed temperature inside the box with an inner volume of ca. $4 \times 4 \times 10$ cm^3, a ventilator is constantly circulating the air.

The explained setup provides, depending on the experiment, a constant temperature or a linear temperature change until reaching a target temperature. This exactly enables the freezing and melting of a water under controlled environmental conditions.

2.3. Sensor Setting

Figures 2 and 3 show a sketch and a photograph of the mechanical setup inside the temperature-controlled box respectively. This concentrates on a good acoustic coupling between the PZT sensor and the probe and an equally distributed air flow throughout the experiment.

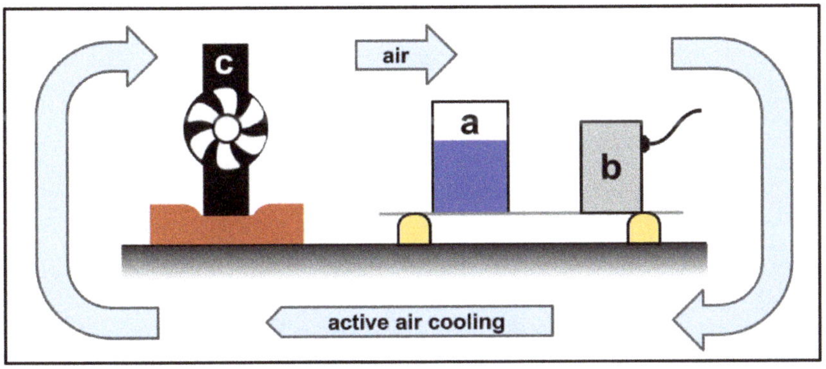

Figure 2. A sketch of the experimental setup inside the temperature-controlled box. An 1 mm thick aluminum plate is holding the sample of water (**a**) as well as the PZT sensor (**b**) and therefore couples the two acoustically. The internal van (**c**) ensures sufficient air movement and a uniform temperature inside the box.

Figure 3. A photo of the sensor setup. The specimen and the PZT sensor are positioned on an aluminum plate which acts as a wave-guide. The plastic cylinder holds the water (for demonstration purposes colored in blue) and has no bottom so that the water is directly touching the aluminum plate.

The sample holder for the water/ice and the PZT sensor were both positioned on a 1 mm thick aluminum plate that serves as a wave-guide for the AE signals. To guarantee a good coupling between the PZT and the plate, some vacuum grease (Dow Corning, High vacuum grease) is used as couplant. As the sample holder has no bottom, the water/ice is directly touching the aluminum plate and no additional couplant is needed. The hollow sample holder, thus sealed with vacuum grease, allows the initially liquid water directly touching the plate. The PZT sensor is placed close to the sample (ca. 2 cm) and acoustically coupled to the plate with some vacuum grease.

Small foam pieces are placed under the aluminum plate to damp mechanical and acoustic vibrations from the outer housing. The aluminum plate is providing a good acoustic coupling of the ice and the PZT sensor and reproduces realistic conditions as in a real application (e.g., in an aircraft fuel tank), where signals will also be transmitted via sheet-structures comparable to the aluminum plate. Although later applications monitor melting ice systems that that are a factor of 10–100 larger in size, a down-scaled experimental setup was designed to ensure an equal and thorough controlled temperature during the measurements with in-house equipment.

As described in Section 2.2, for an optimal performance of the temperature control system, a good air circulation inside the box is required. The drawback of the internal ventilator is its vibration that interferes with the AE measurement and is therefore reliably damped with a piece of foam between the ventilator and the outer housing of the box.

2.4. Measurement

The explained measurement setup allows AE measurements on freezing water and melting ice under constant environmental conditions in which only the temperature changes and the AE setup stays untouched.

For the experiments explored in this paper, the water experiences temperature profile as discribed in Table 1.

Table 1. Temperature profile during the measurement in the temperature–controlled box.

Phase	Time	Temperature
constant	1 h	$T = 10\,°\mathrm{C}$
freezing	3 h	$\Delta T_{\downarrow} = -8\,°\mathrm{C\,h^{-1}}$
constant	2 h	$T = -14\,°\mathrm{C}$
melting	3 h	$\Delta T_{\uparrow} = 8\,°\mathrm{C\,h^{-1}}$
constant	2 h	$T = 10\,°\mathrm{C}$

3. Results

The gray histograms in Figure 4 show the number of acoustic signals during the melting of a water sample with different weights between m = 1.0 ± 0.1 g and m = 4.0 ± 0.1 g. The black lines show the air temperature in the isolated box.

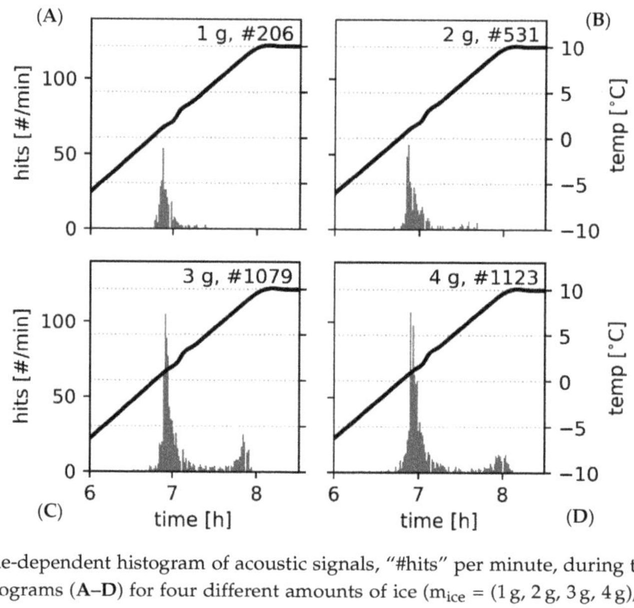

Figure 4. Time-dependent histogram of acoustic signals, "#hits" per minute, during the melting of ice. Four histograms (**A–D**) for four different amounts of ice (m_{ice} = (1 g, 2 g, 3 g, 4 g), upper left to lower right) plotted against the time (grey). On the second x-axis (right), the air temperature inside the experimental setup is plotted (black line).

All four histograms show a high acoustic activity when the temperature reaches the melting temperature. Two peaks of acoustic activity can be identified. The first peak appears at the beginning of the melting process where the Celsius temperature just reaches a positive value. After an "activity valley", another, lower peak appears at the end of the melting phase.

Prior to the increase of temperature in Figure 4 at $t < 6\,\mathrm{h}$, the temperature is constant for approximately two hours. This time included, none of the performed measurements show more than five detected events until the temperature reaches $T \approx -2.5\,°\mathrm{C}$.

The temperature curve confirms the reliable performance of the temperature controlling facility. The experimental setup reaches the target temperature within its limits and stabilizes the temperature at the lower plateau at $T = -14.00 \pm 0.05$ °C. Taking a possible systematic offset from the NTC sensor into account, an air temperature uncertainty of less than $\Delta T = 0.1$ K can be assumed . During the constant raise of target temperature, the temperature shows a small variation of about $\Delta T \sim 30$ mK around the target temperature at $t \sim 7$ h. The registered time-dependent power that drives the Peltier-cooling element confirms this fluctuation around the constantly rising target temperature (not shown in Figure 4). Both can be explained with the latent heat of the melting process which results an additional "energy consumption" in the system. The temperature control unit reacts with a certain delay to this, which results in a slower temperature increase. This is followed by an "over-steering" of the temperature control unit and an increase of the temperature above the target temperature. In the upper right corner of each plot, the mass of water and total sum of acoustic events is indicated.

As described in Section 2.1, common AE parameters such as AMPL, D, RT and E are calculated for each acoustic event. This allows a later filtering and detailed parameter-based analysis. However, in this experiment, only the RT was used to distinguish between AE signal from the ice-melting and background noise. For this, a RT-filter with a threshold of RT $\geq$ 0.1 ms was implemented. A different representation of the data can be seen in Figure 5 where all detected AE events are plotted in a scatter plot showing their AMPL and the time of occurrence (gray dots). The histograms in Figure 5 show the accumulated energy per time unit during the melting (black line). The single events, as well as the histograms which are very similar to the histograms in Figure 5 indicate that there is no significant change in the AMPL or acoustic energy of the events along the melting process. Also, the other parameters (D, RT) that are implicitly influencing the acoustic energy do not show any significant change during the melting process.

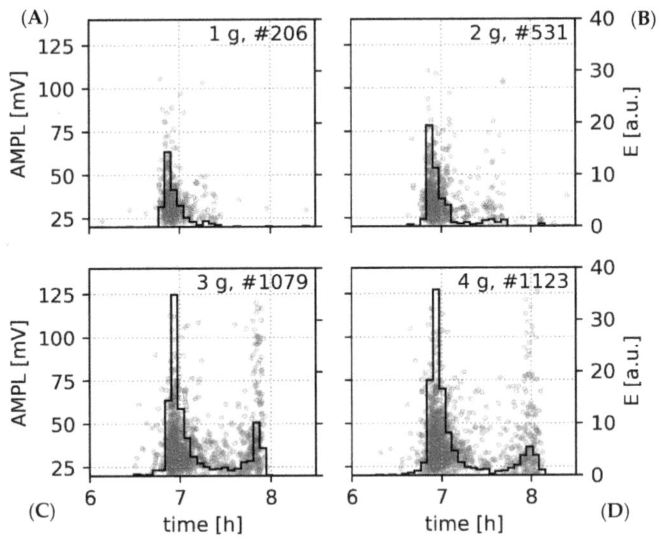

Figure 5. To show the temporal distribution of the acoustic energy, four plots (**A–D**) of different measurements (m_{ice} = (1 g, 2 g, 3 g, 4 g), upper left to lower right) show every AE event plotted with its amplitude AMPL verses the time of occurrence (grey dots). Taking the energy of each event into account, the temporal distribution of accumulated energy per time unit (black line) shows, similar to Figure 4, two peaks at the start and end of the melting phase.

Figure 6 shows the acoustic activity during the melting of different samples with varying water quantities. The solid markers indicate the number of events measured during the complete melting process. As seen in Figure 4, the AE events in the initial melting phase dominate the total number of AE events. For the further analysis, the time between the two peaks divide the histograms in two parts ("first peak", "second peak"). Besides all events ("all" = "first peak"+"second peak"), the number of events in the first peak ("peak") are separately plotted in Figure 6 as circles. Only the missing "second peak" the measurement with the smallest quantity makes this distinction impossible and it is assumed that all AE events are in the "first peak".

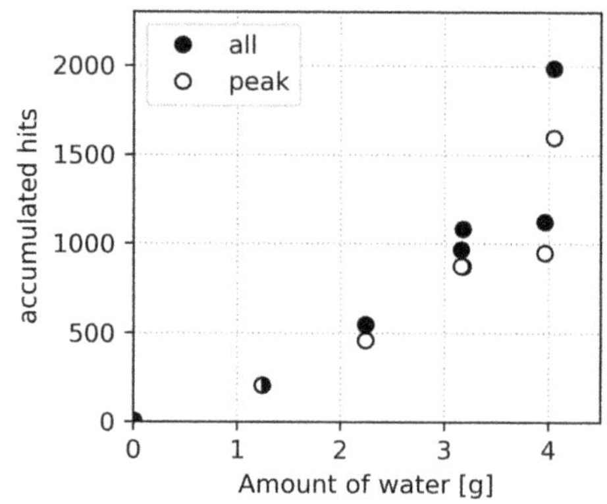

Figure 6. The total amount of acoustic signals during the melting of ice plotted against the quantity of ice. The black and white data points show the complete sum of events and the number of events in the first peak respectively.

The usage of a weight scale reduces the uncertainty of the quantity of water per measurement that is assumed to be below $\Delta m_{ice} < 0.1$ g.

4. Discussion

The measurements presented above show that the AE technique can monitor melting processes of ice quantitatively as well as qualitatively. The acoustic activity of ice in an environment with a constantly increasing temperature starts when the temperature reaches the melting temperature of water. Independent of the amount of water, the acoustic activity is the highest at the beginning of the melting process where rates of up to 125 AE events per minute where measured with the described setup. This number is however strongly dependent on the physical measurement setup, used hardware and signal filters.

Wentzell et al. [13] already reported a peak in the acoustic activity at the beginning of the melting process of ice. They identified the temperature gradient between the surrounding and the frozen sample as a root cause for crystal fractures that cause this peak. In the presented work, the very slow temperature increase ($\Delta T_{\uparrow} = 8\,\mathrm{K\,h^{-1}}$) prevents high temperature gradients within the sample and crystal fractures should not be a relevant source of acoustic events.

After this peak, the rate of the acoustic events decreases and a "valley" with significantly lower AE activity appears. For the measurements with amounts of water $m_{ice} > 2.0$ g, the acoustic activity again increases towards the end of the melting process.

This behavior can have multiple reasons that include phenomena related to the intrinsic acoustic activity during the melting process or the acoustic coupling between source and sensor.

In Figure 7 a schematic, simplified of the temporal histogram of AE events during the melting process is shown and divided into three parts (A, B, C) that might correspond to different situations inside the sample holder.

During the first part of the melting (A), the sample is dominated by ice. At a certain point in time, the constantly increasing temperature results in the melting of ice and stress in the crystal due to the sudden change in volume during the melting. Therefore, a significant amount of acoustic energy is released in this initial stage. As seen in Figure 6, the number of acoustic events at the beginning of the melting process is dominated by the first peak and linear dependent on the amount of ice.

After the first peak, the acoustic activity decreases and the second phase (B) with less acoustic activity sets in. Therefore, the coupling of the AE source (melting ice) and the sensor changes. An additional film of water between the sensor and the ice results in an additional surface that reflects a part of the acoustic energy.

Finally, as described above and seen in all additional measurements, a remarkable increase of observed acoustic activity is measured at the end of the melting process. In this phase (C), only a little ice amount is left swimming at the surface of the water reservoir. This ice is therefore in contact with the air which is warmer than the water in the sample holder resulting in a higher temperature gradient. This could be one reason for a higher melting rate of the last bits of ice resulting in an increase of the acoustic activity.

To investigate this effect, measurements with a bucket that was closed with a cover and some vacuum grease were done. In this case, the air inside the bucket was trapped and not interchangeable with the external air. Therefore, it can be assumed that the effect of the ice touching "warm" air was reduced. However, also these measurements show the three phases of melting including the second peak at the end of the melting phase which suggests that the ice touching the "warm" air has no or only a minor effect.

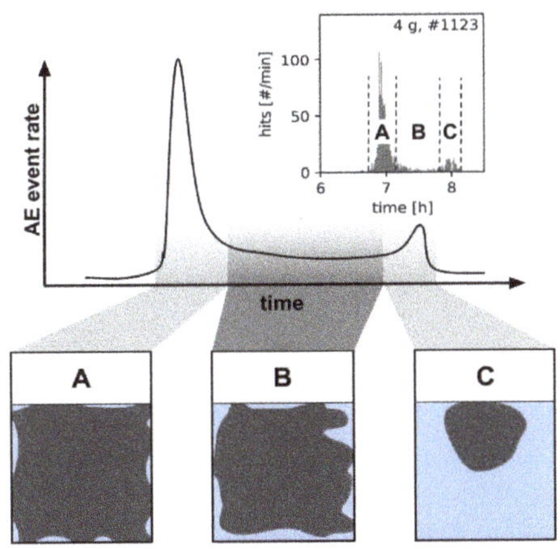

Figure 7. A sketch of a temporal histogram of the acoustic activity during a melting process (above) and the water/ice composition inside the sample holder (below). The physical composition of the ice in the sample holder might explain the three phases in the histogram (**A–C**). For comparison of the sketch with the measured data, measurement of the $m_{ice} = 4.0$ g sample is shown in the upper right corner.

5. Conclusions

The presented work shows that AE can monitor the melting process of ice which is accompanied by acoustic signals observed in the ultrasonic frequency range above 30 kHz.

To conduct measurements in a controlled environment, an actively temperature-controlled box which can be heated and cooled with Peltier elements was build. With this, measurements on melting processes of different amounts of tap water were conducted. Basic signal analysis based on the time of occurrence, amplitude and length of acoustic events enables the monitoring of the melting processes. A systematic pattern of acoustic activity during the melting process with two peaks was measured. Due to these two peaks, which occur at the beginning and the end of the melting phase, the start, as well as the end of the melting, can be clearly estimated. In addition to this, the number of acoustic events in the first peak seems to be correlated with the amount of ice. These results show that AE can monitor melting processes in systems with a good acoustic coupling between the AE sensor and the melting ice. This is especially interesting when melting ice must be identified in areas which are difficult to access, as for instance a fuel tank within an aircraft during the daily maintenance. Here, information about ice in the fuel tank would improve the safety and efficiency of draining processes which remove liquid water from the tanks in the hangar.

6. Outlook

The presented work shows that melting of ice can be monitored with AE. However, in this work only very little information (time of occurrence, AMPL, E, D, RT) were deduced from the acoustic signals and no spectral analysis was performed yet. Thus, additional information that can be deduced from a spectral analysis with well-known methods such as modal analysis or principal component analysis as done by Surgeon and Wevers [20], Johnson [21] respectively. One question is if the two peaks in the histogram originate from different physical phenomena during the melting and if a detailed spectral analysis can differentiate the signals in the first and second peak. However, for this, it must be ensured that the spectral composition of the acoustic transients is not heavily influenced by the experimental setup and the changing physical properties of the sample during each measurement.

In addition, the influence of physical parameters such as the cooling rate during the freezing, the purity of the sample and related freezing temperature and chemical composition of the sample to the acoustic activity have to be investigated. For this, a temperature measurement inside the ice/water mixture during the experiment would provide further insights into the freezing/melting behavior of the sample. Combined with repetitive measurements of the same setting and resulting statistics, fundamental properties of the sample could be derived only by recording and analyzing the acoustic signals during the melting.

Author Contributions: M.S. recorded and analyzed the presented data and was the major contributor in writing this paper. H.P. contributed in the design and the interpretation of the experiment. J.R. was mainly responsible for the technical clarifications and formulations regarding the daily aircraft maintenance. M.W. supervised and reviewed the analysis of the data and contributed to the interpretation of the results. All authors read and approved the final manuscript.

Funding: The Research leading to these results has received funding from the "NDTonAIR" project (Training Network in Non-Destructive Testing and Structural Health Monitoring of Aircraft structures) under the action: H2020–MSCA–ITN–2016–GRANT 722134.

Acknowledgments: Special thanks to Johan Vanhulst (KU Leuven) for the technical support and to the M&E department of Brussels Airlines for daily cooperation.

Conflicts of Interest: The authors declare no conflict of interest. The funders had no role in the design of the study; in the collection, analyses, or interpretation of data; in the writing of the manuscript, or in the decision to publish the results.

Abbreviations

The following abbreviations are used in this manuscript:

AE	acoustic emission technique
NDT	non-destructive testing
PZT	piezoelectric
AC	alternating current
AMPL	amplitude
D	duration
RT	rise time
E	energy
THD	threshold
HDT	hit definition time
HLT	hit lockout time
PDT	peak definition time

References

1. Stamm, M.; Pfeiffer, H.; Reynaert, J.; Wevers, M. Development of a technique for the detection and quantification of water and ice in the fuel tanks of an aircraft. *ECNDT* **2018**, *2018*, 1–9.
2. Baena-Zambrana, S.; Repetto, S.; Lawson, C.; Lam, J.W. Behaviour of water in jet fuel—A literature review. *Prog. Aerosp. Sci.* **2013**, *60*, 35–44. [CrossRef]
3. Li, X. A brief review: Acoustic emission method for tool wear monitoring during turning. *Int. J. Mach. Tools Manuf.* **2002**, *42*, 157–165. [CrossRef]
4. Behnia, A.; Chai, H.K.; Shiotani, T. Advanced structural health monitoring of concrete structures with the aid of acoustic emission. *Constr. Build. Mater.* **2014**, *65*, 282–302. [CrossRef]
5. Zaki, A.; Chai, H.K.; Aggelis, D.G.; Alver, N. Non-destructive evaluation for corrosion monitoring in concrete: A review and capability of acoustic emission technique. *Sensors* **2015**, *15*, 19069–19101. [CrossRef] [PubMed]
6. Verstrynge, E.; Pfeiffer, H.; Wevers, M. A novel technique for acoustic emission monitoring in civil structures with global fiber optic sensors. *Smart Mater. Struct.* **2014**, *23*, 065022. [CrossRef]
7. Schiavi, A.; Niccolini, G.; Tarizzo, P.; Carpinteri, A.; Lacidogna, G.; Manuello, A. Acoustic emissions at high and low frequencies during compression tests in brittle materials. *Strain* **2011**, *47*, 105–110. [CrossRef]
8. Boyd, J.W.; Varley, J. The uses of passive measurement of acoustic emissions from chemical engineering processes. *Chem. Eng. Sci.* **2001**, *56*, 1749–1767. [CrossRef]
9. Sawada, T.; Gohshi, Y.; Abe, C.; Furuya, K. Acoustic emissions arising from the gelation of sodium carbonate and calcium chloride. *Anal. Chem.* **1985**, *57*, 366–367. [CrossRef]
10. Cao, Z.; Wang, B.F.; Wang, K.M.; Lin, H.G.; Yu, R.Q. Chemical acoustic emissions from gas evolution processes recorded by a piezoelectric transducer. *Sens. Actuators B Chem.* **1998**, *50*, 27–37. [CrossRef]
11. Wentzell, P.D.; Wade, A.P. Chemical acoustic emission analysis in the frequency domain. *Anal. Chem.* **1989**, *61*, 2638–2642. [CrossRef]
12. Sawada, T.; Gohshi, Y.; Abe, C.; Furuya, K. Acoustic emission from phase transition of some chemicals. *Anal. Chem.* **1985**, *57*, 1743–1745. [CrossRef]
13. Wentzell, P.D.; Vanslyke, S.J.; Bateman, K.P. Evaluation of acoustic emission as a means of quantitative chemical analysis. *Anal. Chim. Acta* **1991**, *246*, 43–53. [CrossRef]
14. Lube, E.; Zlatkin, A. In-process monitoring of crystal perfection during melt growth. *J. Cryst. Growth* **1989**, *98*, 817 – 826. [CrossRef]
15. Ersen, A.; Smith, A.; Chotard, T. Effect of malic and citric acid on the crystallisation of gypsum investigated by coupled acoustic emission and electrical conductivity techniques. *J. Mater. Sci.* **2006**, *41*, 7210–7217. [CrossRef]
16. Wang, X.; Huang, Y. An investigation of the acoustic emission generated during crystallization process of salicylic acid. *Powder Technol.* **2017**, *311*, 350 – 355. [CrossRef]
17. Grosse, C.U.; Ohtsu, M. *Acoustic Emission Testing*; Springer Science & Business Media: Berlin/ Heidelberg, Germany, 2008.
18. Vallen. *AE-Sensor Data Sheet VS375-M*. Vallen Systeme GmbH. 2018. Available online: www.vallen.de (accessed on 9 December 2019).
19. Olympus. *Olympus Ultrasonic Preamplifiers, Data sheet: Model 5670*; Olympus NDT Inc.: Tokyo, Japan, 2018.
20. Surgeon, M.; Wevers, M. Modal analysis of acoustic emission signals from CFRP laminates. *NDT&E Int.* **1999**, *32*, 311–322.
21. Johnson, M. Waveform based clustering and classification of AE transients in composite laminates using principal component analysis. *NDT&E Int.* **2002**, *35*, 367–376.

Sample Availability: The datasets used and/or analyzed during the current study are available from the corresponding author on reasonable request.

applied sciences

MDPI

Article

The Teager-Kaiser Energy Cepstral Coefficients as an Effective Structural Health Monitoring Tool

Marco Civera [1,*], Matteo Ferraris [2], Rosario Ceravolo [2], Cecilia Surace [2] and Raimondo Betti [3]

1. Department of Mechanical and Aerospace Engineering-DIMEAS, Politecnico di Torino, Corso Duca degli Abruzzi 24, 10129 Turin, Italy
2. Department of Structural, Geotechnical and Building Engineering-DISEG, Politecnico di Torino, Corso Duca degli Abruzzi 24, 10129 Turin, Italy; ferraris.matteo@gmail.com (M.F.); rosario.ceravolo@polito.it (R.C.); cecilia.surace@polito.it (C.S.)
3. Department of Civil Engineering and Engineering Mechanics, Columbia University, New York, NY 10027, USA; rb68@columbia.edu

* Correspondence: marco.civera@polito.it

Received: 22 October 2019; Accepted: 15 November 2019; Published: 23 November 2019

Featured Application: This paper presents a damage sensitive feature, the Teager-Kaiser Energy Cepstral Coefficients (TECCs), which can be used to train a Machine Learning algorithm to perform damage detection and Structural Health Monitoring (SHM) on complex buildings and/or mechanical systems.

Abstract: Recently, features and techniques from speech processing have started to gain increasing attention in the Structural Health Monitoring (SHM) community, in the context of vibration analysis. In particular, the Cepstral Coefficients (CCs) proved to be apt in discerning the response of a damaged structure with respect to a given undamaged baseline. Previous works relied on the Mel-Frequency Cepstral Coefficients (MFCCs). This approach, while efficient and still very common in applications, such as speech and speaker recognition, has been followed by other more advanced and competitive techniques for the same aims. The Teager-Kaiser Energy Cepstral Coefficients (TECCs) is one of these alternatives. These features are very closely related to MFCCs, but provide interesting and useful additional values, such as e.g., improved robustness with respect to noise. The goal of this paper is to introduce the use of TECCs for damage detection purposes, by highlighting their competitiveness with closely related features. Promising results from both numerical and experimental data were obtained.

Keywords: damage detection; cepstral analysis; Structural Health Monitoring; Machine Learning; Teager-Kaiser Energy; gammatone filter

---

1. Introduction

Damage detection and Structural Health Monitoring (SHM) indicate the field of engineering that is involved in the assessment of the structural integrity for civil, mechanical, and aerospace applications, among others [1]. More specifically, vibration-based SHM [2] deals with the structural response, as recorded by means of a sensor network, to extract damage-related features from its free or forced oscillations. In this context, recent years have seen a continuous surge of interest in Machine Learning (ML) algorithms, as they are perfectly suited for damage detection [3], if this is seen from a Pattern Recognition/Novelty Detection standpoint [4]. The concept is simple: by knowing the current state of a system (which is usually undamaged, but can be indistinctly previously damaged, as long as it is stable and not dramatically changing under "normal" conditions), a baseline can be defined. This is performed by training a "normality" model via ML techniques over the features that were extracted from the known set of data. Subsequently, the structural response from other scenarios can be labelled as divergent from this model or not by regarding the value of a distance metric between its current response and the baseline model. One can refer to the previously cited book of Farrar & Worden for a detailed description of the basics of ML for SHM [1].

Some damage-related features (also known as damage sensitive features, DSFs) are needed for training and testing a Machine Learning algorithm. Eventually, DSFs need to be as low in dimensionality as possible, since a very large number of cases is required for the training phase, preserving the relevant information while discarding unrelated effects. However, feature selection is a concern much broader than its SHM applications. One of the fields where it has seen a major development is in speech, speaker, and language recognition.

Virtual assistants, human-like speech synthesis, and audio indexing are all examples of ML applied to Speech Processing for Artificial Intelligence (AI) purposes. Other applications are countless: automatic transcription to text and conversational human-machine interfaces are the most obvious uses of speech recognition. Speaker recognition, i.e., the process of automatically predicting the identity of the speaker by a given utterance, is used for voice biometric authentication, telephone-based services, speaker diarisation, and forensic activities. Finally, language recognition finds its main applications in emergency call routing, spoken language translation, and audio surveillance, to cite a few. Speaker and language recognition rely on similar techniques, as they mostly share the same problems: preprocessing the signal, extracting its relevant features, and representing them by means of statistical models.

There are no conceptual impediments for the application of the related techniques for vibration-based SHM if acceleration time histories (THs) from the structure of interest are considered instead of audio samples. As for speaker recognition, distinctive traits, related to specific phenomena of interest, have to be spotted and discerned from uninteresting effects. On the one hand, mechanical vibrations and biomedical signals show the same issues, such as changing content along time (both frequency- and amplitude-wise) during dynamic excitation, which cannot be properly depicted by means of Discrete Fourier Transform (DFT) [5], and thus require joint time-frequency analysis techniques for a two-dimensional, time-vs-frequency representation [6,7], such as the one obtained via wavelet analysis [8–11]. On the other hand, speech signals present some difficulties—such as different duration between samples, unknown input, and high level of non-stationarity—which are generally much less impelling for civil and mechanical structures rather than in biological systems [12]. Indeed, it is possible to figure the human speech production system out as a control system, where the plant dynamics are related to the vocal tract characteristics, while the constriction and airflow mechanisms are the biological counterparts of the controller. In this sense, the analogy with structural vibrations is straightforward. Some other examples of contamination between the speech processing techniques and structural health monitoring include some recent works on Wavelet Levels (WLs), on Hilbert-Huang Transform (HHT), and on Empirical Mode Decomposition (EMD) [13,14]. Other proposals include well-known techniques, such as the Continuous Wavelet Transform (CWT), the Unscented Kalman Filter (UKF), and the Blind Source Separation (BSS) [5].

By far, the cepstrum [15] has been the main contribution to the field, being often coupled with other techniques, such as autoregressive moving-average (ARMA) models [16] or directly applied for the cepstral editing of signals [17,18]; an extensive review of the uses of cepstral features for the monitoring of mechanical (non-biological) vibrations can be found in [19].

Mel-Frequency Cepstral Coefficients (MFCCs) are an example of low-dimensional, fixed dimension feature vectors defined into a cepstral framework, which are very effective in speech processing and already successfully tested as DSFs for mechanical systems [20,21] and subsequently refined [22]. MFCCs proved to be apt in detecting damage in experimental and numerically simulated data for simpler structural elements and in more complex cases, such as historic masonry buildings. However, this feature is not free of issues; one of the major problems of MFCCs is how they can be swayed by additive noise with relative ease. Indeed, log-mel-amplitudes are influenced by the low-energy component of the signal and this is more strongly evident in the first MF cepstral coefficient.

Similarly to the MFCCs, the Teager-Kaiser Energy Cepstral Coefficients (TECCs) were first introduced in the ambit of Speech Recognition in 2005 by Dimitriadis et al. [23], relying on the Kaiser definition of the Teager Energy Operator [24]. The process for their extraction is two-stepped. Firstly, the signal is filtered through a dense non-constant-Q Gammatone filterbank (details will follow shortly), which replaces the triangular filters utilised in the standard MFCCs approach. This initial step is shared with other closely related alternatives, such as, specifically, the Gammatone Cepstral Coefficients (GTCCs) [25], which have also been tested here in this work. Secondly, the short-time average of the output of the previously mentioned Teager-Kaiser Energy (TKE) Operator is computed. A deeper discussion for both MFCCs and TECCs is postposed to the respective subsections; it is important to remark here how the proposed TECC-based method relies on the TKE operator, which, in turn, is the basis of several time-frequency analysis approaches for the estimation of instantaneous amplitude and frequency [26]. These techniques already proved to be efficient in other fields of signal processing apart speech analysis and very recently they have been successfully applied to Structural Health Monitoring. For instance, the TKE has been combined with deep belief networks for the fault diagnosis of reciprocating compressor valves [27] and with least-squares support vector machine (LS-SVM) classifiers to detect bearing fault [28]. Nevertheless, its use is still very limited and few applications are reported in the literature.

The proposed TECC-based approach has been tested on two numerical and one experimental case studies. The first numerical case is a Finite Element (FE) model that represents the Santa Maria and San Giovenale Cathedral bell tower in Fossano, Italy, which is a structure of relevant importance for cultural heritage and it has been deeply investigated in the literature (for instance, see [29]). This is representative of historical masonry buildings, which noteworthy often present several structural problematics and require proper monitoring [30], especially in the case of high-rise medieval structures [31,32]. The second case study regards the spar of the high-aspect-ratio (HAR) prototype wing XB-1 studied in [33]. This case also presents its own difficulties, arising from the high flexibility of the cantilevered structure. The experimental data that are described in [21] have also been used for comparison. In this last case study, the damage is modelled as a breathing crack mechanism, introducing a source of nonlinearities often encountered in damaged structures, where the presence of crack often results in nonlinear behaviour [34,35]. As a benchmark, the original five alternative methods that were proposed in [21] and the six new closely-related variants described in [22] have also been applied to the numerical and experimental data; the results show interesting improvements.

It is important to stress how the choice of different structural systems, materials, and damage mechanisms for this methodology's validation is intended to show its reliability for novelty detection in a wide range of applications and conditions, as "damage" (in the broader sense possible) might manifest itself in many different ways, also depending on the specific constructional material [36], and these different occurrences may not be all similarly easy to detect—especially in larger structures with complicated geometries.

The outline of this paper is as following: in Section 2, the needed background for a full comprehension of the context is recalled. The algorithm that is utilised for damage detection is described in Section 3. In Section 4, the numerical case studies are presented. Section 5 briefly describes the experimental data used for further validation. A comparison with previous works and similar alternatives is also proposed in the Results Section, and Conclusions follow.

2. Theoretical Background

The novel method here proposed for SHM purposes, based on the Teager-Kaiser Energy Cepstral Coefficients (TECCs), requires some previous knowledge regarding the definitions of cepstrum, quefrency [37], and the melody-frequency scale (also abbreviated as the mel-scale) [38]. This Section is intended to also provide a basic common ground for non-experts in signal and speech processing, not familiar with the jargon, and to establish some definitions before continuing. For the sake of better comprehensibility, the discussion is sequenced here into the following subsections: cepstral analysis, a brief recall about MFCCs, some hints about their (many) variants, and finally the description of the TECCs themselves.

2.1. Cepstral Analysis

A somehow basic but effective description of the cepstrum (that can be applied to both its power and complex definitions) is the "spectrum of the spectrum on logarithmic amplitude axis" [39], as it is computed as the inverse transform of the logarithm of its estimated spectrum. First described in [37], its original and main use is for pitch detection and determination [40], but in the years it has found plenty of applications in speech processing, such as for the estimation of harmonics-to-noise ratios [41]. Resorting to the cepstrum rather than the standard frequency spectrum gives some interesting advantages, mainly the possibility to easily discern and eventually edit off uninteresting harmonics. Some slightly different definitions exist, such as e.g., the discrete complex cepstrum with unwrapped phase, which are useful for time-frequency analysis purposes [42]. Preserving the phase information, the complex extension of the cepstrum allows for reconstructing the original time-dependent signal; however, for the purpose of this research, the definition of the power cepstrum is more convenient. Analytically, this can be defined by resorting to the direct and inverse z-trasform definitions, as ([43], Chapter 24)

$$\hat{f} \triangleq \left[\mathcal{Z}^{-1}\left\{\log\left(|H(z)|^2\right)\right\}\right]^2 = \left[\frac{1}{2\pi j}\oint_{\Gamma_c}\log\left(|H(z)|^2\right)z^{n-1}dz\right]^2 \tag{1}$$

where '^' denotes the power cepstrum of function f, $\mathcal{Z}^{-1}(\cdot)$ is the inverse z-transform operator, $H(z) = \mathcal{Z}(f)$ represents the z-transform of an ordered, discretised time sequence $f = x[n]$, and n indicates the specific time instant; the contour of integration, Γ_c, lies within an annular region where $H(z)$ is defined as analytic and single-valued. Evaluating Equation (1) on the unit circle, the problem is reconducted to the Fourier domain, and it can be restated as

$$\hat{f} = \left[\frac{1}{2\pi}\int_{-\pi}^{\pi}\log\left(|H(\omega)|^2\right)e^{j\omega t}dt\right]^2 \tag{2}$$

For an angular frequency $\omega = 2\pi f$ and the Discrete Fourier Transform $H(\omega)$. The basic idea exploited is the concept of homomorphic deconvolution, i.e., separating different components of one given signal, if f itself is made up by a convolution of some distinct underlying signal components. With respect to the simple, more common Fourier Transform in the frequency domain, this process allows for better discerning between the content of interest and unnecessary/unrelated influences, disturbances or echoes, as convolution in time domain becomes summation in cepstral domain. Thus, thanks to the frequency-warping, the cepstral response simultaneously incorporates properties from both the time and the frequency domains. Please note that, due to direct and inverse transform, the measurement unit is the same as the departing one i.e., time expressed in seconds; nevertheless, $\hat{f}$ cannot be considered to be an ordered time series, because of the warping. Further details can be found in [15]. However, the cepstrum of a signal maintains its same dimensionality, without any gain in conciseness with respect to raw THs. Thus, specific features, i.e., the Cepstral Coefficients (CCs), have to be extracted from it.

2.2. Cepstral Coefficients (CCs) and Mel-Frequency CCs

The Cepstral Coefficients (CCs) can be extracted from the cepstrum of a signal. In the field of speech and sound processing, the mel-scale is intensively used for this purpose, by rearranging the cepstrum in a nonlinear log-like scale of frequency. The idea is generally credited to the works of Mermelstein [44] and Bridle and Brown [45], and it has since then been developed to become a standard in speech processing.

The mel-scale is a purely empirically definition, being intended to mimic the auditory system of the human hear, which is a nonlinear sensing device with almost logarithmically spaced intervals of equivalent perception. The definition of its base unit, the Mel, is one-thousandth of the pitch (℘) of a simple tone with a 40 dB amplitude above the auditory threshold and a frequency that is equal to 1000 Hz [38,46]. The reason behind the thresholding lies in several experimental investigations (such as [47] to name one), which proved that speech processing linear-like for small spectral values and logarithmic-like for large ones yield to more robust identifications with data affected by additive or convolutional artificial noise. Obviously, this definition is very specific, perceptual, and suited for the aim of emulating the human ear. Nevertheless, this was found from previous studies to be apt for SHM purposes; this might be explained by its logarithmic-like spacing, which is recurrent in many natural events not limited to speech production and perception [21]. Another point of relevance is that the definition itself is perceptual, psychoacoustical, and based on perceived equidistance between subsequent pitches; thus, it is not directly provided with a mathematical formulation. Of the many concurrent analytical definitions for the Mel–Hertz conversion, the most commonly accepted one is the variant proposed by Fant in [48]. Other alternative definitions exist; a comparative study was performed by Umesh et al. in [49].

The procedure to extract MFCCs from a non-stationary recorded signal of interest are the following: (i) framing of the signal into more stationary tracts; (ii) windowing of the frames; (iii) computation of the Power Spectral Density (PSD) in the frequency domain; (iv) frequency warping to re-order the discrete power spectrum in the melody-frequency scale (Mel-scale); and finally, (v) L-points inverse Discrete Cosine Transform (iDCT). This last passage can be also performed by means of inverse Discrete Fourier Transform (iDFT), but iDCT is more technically convenient as more adherent to the Karhunen–Loève Transform. When considering (arbitrarily) $L-1$ coefficients to be extracted and M critical bands, the resulting array of MFCCs, $\boldsymbol{c} \in \mathbb{R}^{L\times 1}$, can be defined as

$$\begin{gathered} c[l] = \sum_{m=0}^{M-1} a_m \log(H_{\text{mel}}[m]) \cos\left(\frac{\pi(2l+1)m}{2M}\right), \\ l = 0,1,\ldots,L-1,\ a_m = \begin{cases} \frac{1}{M}\ if\ m=0 \\ \frac{2}{M}\ if\ m \geq 1 \end{cases} \end{gathered} \tag{3}$$

where a_m is the amplitude multiplying the m-th point of the mel-spectrum $H_{\text{mel}}[m]$. The procedure is illustrated for the example of frame #5 in Figure 1.

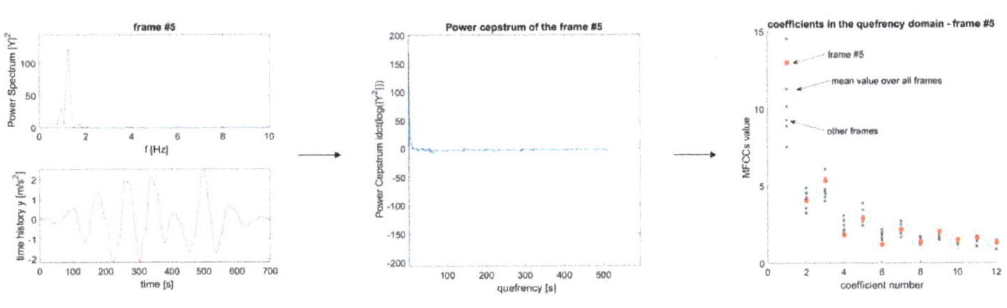

Figure 1. The Mel-Frequency Cepstral Coefficients (MFCCs) extraction procedure; frame #5 is used as an example. From left to right: the acceleration time history and its corresponding power spectrum, the power cepstrum of the same, and the resulting MFCCs in the quefrency domain.

2.3. Similar Scales and Features Investigated

While vastly predominant in the common use, the mel-scale is not the sole nonlinear scale viable. Other options proposed in the literature include the Bark Scale [50] and the Equivalent rectangular bandwidth (ERB)-rate scale [51], to cite the two main competitors. In both cases, several Hertz–Bark/Hertz–ERB relationships were proposed in an analytical form, similarly to the Mel-Scale, as defined in [52] and in [53], respectively. Some of these alternatives, such as the Bark-Frequency Cepstral Coefficients (BFCCs) [54], have also been recently tested by the Authors for SHM purposes. In this context, three novel scales, namely, the Energy COntent-based MMFCCs (ECO-MMFCCs), the MOdal Frequency-centred MMFCCs (MO-MMFCCs), and the LOw-Frequency Restricted MMFCCs (LO- MMFCCs) have been recently proposed [22].

Apart from different psychoacoustical, Mel-like scales as the ones just mentioned, non-triangular weighting functions can also be used, such as gammatone filters [55]. Additionally, the acceleration THs can be pre-processed before applying the extraction procedure. An option is the mentioned Teager-Kaiser Energy (TKE) Operator. This, combined with the Gammatone filters, is the core of the method proposed in this work and will be better detailed in the two next subsections.

To sum up and for comparison with the previous studies, Figure 2 reports a pictorial representation of the tested approaches.

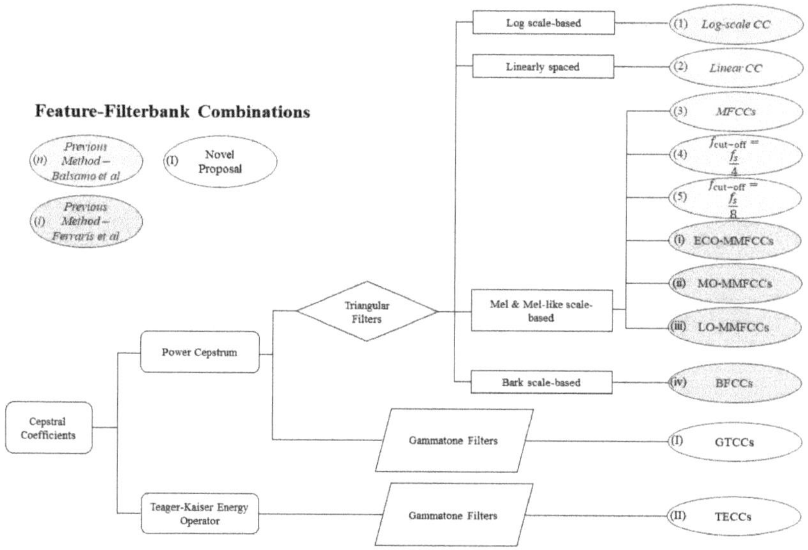

Figure 2. Conceptual relationships between the proposed methodology and the other tested approaches.

2.4. Advantages Over the Auto-Regressive Model Coefficients

The MFCCs, TECCs, and similar cepstral-based features can be compared to other DSFs to better describe their advantages. As done in [21], the discussion is here focused on the Auto-Regressive Model (AR) Coefficients, which are a classic feature widely used for SHM purposes [3,4]. An AR model of order p can be defined (for a time-discrete signal $x[n]$), as

$$x[k] = \sum_{j=1}^{p} a_j x[k-1] + \epsilon_k \quad (4)$$

With $x[k]$ being the signal value at the k-th integer multiple of the sampling interval Δt, a_j the j-th AR coefficient, and ϵ_k the modelling error at $k\Delta t$. The Akaike Information Criterion (AIC) can be used to select a proper value of p to both prevent over- and under-fitting (i.e., to have a model complete enough to describe properly the observed data, but not unnecessarily overcomplicated), as

$$AIC(p) = n\left(\ln\left(\sigma_p^2\right) + 1\right) + 2p \quad (5)$$

where n is the total amount of estimated data point, p is the specific order under investigation, and $\ln\left(\sigma_p^2\right)$ is the natural logarithm of the mean of the sum of squared residual errors ϵ_k between the model and the data at any data point.

However, with respect to AR coefficients, the cepstral features have the following useful qualities: (1) they do not require a somehow arbitrary definition of the model order (even if some arbitrariness does still exist in the definition of the $L-1$ coefficients to be extracted from the M critical bands); (2) their results are much less affected by the choice of L and M with respect to AR's results accordingly to the order p; and, (3) the effects of noise, echoes, and nonlinearities can be easier discerned and eventually removed in the quefrency domain, as they affect separate and distant regions in the logarithmic scale.

A proper comparison between cepstral features and AR coefficients over some experimental data is postponed to Section 6.2.1.

2.5. *Teager-Kaiser Energy Cepstral Coefficients (TECCs)*

The TECCs are known in speech recognition and similar applications to perform better than MFCCs when noise is artificially added, and comparably to them for convolutional (non-additive) noise and in clean conditions [23]; the extraction procedure of these coefficients is made up by two main components that distinguish them from the MFCCs. These steps are the definition of the Teager-Kaiser Energy (TKE) and the convolution with Gammatone filters, which are described here below. The whole extraction procedure is graphically displayed in the flowchart of Figure 3.

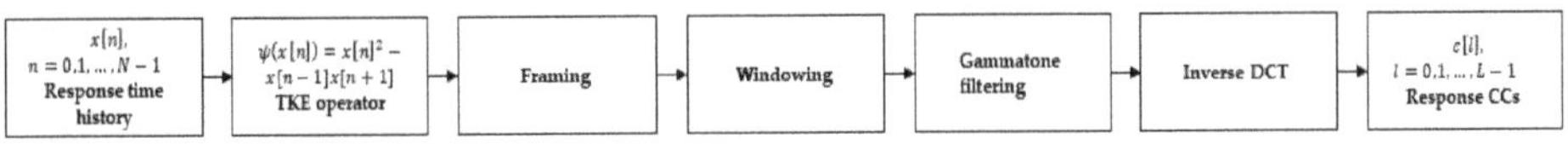

Figure 3. Flowchart of the Teager-Kaiser Energy Cepstral Coefficients (TECCs) extraction procedure.

2.5.1. Teager-Kaiser Energy (TKE)

The Teager-Kaiser Energy (TKE) Operator, also simply known as the Teager Energy Operator, was firstly mentioned by Teager & Teager [56] in 1990. However, this first occurrence did not report an analytical definition of it, but only a vaguely defined as the "energy creating the sound". Thus, it was Kaiser [24], from which the complete name, to propose a proper definition of it as it is currently used. An extensive description of it can be found in [57]; the basic concepts are recalled here to provide a minimal context.

Rather than directly relying on the discretised signal $x[n]$, extracting the CCs from the nonlinear TKE operator allows for more correctly approximating the actual signal energy. The rationale is that power spectrum/cepstrum approximate the total energy of the system as the square of its amplitude, x^2, while the total energy of a vibrating, single degree-of-freedom undamped mechanical system is given by the sum of kinetic and potential energies, that is to say,

$$E = \frac{1}{2}kx^2 + \frac{1}{2}m\dot{x}^2 \tag{6}$$

Unfortunately, this also means that the TKE is limited by its implicit assumption that the source of the analysed signal can be approximated to a single undamped SDoF oscillator, which is a clearly incomplete model. In its discrete form, the TKE of a given signal can be defined as

$$\psi(x[n]) = x[n]^2 - x[n-1]x[n+1] \tag{7}$$

From this definition, the magnitude of the frequency transform is used for the extraction of the CCs. The process is the same as described in Section 2.2. and represented in Figure 1, except for the use of TKE operator and of the Gammatone filter in lieu of acceleration time histories and triangular filters. Noteworthy, the TKE operator can be approximately seen as a high-pass Volterra system, in both continuous- and discrete-time. In the latter case, as used in this work, the operator might be further approximated as the product of an appropriate finite impulse response (FIR) filter and of the signal local (along time) mean. This approximation is only legitimate if the local mean is much larger than the variance of the signal system; however, this error is much more evident in speech signals, which are commonly strongly non-stationary, rather than in the mechanical vibration of structures.

2.5.2. Gammatone Filter

The Gammatone is a linear filter that is obtained from the impulse response of a sinusoid modulated by a gamma distribution function, that is to say,

$$g(t) = at^{n_f-1}e^{-2\pi b\mathrm{ERB}(f_c)t}\cos(2\pi f_c t + \phi) \tag{8}$$

for a given centre frequency f_c. In (8), a is the impulse response function amplitude, n_f is the desired filter order, b the filter bandwidth, and $\mathrm{ERB}(f_c)$ is the Equivalent Rectangular Bandwidth defined by [51]; ϕ indicates the phase of the carrier. Therefore, the filter frequency response becomes

$$G(\omega) = \frac{A}{2}\frac{6}{(2\pi b\mathrm{ERB}(f_c) + j(\omega - \omega_c))^4} + \frac{A}{2}\frac{6}{(2\pi b\mathrm{ERB}(f_c) + j(\omega + \omega_c))^4}) \tag{9}$$

with filter gain A and non-dimensionalised $\omega_c = 2\pi f_c$. The main advantages of Gammatone filters, with respect to the symmetric triangular ones originally employed for MFCCs extraction, are that (i) they are broader and smoother and (ii) they are asymmetrical and non-constant-Q, i.e., the filter frequency response overlap is not constantly 50%. This second point allows for them to emphasise the lower frequencies content, which is of greater relevance in both speech processing and Structural Health Monitoring, while larger bandwidths and less sharp transitions improve the noise robustness [58]; moreover, a Gammatone filterbank is denser in the frequency domain than the mel-frequency-based triangular filterbank. These advantages with respect to the MFCCs and similar alternatives are documented for making the TECCs more effective in the ambit of speech and sound processing.

3. Damage Detection Algorithm

The chosen feature, the TEC coefficients, has been used to train a machine learning (ML) algorithm, obtaining a model of the structural response of interest on unaltered conditions. The algorithm applied is essentially the one that is described in [21], being the proposed DSF, and not the ML procedure utilised, the novelty of this paper; the procedure is only briefly restated for completeness.

As for any ML approach, two phases exist—training and testing—where the former can be executed offline, while the latter may be performed online. During the training phase, the pattern of the DSF coming from the baseline model is 'learnt' by the algorithm, which builds a statistical model out of it. This model is then used as a comparison for the same feature when extracted from the structural response under unknown conditions. A metric of distance between any test case and the reference model is then utilised to discern whether these conditions under inspection correspond to a structural change that is substantial enough to be linked with damage. This threshold value, as will be better explained later, is derived via statistical means. This classic statistical framework is well described in [59].

Signal pre-processing on the acceleration THs was executed, as indicated in [22]. For s acquisition channels, n_{tr} realisation and thus $n_{tr}{\cdot}s$ THs for training the ML algorithm, it is then possible to extract n_{tr} feature vectors, i.e., $\boldsymbol{c}^{(i)} \in \mathbb{R}^{s\cdot(L-1)\times 1}$, where $i = 1, \dots, n_{tr}$ represents the i-th vectors of L CCs, where L has been arbitrarily chosen and set equal for all of the alternatives tested here (both numerically and experimentally) to enforce comparability; note that the size $L-1$ derives from the first CC, $\boldsymbol{c}^{(i)}[0]$, being always discarded to mitigate the DC component and other input-specific effects [20].

The following hypotheses have been made. First, the cepstral features are assumed to follow a Gaussian multi-variate distribution (which is reasonable for a large enough training set and very commonly used for ML); therefore, all $\boldsymbol{c}^{(i)}$ are considered as being practically uncorrelated and thus independent identically distributed (i.i.d.) vectors. This is necessary to completely define the 'normality' model (i.e., the baseline model of the unaltered reference situation) by means of its sample covariance matrix $\mathbf{S_{tr}}$ and the sample mean vector $\mathbf{m}_{tr}$ alone. These can be computed, respectively, as

$$\mathbf{S_{tr}} = \frac{1}{n_{tr}} \sum_{i=1}^{n_{tr}} \left(c^{(i)} - \mathbf{m}_{tr}\right)\left(c^{(i)} - \mathbf{m}_{tr}\right)^T \tag{10}$$

and

$$\mathbf{m}_{tr} = \frac{1}{n_{tr}} \sum_{i=1}^{n_{tr}} c^{(i)} \tag{11}$$

The Squared Mahalanobis Distance (SMD) of a d-dimensional point $\mathbf{x}$ from the baseline model is used as a damage metric; $\mathbf{x}$ corresponds to one element of the test set, i.e., to a single test feature vector $\widetilde{c}$. The physical meaning behind the use of SMD is that features coming from more different structural conditions will be more distant in the feature space; the advantages of this approach, which is the favourite (and often the default) option for outlier detection, are well-known and properly described in [21]. Basically, it relies on the sample covariance matrix $\mathbf{S_{tr}}$; this allows for accounting for feature variability under confounding factors, such as temperature variation, wind speed, or operational conditions, as long as the given samples are measured during these different, non-damage-related external conditions. The SMD can be analytically defined as

$$\mathcal{D}^2(\mathbf{x}) = (\mathbf{x} - \mathbf{m}_{tr})^T\, \mathbf{S_{tr}}^{-1} (\mathbf{x} - \mathbf{m}_{tr}) \tag{12}$$

and the procedure can be easily extended and generalized to multiple elements in the test set. As a Damage Index (DI), $\mathcal{D}^2(\mathbf{x})$ needs a threshold (hereinafter, Γ) to discern actual outliers—hopefully, linked to damaged conditions—to statistical fluctuations of the normality model; it is, essentially, a lower limit of discordancy to call for novelty. Γ is defined here by exploiting a known property of the inspected distribution. The distribution of the SMD of a d-variate point $\mathbf{x}$ follows a scaled F-distribution [60], defined by two degrees of freedom, the dimension of the scalar $\mathbf{x}$ and the number of observations in the statistical population used to define the model. In this specific case, they take the form of $d = s{\cdot}(L-1)$ and $n = n_{tr}$, which leads to

$$\frac{n_{tr}(n_{tr} - s{\cdot}(L-1))}{\left(n_{tr}^2 - 1\right){\cdot}s{\cdot}(L-1)} \mathcal{D}^2(\mathbf{x}) \sim F_{s\cdot(L-1),n_{tr}-s\cdot(L-1)} \tag{13}$$

where $\mathcal{D}^2(\mathbf{x})$ can be directly compared to the F-distribution after being properly scaled. Noteworthy, this is valid as long as $\mathbf{x}$, which belongs to the test set, has not been used to define the sample statistics' estimators. Γ is then simply defined correspondingly to the $(1-\alpha)-$ quantile of F, with α arbitrarily set to 1%. Moreover, it is mandatory that the total number of training data is larger than the degrees of freedom of the system analysed (i.e., of the product between the number of cepstral coefficients per channel and the number of channels) to produce the needed statistics.

4. The Numerical Case Studies

Two numerical case studies are introduced here. These are representative of several cases that are frequently encountered in Aerospace, Mechanical, and Civil Engineering. In both cases, a 10% root mean square (RMS) white Gaussian noise (WGN) has been added to the simulated acquisitions; response THs have been normalised and standardised (by subtracting the mean and dividing by the standard deviation) at any frame before the feature extraction.

4.1. Santa Maria and San Giovenale Cathedral Bell Tower

The Finite Element Model (FEM) of the Santa Maria and San Giovenale Cathedral bell tower (as visible in Figure 4) has been used as a numerical case study for the proposed damage-sensitive feature, as previously proposed for similar alternatives in [22]. The baseline FE Model of the structure 'as is' has been calibrated accordingly to the results of a campaign of ambient vibration tests [29], aimed to establish the mechanical properties of the deteriorated (or cracked) masonry walls. Table 1 provides material properties, with Young's moduli defined accordingly to the story level, as also pictorially described in Figure 2. The story levels are labelled as follow. Level 0: from ground floor to 9.9 m; level 1: from there to 28.2 m; level 2: from the previous level top m to 35 m, and level 3 corresponds to the belfry (35 m to 46 m high). The octagonal belfry at the top of the bell tower has a wooden rooftop and 0.5 m thick masonry walls, while the thickness of the external perimeter is 1.5 m for the lower levels. By assuming a Rayleigh's viscous model, the damping parameters α and β were defined from the experimentally determined first two flexural eigenfrequencies, $f_1 = 1.297$ Hz and $f_2 = 4.253$ Hz.

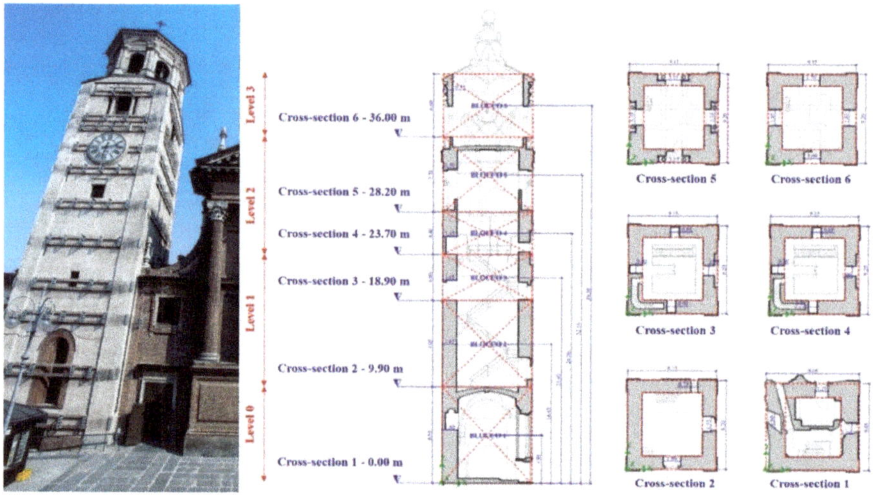

Figure 4. Structural scheme of the bell tower (to the right in the image) and picture of its front view (to the left); autofrettage interventions are clearly visible on the external façades.

Table 1. Fine-tuned Bell Tower baseline Finite Element Model's (FEM's) mechanical properties.

Structure-Wise Properties		Story-Wise Properties	
Rayleigh's Damping Parameter [-]		**Young's Modulus E [MPa]**	
α	0.0993	Level 0	2.690 10^3
β	0.0180	Level 1	1.320 10^3
Density [kg/m^3]		Level 2	1.250 10^3
ρ	2000	Level 3	2.470 10^3
Poisson's ratio [-]			
ν	0.3		

8-noded, 6 DoF-per-node rectangular shell elements were used everywhere in the FEM. The mesh size was 0.5 × 0.5 m wherever possible. The distributed connections with the Cathedral were modelled by means of three-dimensional (3D) spring elements at the contact points [29]. For all simulations (both on the baseline model and on the damaged scenarios), the acceleration THs were recorded at the nodes closest to the actual sensor positions visible in Figure 5, along with the 20 directions of the corresponding acquisition channels. The sampling frequency $f_{s,num\ 1}$ was set to 100 Hz. Seismic strong ground motions were applied as input; 300 spectrum-compatible earthquakes were artificially generated for this aim accordingly to the Italian normative requirements for earthquake engineering at the time of the research (*Norme Tecniche per le Costruzioni* - NTC, D.M. 14th January 2008, Chapter 7) and in compliance with the Eurocodes. The limit State of safeguard of life was considered. The seismic intensity parameters are reported in Table 2. Some of the values were fixed, while the remaining parameters were left free to vary in between realistic ranges, generating 10 combinations. Additive white Gaussian noise (AWGN) was then inserted to obtain 10 slightly different dynamical inputs from any of these 30 cases. Elastic spectra were employed, with the smallest period of the response being equal to 0.02 s, the largest period 4.00 s, the beginning of the stationary part of the accelerogram at 2 s, and the duration of the stationary part TLVL (NTC §3.2.3.6) free to float in a given range. The baseline FE model was modified to obtain several damage scenarios. These are, again for comparison' sake, the same as utilised in [22]. In all cases, the damage was simulated as a reduction of the Young's modulus E at given locations to emulate damage in the masonry external walls. Table 3 lists all the scenarios; the corresponding first natural frequency is also reported to highlight the frequency shift induced by the presence of damage.

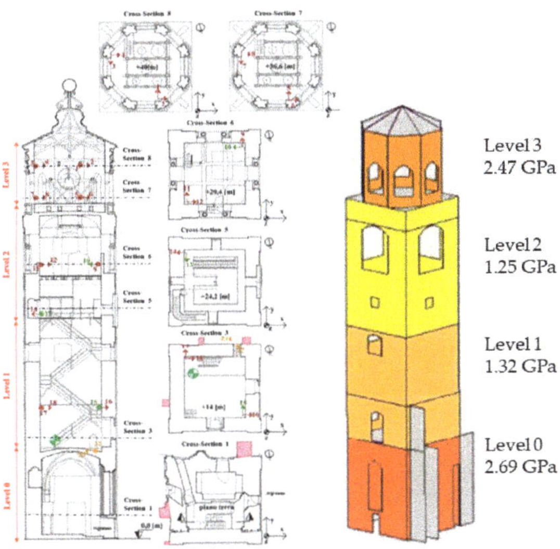

Figure 5. Left: scheme of the sensor placement and direction of the acquisition channels (red, green, and yellow arrows). **Right:** the updated Finite Element (FE) model.

Table 2. Seismic intensity parameters.

Fixed Parameters		Constrained Parameters	
return period T_R	475 years	TLVL	10–25 s
rated life V_N	50 years	Damping coefficient	0.05–0.07
Probability of exceedance P_{V_R}	0.9		
Undrained shear strength C_U	1.0		
topographic category	T1		
stratigraphic category	C		
peak ground acceleration a_g	0.109 g		
seismic input duration	35 s		

Table 3. Baseline model and other scenarios for numerical validation (Fossano bell tower).

Case	Name	Description
00	Undamaged	As it is, no Young's modulus reductions
01	1st Damaged	5.00% reduction of the E modulus at Level 0 (all four façades)
02	2nd Damaged	5.00% reduction of the E modulus at Level 1 (all four façades)
03	3rd Damaged	5.00% reduction of the E modulus at Level 2 (all four façades)
04	4th Damaged	5.00% reduction of the E modulus at Level 3 (all four façades)
05–10	5th–10th Damaged	15.00% local reduction of the E modulus at specific, realistic points (as reported in Figure 6)
11	11th Damaged	10.00% global reduction of the E modulus of all levels, all façades
12	12th Damaged	50.0% reduction of spring stiffness at the linking with the Cathedral, x- and y- directions
13	13th Damaged	50.0% reduction of spring stiffness at the linking with the Cathedral, x-direction only
14	14th Damaged	50.0% reduction of spring stiffness at the linking with the Cathedral, y-direction only
15	1st Unaltered	+0.25% global increase of the E modulus of all levels, all façades
16	2nd Unaltered	+1.00% global increase of the E modulus of all levels, all façades
17	3rd Unaltered	−0.25% global reduction of the E modulus of all levels, all façades
18	4th Unaltered	−1.00% global reduction of the E modulus of all levels, all façades

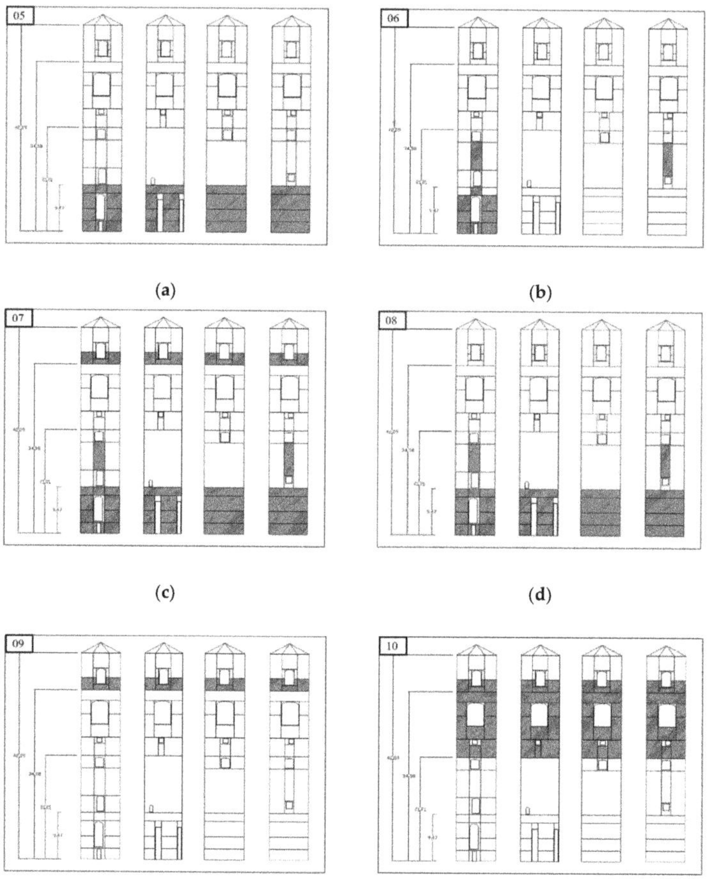

(a) (b) (c) (d) (e) (f)

Figure 6. (**a**) to (**f**): Damage Scenarios 05–10, in the same order. The scenarios have been derived by observation of similar structures damaged by earthquakes; areas with reduced Young's Modulus are portrayed in grey.

Cases 01 to 04 only present a 5% reduction of E along one of the four stories of the structure, on all façades. Cases 05–10 are realistic post-earthquake scenarios, which are based on the Authors' previous experience and intended to model typical damage pattern encountered in similar cases; they are portrayed in Figure 6. Case 11 is a global stiffness reduction of 10% throughout the whole structure. Cases 12–14 emulate damage to the linking with the adjacent cathedral, with 50% of the spring stiffness. Finally, four 'unaltered' scenarios were also accounted for: as for case 11, a global variation of E was applied; specifically, a reduction/increase equal to 1% and 0.25%. This was intended to simulate statistical fluctuations of the material properties, which cannot be unequivocally related to the presence of occurring damage.

4.2. High Aspect Ratio Flexible Wing

The second case study concerns the spar of a prototype wing, the linear and nonlinear dynamics of which have been investigated in recent studies [61,62]. Importantly, the prototype skin is supposed to completely transfer the aerodynamic loads to the spar, making the structural behaviour of the latter a good approximation of the dynamical response of the whole spar-skin ensemble. The HAR, highly flexible wing is cantilevered, its geometry narrowing toward the tip with taper ratio changing at $l = 258$ mm (hereinafter, the mid-length section). Table 4 reports other geometrical and mechanical properties. The wing planform presents a tapered swept configuration, with trapezoidal planform, swept leading edge, and inward-kinked trailing edge. In this case, the fatigue damage that is induced by subsequent load cycles has been modelled as a reduction of the Young's modulus at the two most probable locations, i.e., at the clamped end and at the mid-length section, as portrayed in Figure 7 and enlisted in Table 5. The input, a perpendicular acceleration applied on the wingtip, was defined as a white Gaussian noise of peak value 1.5 g, sampled at $f_{s,num\ 2} = 256$ Hz for 5 s. 50 realisations were numerically simulated on the undamaged case for training; 10 others per case were used for testing the trained model. Five output channels were considered, mimicking the actual sensors array (plus the point of application of the laser Doppler velocimeter) actually used in the previous experimental studies and depicted in Figure 7.

ECO- and MO-MMFCCs, firstly defined in [22] for the specific case of the Santa Maria and San Giovenale Cathedral bell tower, were adapted to this second numerical case. For the ECO-MMFCCs, 90% of the energy content was found at 43.76 Hz, by taking the mean over the five output channels. Therefore, this value has been set as the new cut-off frequency. The centre frequencies for MO-MMFCCs were defined from the modal analysis of the wing spar and benchmarked and validated against experimental data obtained in [61,62]; they correspond, in order of increasing natural frequency, to the 1st, 2nd, 3rd, and 4th flapwise bending mode; the 1st torsional mode; the 5th flapwise flexural; the 1st chordwise bending mode; the 2nd torsional; and, the 6th flexing modes. The tenth mode (the 7th flapwise) fell over the maximum investigable frequency at 333.88 Hz and was therefore replaced by adding '1' to the begin of the array, thus resulting in

$$f_c = [1 \quad 5.49 \quad 23.16 \quad 55.80 \quad 103.99 \quad 125.11 \quad 172.09 \quad 189.15 \quad 219.90 \quad 255.62]$$

To conclude, the low frequency range of LO-MMFCCs was not changed and it remains defined between 0 and $f_s/2$, which here becomes 128 Hz.

For this second case, the major challenge comes from even large reductions in E resulting in minimal differences in terms of frequency shift, as can be seen in the last column of Table 5.

Table 4. Geometrical and mechanical properties of the wing spar.

Density ρ	2850	kg/m^3
Young's Modulus E	7.31×10^{10}	Pa
Poisson Ratio v	0.33	-
Free length (clamp to tip) l_{tip}	706	mm
Thickness t	2	mm
Max width at clamped section b_{max}	180.00	mm
Mid-length width at the section of changing tampering (l = 258 mm) $b_{l=258}$	56.10	mm
Min width at the tip section b_{min}	17.04	mm

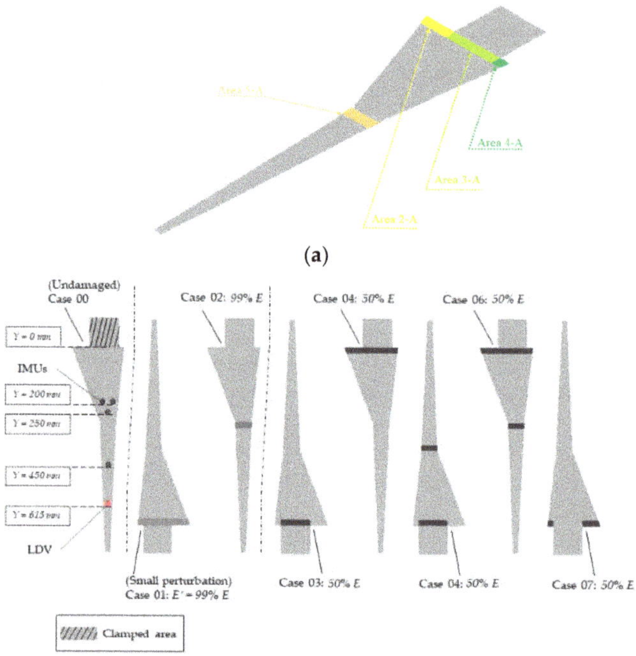

Figure 7. (**a**) Areas subject to decreased Young's modulus (**b**) Damage scenarios.

Table 5. Baseline model and other scenarios for numerical validation (HAR wing).

Case	Name	Description	1st Natural Frequency [Hz]
00	Undamaged	As it is, no Young's modulus reductions	5.490
01	1st Unaltered	1.00% reduction of the E modulus in areas 2A, 3A, and 4A	5.488
02	2nd Unaltered	1.00% reduction of the E modulus in area 5A	5.489
03	1st Damaged	50.00% reduction of E in areas 2A, 3A, and 4A	5.270
04	2nd Damaged	50.00% reduction of E in area 3A	5.330
05	3rd Damaged	50.00% reduction of E in areas 2A, 3A, 4A and 5A	5.168
06	4th Damaged	50.00% reduction of E in areas 3A and 5A	5.224
07	5th Damaged	50.00% reduction of E in areas 2A and 4A	5.460

5. The Experimental Case Study (Frame Behaving Nonlinearly)

For the experimental validation, a laboratory three-story frame, which behaves nonlinearly due to damage occurrence, has been investigated; this is the same case studied by [21], again to ensure the comparison of the results, and proposed by [3]. Please note that this is quite different from the two numerical cases, where the structural damage was not a source of nonlinearity. This is due to the goal of this experimental setup being to mimic breathing crack behaviour [63]. A bumper–column bilinear mechanism achieves this: the column, hanging from the third floor, meets the bumper at the second story. This mechanism generates a directional nonlinear response, with stiffness increasing when the two are touching and pushing against each other. A scheme of the test structure is reported in Figure 8; the shaker and the four sensors' locations are highlighted. A detailed description of the experimental setup and procedure is provided in [59]. The severity of the damage is modelled by moving closer or farther the column's tip—when they are closer, the two elements come in touch for lower amplitudes of the driving force, thus causing major nonlinearities in the recorded response. The experimental THs have a total duration of 25.6 s, with $f_{s,exp} = 320$ Hz. The dynamic input is a band-limited excitation (ranging 20–150 Hz), as defined in [3]. The 17 states (nine undamaged and eight damaged, with the first nine configurations being intended to emulate different operational and environmental conditions) were used for this investigation; they are summarized in Table 6. All the output channels were used for the sensor setup considered. The test set was made up of 10 realisations for each scenario, totalling 180 simulated THs.

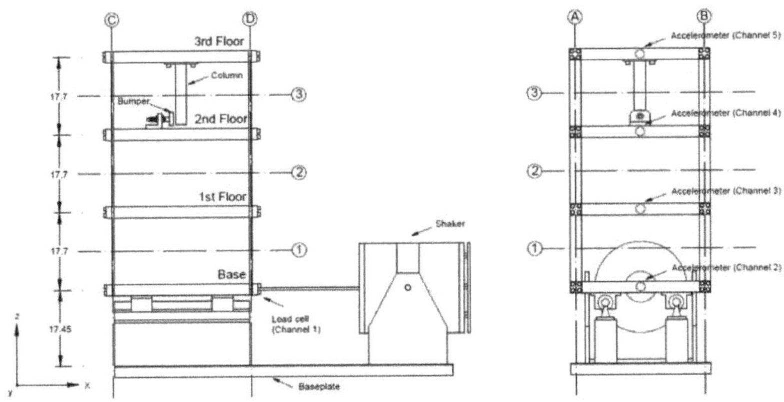

Figure 8. The experimental test setup schematics.

Table 6. Baseline, Undamaged and Damaged Scenarios for experimental validation.

Case	Description
1	Undamaged Baseline
2	Undamaged Added mass of 1.2 kg at the base
3	Undamaged Added mass of 1.2 kg at the 1st floor
4	Undamaged 87.5% stiffness reduction in one column of the 1st inter-story
5	Undamaged 87.5% stiffness reduction in two columns of the 1st inter-story
6	Undamaged 87.5% stiffness reduction in one column of the 2nd inter-story
7	Undamaged 87.5% stiffness reduction in one column of the 2nd inter-story
8	Undamaged 87.5% stiffness reduction in one column of the 3rd inter-story
9	Undamaged 87.5% stiffness reduction in two columns of the 3rd inter-story
10	Damaged Distance between bumper and column tip 0.20 mm
11	Damaged Distance between bumper and column tip 0.15 mm
12	Damaged Distance between bumper and column tip 0.13 mm
13	Damaged Distance between bumper and column tip 0.10 mm
14	Damaged Distance between bumper and column tip 0.05 mm
15	Damaged Bumper 0.20 mm from column tip, 1.2 kg added at the base
16	Damaged Bumper 0.20 mm from column tip, 1.2 kg added on the 1st floor
17	Damaged Bumper 0.10 mm from column tip, 1.2 kg added on the 1st floor

As for the second numerical case study, the definitions of ECO- and MO-MMFCCs have been changed to fit the structure of interest. For ECO-MMFCCs, the 90% bound of f_{co} is approximately 78 Hz (averaging over the channels 3, 4, and 5 for the 50 realisations of Case 1); the $f_{s,exp}/2$ limit of LO-MMFCCs becomes 160 Hz. The array of the centre frequencies for MO-MMFCCs underwent some major rethinking. Being the structure basically assimilable to a three degrees-of-freedom oscillator (floor level does not count as it is subject to the input-induced rigid motion), only the first three frequencies have been retained. The remaining seven filters have been linearly spaced up to $f_s/2$, i.e., 160 Hz, thus resulting in

$$f_c = [31\ 54\ 71\ 83\ 96\ 109\ 122\ 135\ 148\ 160]$$

6. Results

The results from previous works are reported for direct comparison. The methods applied are, by following the same nomenclature (see also Figure 2), (1) the log-scale CCs; (2) the linear CCs; (3) the classic definition of MFCCs; (4) the MFCCs with a cut-off frequency $f_{co} = \frac{f_s}{4}$; and, (5) the MFCCs with a cut-off frequency $f_{co} = \frac{f_s}{8}$ from [21]; the three proposed Mel-Modified Cepstral Coefficients (MMFCCs), i.e., the so-called (i) ECO-MMFCCs; (ii) MO-MMFCCs; and, (iii) LO-MMFCCs, from [22], plus the Bark-scale based (iv) BFCCs; finally, the novel features here presented, (I) the GTCCs and (II) the TECCs.

For better readability, this Section is split between the numerical and the experimental data. All the results are expressed in terms of type 1 (false positives, i.e., false alarms) and type 2 (false negatives or false acceptance) errors. In the latter case, the system is declared healthy while not being so; the opposite happens for false alarms. Being life-safety the main aim of any SHM procedure, type 1 errors are generally overlooked respect to their type 2 counterparts. Nevertheless, economical, practical, and psychological concerns make these as valuable as the other ones. A simple yet effective reasoning is that an alarm system constantly affected by false alarm will most probably be ignored when actual damage is effectively spotted. This nullifies any possible gain from the deployment of the SHM apparatus.

6.1. Results from the Numerical Simulations

Regarding the first numerical case study (the Fossano bell tower), cases 00 to 04—i.e., the baseline and the story-uniform damage cases—are visually compared in Figure 9 to prove the algorithms capability to discern simpler configurations and widely extended damage from the normality model. Figure 10 shows the results for cases 05 to 10 (realistic damage scenarios). To estimate the robustness of each MFCC-based variant to false alarms, cases 15 to 18 (i.e., the scenarios with small fluctuations of the Young's modulus respect to the baseline) are presented in Figure 11. Apart from the Teager-Kaiser Energy Cepstral Coefficients, the standard definition of MFCCs is also shown for comparison. Table 7 reports all results. In all figures, the black dashed line represents the defined threshold; bars exceeding it are coloured in red and they stand for the realisations labelled as damaged, while blue bars indicate values of DI under the prescribed threshold. From a statistical pattern recognition standpoint, this means that the cepstral features that correspond to that specific realisation have not departed considerably enough from their respective values for the reference baseline and are therefore labelled as undamaged.

As can be inferred from the first two columns of Table 7, the GTCCs, the Logarithmic Scale CCs, and the three Mel-Modified Scales (ECO-, MO-, and LO-MMFCCs) have all relatively low type 1 errors. The Damage Index values inside any given damage case are quite similar; that is a valuable proof of the robustness of the method. Bark-scale based CCs wrongly labelled as damaged 60% of the undamaged inputs coming from the same scenario used for the training. The TECCs were the most performing feature; in Figure 9.a one can clearly see that two out of the three mislabeled realisations for TECCs fall very short of the threshold. The third column in Table 7 (cases 05–14) shows that all algorithms work effectively in terms of damage detection. The classic definition of the Mel-Scale, the Bark-Scale, and the linear Scale nevertheless present some relatively larger fluctuations along the ten realisations of the same damage scenario with different inputs. The Log-Scale and the Mel-modified Scales provide more stable results, with a flat trend inside each damage scenario. This seems to indicate that a kind of nonlinear scale performs better than the constantly spaced filters. The filter shape might be even more incisive, as GTCCs and TECCs are all spaced according to the Bark Critical Bands yet perform better than triangular filters-based options both in terms of type I and II errors. Most of the algorithms recognise substantial gaps (Figures 8 and 9) between different scenarios, which reflect the difference of damage extension; this seems to indicate a possibility to also exploit them for instances of damage severity assessment. Finally, the last column of Table 7 (cases 15 to 18) evidently portrays how only the TECCs correctly label all the proposed cases as unmodified respect to the baseline; the other features proved to be quite unsatisfactory, especially the Mel-Scale and Bark-Scale. That seems to once more validate the assumption that the straight application of the original definition of the mel-scale, tuned for speech processing aims, is not well-performing and proper corrections are needed. In particular, the scenarios with a variation in the material properties of ±0.25% are most often correctly labelled as healthy; otherwise, the ±1.00% deviation of the Young's modulus provides DI values above the threshold. This phenomenon could be justified by the fact that even a relatively small difference between the pristine and modified mechanical properties of the model's material cause a greater variation in the power spectrum of the structure, which is also reflected in the power and is wrongly seen as damage-induced. This aspect is not helpful in the damage detection process; the TKE operator seems to be more resilient to these slight changes. By summing up all of the damage scenarios of this first case, it turns out that of all features, the TECCs are the only one always correctly labelling the cases with realistic damage patterns (05–14) and the ones with global, small fluctuations of E, unrelated to damage (15–18). Indeed, the TECCs fall short of a perfect score for only three mislabeled realisations (out of 10) in case 01, which was also the most demanding scenario due to the proximity of the damage to the fixed end.

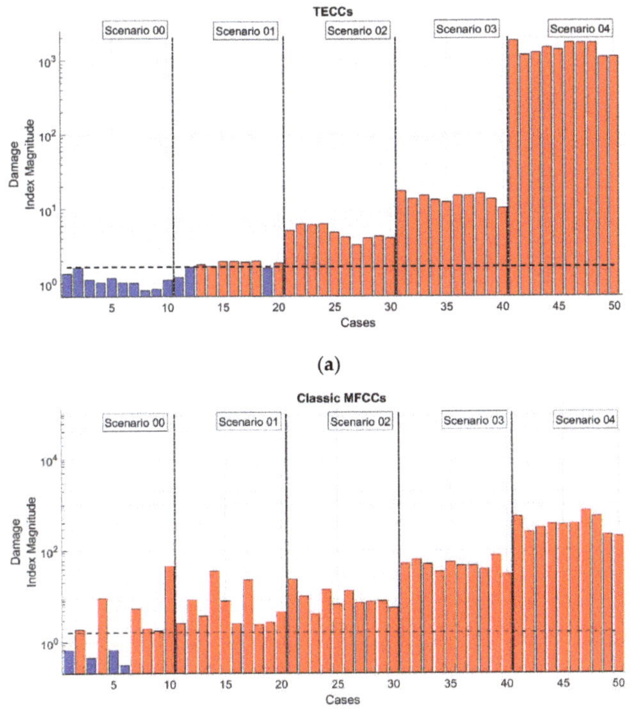

(a)

(b)

Figure 9. Damage indexes for TECCs (**a**) and MFCCs (**b**); first numerical case study (Fossano bell tower), scenarios 00–04 (story-level damage).

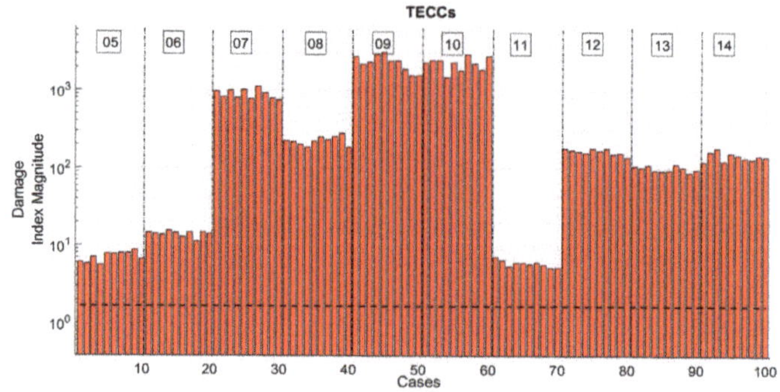

(**a**)

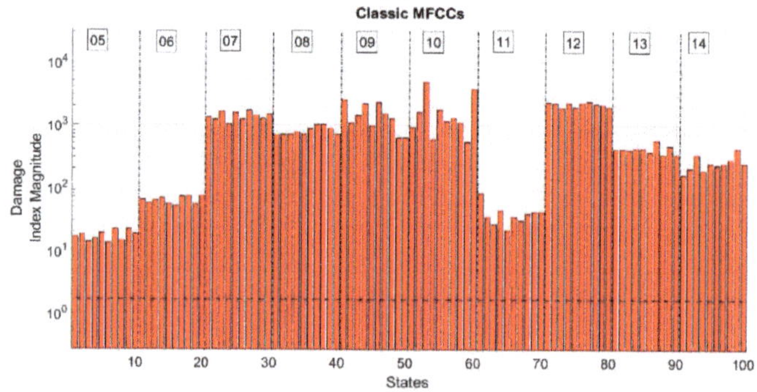

(**b**)

Figure 10. Damage indexes for TECCs (**a**) and MFCCs (**b**); first numerical case study (Fossano bell tower), scenarios 05–14 (realistic damage).

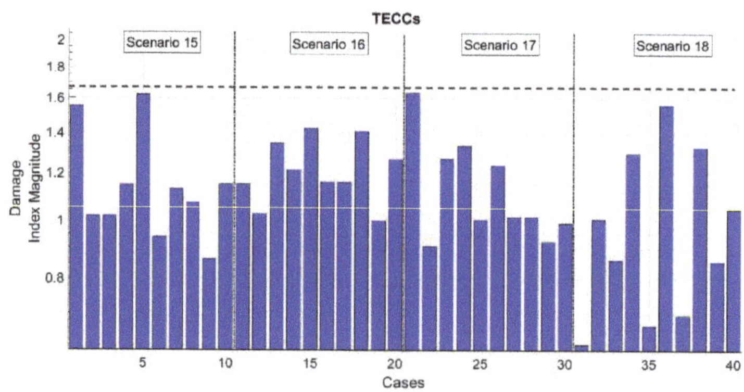

(**a**)

Figure 11. *Cont.*

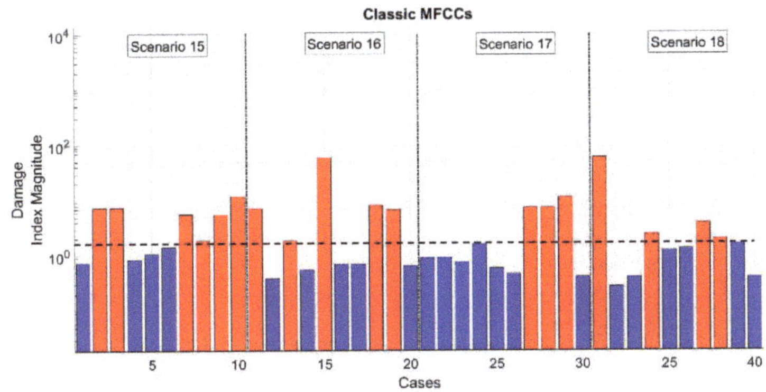

(b)

Figure 11. Damage indexes for TECCs (a) and MFCCs (b); first numerical case study (Fossano bell tower), scenarios 15–18 (no damage).

Table 7. Results for the first numerical case study (bell tower).

	Case 00	Cases 01–04	Cases 05–14	Cases 15–18
	Type 1 Errors [%]	**Type 2 Errors [%]**	**Type 2 Errors [%]**	**Type 1 Errors [%]**
MFCCs	60.00	0.00	0.00	45.00
$f_{\text{cut-off}} = \frac{1}{4} f_s$	30.00	0.00	0.00	45.00
$f_{\text{cut-off}} = \frac{1}{8} f_s$	30.00	0.00	0.00	45.00
Log-Scale	20.00	0.00	0.00	65.00
Linear	60.00	0.00	0.00	45.00
ECO-MMFCCs	20.00	0.00	0.00	62.50
MO-MMFCCs	10.00	0.00	0.00	62.50
LO-MMFCCs	20.00	0.00	0.00	65.00
BFCCs	60.00	0.00	0.00	47.50
GTCCs	20.00	0.00	0.00	60.00
TECCs	0.00	7.50	0.00	0.00

Table 8 reports the results for the second case study (the highly flexible prototype wing); Figure 12 graphically illustrates them. It can be inferred that, in this case, the exact detection of damage was more difficult. MFCCs and similar variants proved to be able to spot everywhere the damage but at the cost of not actually discerning it from the two cases with small perturbations. The TECCs, on the other hand, did not perform perfectly, yet they avoided all but one false alarm and were able to discern the damaged conditions decently in the inspected scenarios. If one does not take into account Case 07, which presents an imperceptible decrease of 0.03 Hz and went completely unnoticed by the trained algorithm, the type 2 error percentage drops to 7.50%, with two errors in Case 04, no one in Case 05, and a single mislabeled element in Case 06.

Table 8. Results for the second numerical case study (HAR wing).

	Cases 00–02	Cases 03–07
	Type 1 Errors [%]	**Type 2 Errors [%]**
MFCCs	66.67	0.00
$f_{\text{cut-off}} = \frac{1}{4} f_s$	66.67	0.00
$f_{\text{cut-off}} = \frac{1}{8} f_s$	66.67	0.00
Log-Scale	70.00	0.00
Linear	66.67	0.00
ECO-MMFCCs	66.67	0.00
MO-MMFCCs	66.67	0.00
LO-MMFCCs	66.67	0.00
BFCCs	66.67	0.00
GTCCs	66.67	0.00
TECCs	3.33	26.00 (*)

(*) 7.50% on Cases from 03 to 06.

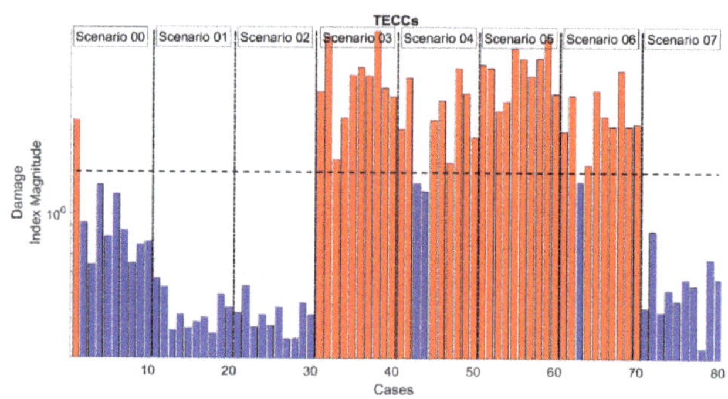

(**a**)

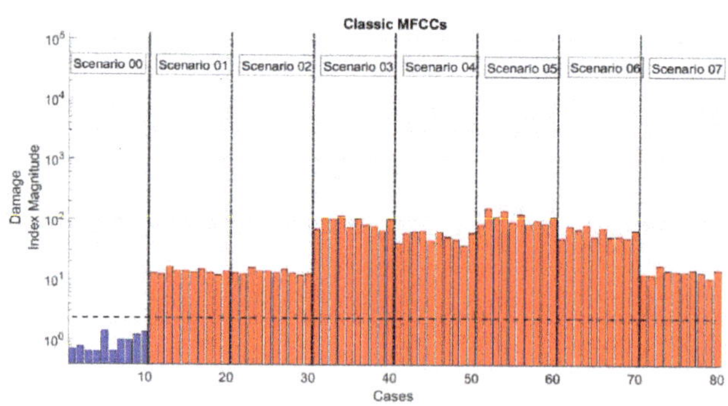

(**b**)

Figure 12. Damage indexes for TECCs (**a**) and MFCCs (**b**); second numerical case study (highly flexible prototype wing).

As in the previous case of the Fossano bell tower, this seems to point out the larger robustness of the TECCs feature to small changes, assumed here as statistical variations of the structural parameters and not directly linked to damage occurrence. A possible explanation comes from the TKE Operator being notably sensitive to frequency and amplitude changes, while much more robust than MFCCs to noise and minimal variations in the field of speech recognition accuracy.

6.2. Results from the Experimental Data

The results from experimental data are displayed here in Figure 13. As before, TECCs and classic MFCCs are only depicted due to space constraints. Table 9 summarises the results, in terms of percentage of type 1 and 2 errors.

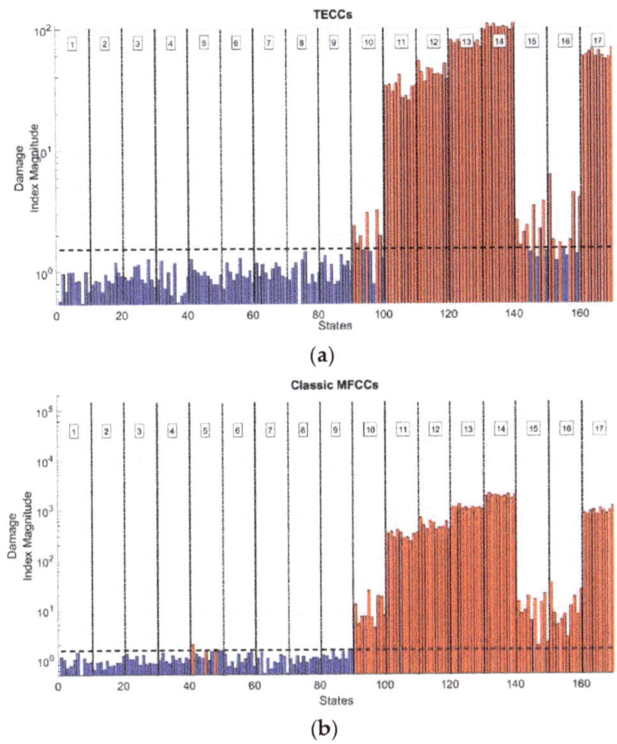

Figure 13. Damage indexes for TECCs (**a**) and MFCCs (**b**); and, experimental data.

Table 9. Results for the experimental case study.

	Cases 1–9	**Cases 10–17**
	Type 1 Errors [%]	**Type 2 Errors [%]**
MFCCs	3.33	0.00
$f_{\text{cut-off}} = \frac{1}{4} f_s$	5.56	0.00
$f_{\text{cut-off}} = \frac{1}{8} f_s$	7.78	1.25
Log-Scale	4.44	1.25
Linear	5.56	0.00
ECO-MMFCCs	4.44	0.00
MO-MMFCCs	2.22	22.50
LO-MMFCCs	12.22	25.00
BFCCs	3.33	0.00
GTCCs	4.44	0.00
TECCs	0.00	13.75

In interpreting the data, one must consider the different definition of 'damage' between the two numerical and the experimental case studies. The Fossano bell tower and the HAR prototype wing behave linearly in both damaged and pristine conditions; again, the difference between the two is only a matter of shifting natural frequencies. The experimental data have a completely different 'meaning' of damage, as here, differences in mass and stiffness are accounted as different operational conditions and the distinction between damaged and undamaged response lies solely on the presence of nonlinearities. In this latter case, the TECCs proved to be the only feature able to reach a 0% type 1 error in the test set, but only in spite of a relatively high type 2 error percentage, only surpassed by the recently proposed MO- and LO-MMFCCs. Nevertheless, a closer inspection of the several damaged and undamaged configurations allows for highlighting that this error is clustered in only three scenarios, respectively, cases 10, 15, and 16. Case 10 is the one with the slightest nonlinearity and, thus, is the most challenging to detect. The algorithm also struggled for cases 15 and 16, where changes in the frame's mass are involved. On the other hand, all of the undamaged scenarios are recognised as such by means of the TECCs, while classic MFCCs generate some misclassifications for case 5 (which is the one with the largest stiffness reduction close to the base of the frame, where such structural changes have the most impact in the vibrational response).

Therefore, experimental results seem to validate what is expected from numerical simulations, that is to say: (i) a greater generalisation capability of the TECCs feature over closely-related alternatives; (ii) a strong robustness to false alarms but at the cost of some false negatives; and, (iii) a sensibility to shifts in the frequency content higher or at least comparable to MFCCs in its original form and proposed variants. On the other hand, the TECCs proved to be less sensitive to the presence of limited nonlinearities during changing operational conditions than to mass changes respect to the MFCCs and similar features; but this outcome might be specific of the experimental data utilised here.

6.2.1. Further Analysis of the Experimental Data

Some further tests were performed to better define the advantages of the proposed TECCs feature. For brevity sake, only the results for experimental data are reported; the same findings were observed for numerical data.

Firstly, the number and distribution of output channels were investigated. Four setups have been considered: setup #0, including all channels (numbered 2 to 5, see Figure 8), for which the results are already reported in Figure 13 and Table 9; setup #1 (channels 3, 4, and 5); setup #2 (channels 3 and 4); and finally, setup #3 (constituted by channels 4 and 5). Table 10 enlists the results for these last three setups. Remarkably, this test evidenced what was already observed in [21], that is to say, MFCCs and similar features produce less accurate results when all of the acquisition channels are considered. This point was confirmed here for type 1 errors in two out of the three new setups considered. The same effects were encountered here for TECCs as well, with the same conditions: MFCCs systematically fail in labelling case 5 as undamaged, due to its relatively large shift in frequency, while TECCs struggle in classifying cases 10, 15, and 16 as damaged due to their lower level of nonlinear distortion.

Secondly, a comparison with a non-cepstral feature, the aforementioned AR coefficients, was also performed and reported in Table 10. The AIC approach was applied similarly to what done in [21]. An order of 12 was used for sensor setup #1 and #2; order 10 was applied instead for the sensor setup #3. It can be seen that the percentage of type 1 error is generally higher than most of the cepstral features that are proposed here.

Table 10. Results according to the sensor setup considered.

	Sensor Setup #1		**Sensor Setup #2**		**Sensor Setup #3**	
	Cases 1–9	**Cases 10–17**	**Cases 1–9**	**Cases 10–17**	**Cases 1–9**	**Cases 10–17**
	Type 1 Errors [%]	**Type 2 Errors [%]**	**Type 1 Errors [%]**	**Type 2 Errors [%]**	**Type 1 Errors [%]**	**Type 2 Errors [%]**
MFCCs	4.44	0.00	2.22	1.25	2.22	0.00
$f_{\text{cut-off}} = \frac{1}{4} f_s$	3.33	0.00	1.11	0.00	1.11	0.00
$f_{\text{cut-off}} = \frac{1}{8} f_s$	5.56	0.00	3.33	1.25	3.33	0.00
Log-Scale	5.56	1.25	3.33	2.50	1.11	1.25
Linear	3.33	0.00	1.11	1.25	1.11	0.00
ECO-MMFCCs	3.33	0.00	1.11	0.00	1.11	0.00
MO-MMFCCs	2.22	25.00	3.33	30.00	1.11	27.50
LO-MMFCCs	10.00	25.00	0.00	23.75	2.22	25.00
BFCCs	1.11	0.00	0.00	2.50	1.11	0.00
GTCCs	2.22	0.00	2.22	1.25	2.22	1.25
TECCs	0.00	7.50	0.00	11.25	0.00	6.25
AR coefficients	13.33	0.00	12.22	0.00	6.67	0.00

The Central Processing Unit (CPU) time that was required for the training phase was then evaluated for all types of DSFs on non-optimised (yet comparable) versions of the code. This was tested by means of the Matlab® stopwatch timer tic–toc; Table 11 reports the results. One can clearly see that the computational time is comparable over the several sensor setups considered. Note that the timing accounts for training operations (feature extraction and population statistics estimation) and for threshold definition; the time required for data preprocessing, differently from what done in [21], was not considered, as it is identical for all features. This returns some faster results with respect to what is reported there. Moreover, the codes were run 10 times to avoid any computational variability; the average result is reported. As expected, the difference between MFCCs, similar variants, GTCCs and TECCs are minimal. It can be observed that GTCCs takes slightly longer than MFCCs (in the order of some fractions of a second), since there it is needed to build up the more complexly shaped gammatone filterbank. The TECCs require that additional time plus a bit more to compute the TKE operator out of the provided time arrays. In the case with all acquisition channels considered, these two operations make the code run in almost 30% more time (0.43 s on average). However, this delay is basically irrelevant if compared to the one that is needed to extract other features, such as the Auto-Regressive (AR) model coefficients [21].

Lastly, the standard deviation is proposed as a metric of dispersion for the results of the several realisations that belong to the same damage case, to quantitatively express the robustness of the classification. Table 12 enlists the results for all 17 cases. Only the values for the sensor setup #0 and for MFCCs, log-scale CCs, linear CCs, BFCCs, GTCCs, and TECCs are reported here due to space constraints. The same behaviour was observed for the other setups as well. By considering ten realisations per investigated case, it can be observed that the use of TECCs as a DSF produce a much smaller scattering of the results, with the standard deviation never exceeding $\sigma = 6$, while the same measure goes well over 50 or even 70 for MFCCs in some cases.

Table 11. CPU time accordingly to the sensor setup considered (training phase).

	Sensor Setup #0	Sensor Setup #1	Sensor Setup #2	Sensor Setup #3
	CPU Time [s]	CPU Time [s]	CPU Time [s]	CPU Time [s]
MFCCs	1.60	1.31	1.00	1.03
$f_{\text{cut-off}} = \frac{1}{4} f_s$	1.63	1.31	1.01	1.03
$f_{\text{cut-off}} = \frac{1}{8} f_s$	1.62	1.35	1.04	1.04
Log-Scale	1.62	1.37	1.01	1.02
Linear	1.56	1.30	0.99	0.99
ECO-MMFCCs	1.56	1.31	1.00	1.02
MO-MMFCCs	1.58	1.29	1.00	0.99
LO-MMFCCs	1.59	1.29	1.01	0.99
BFCCs	1.58	1.32	1.02	1.00
GTCCs	1.78	1.38	1.06	1.07
TECCs	2.03	1.62	1.15	1.16

Table 12. Standard deviation of Damage Index (DI) (sensor setup #0).

	σ(DI(n))-Cases 1 to 17
MFCCs	0.32, 0.20, 0.28, 0.19, 0.28, 0.32, 0.26, 0.21, 0.22, 6.21, 41.74, 62.64, 52.52, 72.21, 6.16, 8.63, 81.06.
Log-Scale	0.24, 0.29, 0.15, 0.22, 0.28, 0.22, 0.27, 0.22, 0.23, 3.29, 22.23, 33.42, 31.63, 32.26, 3.45, 5.39, 35.00.
Linear	0.29, 0.30, 0.22, 0.19, 0.27, 0.22, 0.24, 0.31, 0.24, 6.55, 39.27, 67.89, 46.22, 52.90, 5.50, 7.85, 72.59.
BFCCs	0.31, 0.20, 0.26, 0.19, 0.26, 0.35, 0.22, 0.21, 0.19, 6.73, 44.16, 68.87, 47.46, 64.38, 6.72, 9.36, 90.09.
GTCCs	0.20, 0.28, 0.15, 0.16, 0.20, 0.31, 0.32, 0.19, 0.24, 4.19, 25.88, 45.77, 51.07, 73.55, 3.80, 6.54, 41.64.
TECCs	0.18, 0.14, 0.17, 0.23, 0.14, 0.17, 0.13, 0.26, 0.22, 0.75, 4.78, 5.57, 4.25, 4.64, 0.84, 1.70, 4.35.

7. Conclusions

Any Structural Health Monitoring system relies on data, almost always pre-processed, and on features extrapolated from them. Mel-Frequency Cepstral Coefficients (MFCCs) have recently been proven to be effective in damage detection, relying on the cepstrum of the recorded structural response, even if margins for improvements were evident. The Teager-Kaiser Energy (TKE) operator has been proposed here as the basis for a similar feature, less subject to false positive mislabeling when used for Machine Learning. That is adherent to what is encountered for speech signals and well-known in the field of speaker and speech recognition. The investigation reported here spans over different structures, with different input and very different setups—varying acquisition parameters, such as the number of the output channels, the sampling frequency, and so on. Moreover, in the first and in the second numerical study, the damage was modelled as a linear reduction of stiffness inserted in a linear system, while the experimental case emulated the damage as a pointwise source of nonlinearity in an otherwise linear system and stiffness local reduction was intended as a change in the operational conditions that were unrelated to damage. This shows the excellent capability of generalisation of this approach.

The proposed damage sensitive feature, the Teager-Kaiser Energy Cepstral Coefficients (TECCs), has been benchmarked against the MFCCs and some similar variants. Interesting numerical and experimental results were achieved for both the linear and nonlinear models of damage.

The main conclusions are the following:

1. the TECCs, MFCCs, and similar features perform efficiently both with damage emulated by a reduction of the Young's modulus E or by the presence of nonlinearities, even if the signal is noisy;
2. the Mel-Scale performed similarly to other nonlinear scales such as the Bark Scale or logarithmic spacing, with no clear best option among all the investigated case studies;
3. the algorithms that resort to a Gammatone filterbank generally produced better results than the ones with triangular filters; even if the TECCs still outperforms the GTCCs;
4. the TECCs present a strong reduction of the order of magnitude of the Damage Index (DI) with respect to MFCCs and similar options, in absolute terms; this does not affect the damage detection

algorithm, as the relative distance between the 'normality' model and the damaged cases remains and seems even to increase;

5. the Teager-Kaiser Energy Cepstral Coefficients outperforms all the competitors in terms of little or no type 1 errors, but it is slightly more prone to type 2 errors respect to MFCCs and derived features;
6. with respect to non-cepstral features, such as the AR coefficients, the main benefits lie in the less computational cost and greater robustness to noise and to confounding influences, such as environmental and operational effects, unrelated to damage.

The outcomes of this research leave plenty of room for improvements. Indeed, apart from the TECCs and other cepstral-based alternatives, wavelet-based alternatives are of great interest. The Mel-Frequency Discrete Wavelet Coefficients may be a relevant alternative. Thanks to the Discrete Wavelet Transform, the filter spacing issue is by-passed; filter shaping is reduced to the selection of the mother wavelet. The Authors are committed to pursuing further studies in this direction.

Author Contributions: Conceptualization, M.C., M.F., R.C., C.S. and R.B.; Data curation, M.C. and M.F.; Formal analysis, M.C. and M.F.; Funding acquisition, R.C. and C.S.; Investigation, M.C. and M.F.; Methodology, M.C., R.C., C.S. and R.B.; Project administration, R.C., C.S. and R.B.; Resources, R.C., C.S. and R.B.; Software, M.C., M.F. and R.B.; Supervision, R.C., C.S. and R.B.; Validation, M.C. and M.F.; Visualization, M.C.; Writing—original draft, M.C. and M.F.; Writing—review & editing, M.C., R.C., C.S. and R.B.

Funding: This research was funded by the INTE project by Compagnia di San Paolo (2017-18) "System Identification, model updating and damage assessment of heritage buildings and structures", as a joint partnership of Politecnico di Torino (IT) and Columbia University (USA).

Acknowledgments: The Authors wish to thank Luciana Balsamo and Marica Leonarda Pecorelli for their precious help and advice. The Authors also wish to acknowledge the Engineering Institute at Los Alamos National Laboratory for making available to the public domain the experimental data used in this work.

Conflicts of Interest: The authors declare no conflict of interest. The funders had no role in the design of the study; in the collection, analyses, or interpretation of data; in the writing of the manuscript, or in the decision to publish the results.

References

1. Farrar, C.R.; Worden, K. *Structural Health Monitoring: A Machine Learning Perspective*; John Wiley & Sons: Hoboken, NJ, USA, 2013.
2. Doebling, S.W.; Farrar, C.R.; Prime, M.B. A Summary Review of Vibration-Based Damage Identification Methods. *Shock Vib. Dig.* **1998**, *30*, 91–105. [CrossRef]
3. Figueiredo, E.; Park, G.; Farrar, C.R.; Worden, K.; Figueiras, J. Machine learning algorithms for damage detection under operational and environmental variability. *Struct. Health Monit.* **2011**, *10*, 559–572. [CrossRef]
4. Sohn, H.; Farrar, C.R.; Hunter, N.F.; Worden, K. Structural Health Monitoring Using Statistical Pattern Recognition Techniques. *J. Dyn. Syst. Meas. Control* **2001**, *123*, 706–711. [CrossRef]
5. Amezquita-Sanchez, J.P.; Adeli, H. Signal Processing Techniques for Vibration-Based Health Monitoring of Smart Structures. *Arch. Comput. Methods Eng.* **2016**, *23*, 1–15. [CrossRef]
6. Bonato, P.; Ceravolo, R.; Stefano, A. De Time-Frequency and Ambiguity Function Approaches in Structural Identification. *J. Eng. Mech.* **1997**, *123*, 1260–1267. [CrossRef]
7. Bonato, P.; Ceravolo, R.; De Stefano, A.; Molinari, F. Cross-Time Frequency Techniques for the Identification of Masonry Buildings. *Mech. Syst. Signal Process.* **2000**, *14*, 91–109. [CrossRef]
8. Law, S.S.; Zhu, X.Q.; Tian, Y.J.; Li, X.Y.; Wu, S.Q. Statistical damage classification method based on wavelet packet analysis. *Struct. Eng. Mech.* **2013**, *46*, 459–486. [CrossRef]
9. Liu, J.; Wang, Z.; Liu, J.-L.; Wang, Z.-C.; Ren, W.-X.; Li, X.-X. Structural time-varying damage detection using synchrosqueezing wavelet transform. *Smart Struct. Syst.* **2015**, *15*, 1738–1991. [CrossRef]
10. Wang, C.; Ren, W.X.; Wang, Z.C.; Zhu, H.P. Time-varying physical parameter identification of shear type structures based on discrete wavelet transform. *Smart Struct. Syst.* **2014**, *14*, 831–845. [CrossRef]
11. Pnevmatikos, N.; Blachowski, B.D.; Hatzigeorgiou, G.; Pnevmatikos, N.G.; Blachowski, B.; Hatzigeorgiou, G.D.; Swiercz, A. Wavelet analysis based damage localization in steel frames with bolted connections. *Smart Struct. Syst.* **2016**, *18*, 1738–1991. [CrossRef]

12. Civera, M.; Zanotti Fragonara, L.; Surace, C. Nonlinear Dynamics of Cracked, Cantilevered Beam-like Structures Undergoing Large Deflections. In Proceedings of the 2019 IEEE 5th International Workshop on Metrology for AeroSpace (MetroAeroSpace), Torino, Italy, 19–21 June 2019.
13. Pines, D.; Salvino, L. Structural health monitoring using empirical mode decomposition and the Hilbert phase. *J. Sound Vib.* **2006**, *294*, 97–124. [CrossRef]
14. Civera, M.; Filosi, C.M.; Pugno, N.M.; Silvestrini, M.; Surace, C.; Worden, K. Assessment of vocal cord nodules: A case study in speech processing by using Hilbert-Huang Transform. *J. Phys. Conf. Ser.* **2017**, *842*, 012025. [CrossRef]
15. Childers, D.G.; Skinner, D.P.; Kemerait, R.C. The cepstrum: A guide to processing. *Proc. IEEE* **1977**, *65*, 1428–1443. [CrossRef]
16. Zheng, H.; Mita, A. Damage indicator defined as the distance between ARMA models for structural health monitoring. *Struct. Control Health Monit.* **2008**, *15*, 992–1005. [CrossRef]
17. Kim, J.T.; Lyon, R.H. Cepstral analysis as a tool for robust processing, deverberation and detection of transients. *Mech. Syst. Signal Process.* **1992**, *6*, 1–15. [CrossRef]
18. Peeters, C.; Guillaume, P.; Helsen, J. A comparison of cepstral editing methods as signal pre-processing techniques for vibration-based bearing fault detection. *Mech. Syst. Signal Process.* **2017**, *91*, 354–381. [CrossRef]
19. Randall, R.B. A History of Cepstrum Analysis and its Application to Mechanical Problems. *Mech. Syst. Signal Process.* **2017**, *97*, 3–19. [CrossRef]
20. Balsamo, L.; Betti, R.; Beigi, H. Structural Damage Detection Using Speaker Recognition Techniques. In Proceedings of the 11th International Conference on Structureal Safety and Reliability (ICOSSAR), New York, NY, USA, 16–20 June 2013.
21. Balsamo, L.; Betti, R.; Beigi, H. A structural health monitoring strategy using cepstral features. *J. Sound Vib.* **2014**, *333*, 4526–4542. [CrossRef]
22. Ferraris, M.; Civera, M.; Ceravolo, R.; Surace, C.; Betti, R. *Using Enhanced Cepstral Analysis for Structural Health Monitoring*; Springer: Singapore, 2020; pp. 150–165.
23. Dimitriadis, D.; Maragos, P.; Potamianos, A. Auditory Teager Energy Cepstrum Coefficients for Robust Speech Recognition. In Proceedings of the 9th European Conference on Speech Communication and Technology, Lisbon, Portugal, 4–8 September 2005.
24. Kaiser, J.F. On a simple algorithm to calculate the "energy" of a signal. In Proceedings of the International Conference on Acoustics, Speech, and Signal Processing, Albuquerque, NM, USA, 3–6 April 1990; pp. 381–384.
25. Schluter, R.; Bezrukov, I.; Wagner, H.; Ney, H. Gammatone Features and Feature Combination for Large Vocabulary Speech Recognition. In Proceedings of the 2007 IEEE International Conference on Acoustics, Speech and Signal Processing—ICASSP '07, Honolulu, HI, USA, 15–20 April 2007; pp. 649–652.
26. Vakman, D. On the analytic signal, the Teager-Kaiser energy algorithm, and other methods for defining amplitude and frequency. *IEEE Trans. Signal Process.* **1996**, *44*, 791–797. [CrossRef]
27. Tran, V.T.; AlThobiani, F.; Ball, A. An approach to fault diagnosis of reciprocating compressor valves using Teager—Kaiser energy operator and deep belief networks. *Expert Syst. Appl.* **2014**, *41*, 4113–4122. [CrossRef]
28. Henríquez Rodríguez, P.; Alonso, J.B.; Ferrer, M.A.; Travieso, C.M. Application of the Teager-Kaiser energy operator in bearing fault diagnosis. *ISA Trans.* **2013**, *52*, 278–284. [CrossRef]
29. Ceravolo, R.; Pistone, G.; Fragonara, L.Z.; Massetto, S.; Abbiati, G. Vibration-Based Monitoring and Diagnosis of Cultural Heritage: A Methodological Discussion in Three Examples. *Int. J. Archit. Herit.* **2016**, *10*, 375–395. [CrossRef]
30. Carpinteri, A.; Lacidogna, G. Damage monitoring of an historical masonry building by the acoustic emission technique. *Mater. Struct.* **2006**, *39*, 161–167. [CrossRef]
31. Carpinteri, A.; Lacidogna, G. Structural Monitoring and Integrity Assessment of Medieval Towers. *J. Struct. Eng.* **2006**, *132*, 1681–1690. [CrossRef]
32. Carpinteri, A.; Invernizzi, S.; Lacidogna, G. In situ damage assessment and nonlinear modelling of a historical masonry tower. *Eng. Struct.* **2005**, *27*, 387–395. [CrossRef]
33. Pontillo, A.; Hayes, D.; Dussart, G.X.; Lopez Matos, G.E.; Carrizales, M.A.; Yusuf, S.Y.; Lone, M.M. Flexible High Aspect Ratio Wing: Low Cost Experimental Model and Computational Framework. In Proceedings of the 2018 AIAA Atmospheric Flight Mechanics Conference, Reston, VA, USA, 8–12 January 2018.

34. Lanzoni, L.; Tarantino, A.M. Equilibrium configurations and stability of a damaged body under uniaxial tractions. *Z. Angew. Math. Phys.* **2014**, *66*, 171–190. [CrossRef]
35. Tarantino, A.M. Equilibrium paths of a hyperelastic body under progressive damage. *J. Elast.* **2014**, *114*, 225–250. [CrossRef]
36. Bedon, C. Diagnostic analysis and dynamic identification of a glass suspension footbridge via on-site vibration experiments and FE numerical modelling. *Compos. Struct.* **2019**, *216*, 366–378. [CrossRef]
37. Bogert, B.P. The quefrency alanysis of time series for echoes; Cepstrum, pseudo-autocovariance, cross-cepstrum and saphe cracking. *Time Ser. Anal.* **1963**, 09–243.
38. Stevens, S.S.; Volkmann, J.; Newman, E.B. A Scale for the Measurement of the Psychological Magnitude Pitch. *J. Acoust. Soc. Am.* **1937**, *8*, 185–190. [CrossRef]
39. Satyam, M.; Sudhakara, R.; Devy, C.G. Cepstrum Analysis: An Advanced Technique in Vibration Analysis of Defects in Rotating Machinery. *Def. Sci. J.* **1994**, *44*, 53–60. [CrossRef]
40. Noll, A.M. Cepstrum Pitch Determination. *J. Acoust. Soc. Am.* **1967**, *41*, 293–309. [CrossRef] [PubMed]
41. De Krom, G. A Cepstrum-Based Technique for Determining a Harmonics-to-Noise Ratio in Speech Signals. *J. Speech Lang. Hear. Res.* **1993**, *36*, 254–266. [CrossRef] [PubMed]
42. Oppenheim, A.V.; Schafer, R.W.; Buck, J.R. *Discrete-Time Signal Processing*; Prentice Hall: Upper Saddle River, NJ, USA, 1999; ISBN 0130834432.
43. Beigi, H. *Fundamentals of Speaker Recognition*; Springer: Berlin/Heidelberg, Germany, 2016; ISBN 9781489979223.
44. Mermelstein, P. Distance measures for speech recognition, psychological and instrumental. *Pattern Recognit. Artif. Intell.* **1976**, *116*, 374–388.
45. Bridle, J.S.; Brown, M.D.; Chamberlain, R.M. *An experimental automatic word recognition system*; Technical Report JSRU No. 1003; Joint Speech Research Unit: Ruislip, England, 1974.
46. Stevens, S.S.; Volkmann, J. The Relation of Pitch to Frequency: A Revised Scale. *Am. J. Psychol.* **1940**, *53*, 329–353. [CrossRef]
47. Hermansky, H.; Morgan, N.; Hirsch, H.-G. Recognition of speech in additive and convolutional noise based on RASTA spectral processing. In Proceedings of the IEEE International Conference on Acoustics Speech and Signal Processing, Minneapolis, MN, USA, 27–30 April 1993; Volume 2, pp. 83–86.
48. Fant, G. *Acoustic Theory of Speech Production: With Calculations Based on X-ray Studies of Russian Articulations*; Mouton: The Hague, Netherlands, 1970; ISBN 9027916004.
49. Umesh, S.; Cohen, L.; Nelson, D. Fitting the Mel Scale. In Proceedings of the 1999 IEEE International Conference on Acoustics, Speech, and Signal Processing, Phoenix, AZ, USA, 15–19 March 1999.
50. Zwicker, E. Subdivision of the Audible Frequency Range into Critical Bands (Frequenzgruppen). *J. Acoust. Soc. Am.* **1961**, *33*, 248. [CrossRef]
51. Irino, T.; Patterson, R.D. A time-domain, level-dependent auditory filter: The gammachirp. *J. Acoust. Soc. Am.* **1997**, *101*, 412–419. [CrossRef]
52. Traunmüller, H. Analytical expressions for the tonotopic sensory scale. *J. Acoust. Soc. Am.* **1990**, *88*, 97–100. [CrossRef]
53. Moore, B.C.J.; Glasberg, B.R. Suggested formulae for calculating auditory-filter bandwidths and excitation patterns. *J. Acoust. Soc. Am.* **1983**, *74*, 750–753. [CrossRef]
54. Herrera, A.; Del Rio, F. Frequency bark cepstral coefficients extraction for speech analysis by synthesis. *J. Acoust. Soc. Am.* **2010**, *128*, 2290. [CrossRef]
55. Slaney, M. An Efficient Implementation of the Patterson-Holdsworth Auditory Filter Bank; Technical Report Apple Computer No. 35, Perception Group—Advanced Technology Group, Cupertino, California (USA). 1993. Available online: https://engineering.purdue.edu/~{}malcolm/apple/tr35/PattersonsEar.pdf (accessed on 21 November 2019).
56. Teager, H.M.; Teager, S.M. Evidence for Nonlinear Sound Production Mechanisms in the Vocal Tract. In *Speech Production and Speech Modelling*; Springer: Dordrecht, The Netherlands, 1990; pp. 241–261.
57. Maragos, P.; Kaiser, J.F.; Quatieri, T.F. On amplitude and frequency demodulation using energy operators. *IEEE Trans. Signal Process.* **1993**, *41*, 1532–1550. [CrossRef]
58. Skowronski, M.D.; Harris, J.G. Increased mfcc filter bandwidth for noise-robust phoneme recognition. In Proceedings of the IEEE International Conference on Acoustics Speech and Signal Processing, Orlando, FL, USA, 13–17 May 2002; pp. 801–804.

59. Figueiredo, E.; Park, G.; Figueiras, J. *Structural Health Monitoring Algorithm Comparisons Using Standard Data Sets*; Los Alamos National Lab.(LANL): Los Alamos, NM, USA, 2009.
60. Ververidis, D.; Kotropoulos, C. Information Loss of the Mahalanobis Distance in High Dimensions: Application to Feature Selection. *IEEE Trans. Pattern Anal. Mach. Intell.* **2009**, *31*, 2275–2281. [CrossRef] [PubMed]
61. Civera, M.; Zanotti Fragonara, L.; Surace, C. Video Processing Techniques for the Contactless Investigation of Large Oscillations. *J. Phys. Conf. Ser.* **2019**, *1249*, 012004. [CrossRef]
62. Civera, M.; Zanotti Fragonara, L.; Surace, C.; Civera, M.; Zanotti Fragonara, L.; Surace, C. Using Video Processing for the Full-Field Identification of Backbone Curves in Case of Large Vibrations. *Sensors* **2019**, *19*, 2345. [CrossRef] [PubMed]
63. Bovsunovsky, A.; Surace, C. Non-linearities in the vibrations of elastic structures with a closing crack: A state of the art review. *Mech. Syst. Signal Process.* **2015**, *62–63*, 129–148. [CrossRef]

applied sciences

MDPI

Article

The Rapid Detection Technology of Lamb Wave for Microcracks in Thin-Walled Tubes

Shunmin Yang, Mingquan Wang * and Lu Yang

Science and Technology on Electronic Test and Measurement Laboratory, North University of China, Taiyuan 030051, China

* Correspondence: wangmq@nuc.edu.cn; Tel.: +86-139-3454-8995 or +86-0351-355-7406

Received: 25 July 2019; Accepted: 27 August 2019; Published: 1 September 2019

check for updates

Abstract: Thin-walled tubes are a kind of pressure vessel formed by a stamping and drawing process, which must withstand a great deal of sudden pressure during use. When microcrack defects of a certain depth are present on its inner and outer surfaces, severe safety accidents may occur, such as cracking and crushing. Therefore, it is necessary to carry out nondestructive testing of thin-walled tubes in the production process to eliminate the potential safety hazards. To realize the rapid detection of microcracks in thin-walled tubes, this study could be summarized as follows: (i) Because the diameters of the thin-walled tubes were much larger than their thicknesses, Lamb wave characteristics of plates with equal thicknesses were used to approximate the dispersion characteristics of thin-walled tubes. (ii) To study the dispersion characteristics of Lamb waves in thin plates, the detection method of the A_0 mode was determined using the particle displacement–amplitude curve. (iii) Using a multi-channel parallel detection method, rapid detection equipment for Lamb wave microcracks in thin-walled tubes was developed. (iv) The filtering peak values for defect signal detection with different depths showed that the defect detection peak values could reflect the defect depth information. (v) According to the minimum defect standard of a 0.045-mm depth, 100,000 thin-walled tubes were tested. The results showed that the missed detection rate was 0%, the reject rate was 0.3%, and the detection speed was 5.8 s/piece, which fully meets the actual detection requirements of production lines. Therefore, this study not only solved the practical issues for the rapid detection of microcracks in thin-walled tubes but also provided a reference for the application of ultrasonic technology for the detection of other components.

Keywords: Lamb wave; microcracks testing; multi-channel detection; signal-to-noise ratio; thin-walled tube

1. Introduction

1.1. Purpose and Significance

Thin-walled tubes are pressure vessels formed by stamping and drawing. In the production process, due to errors in the processing parameters and faults in the production equipment, microcrack defects form in the inner and outer walls of thin-walled tubes. However, thin-walled tubes must withstand large sudden changes in pressure during use. When the depths of the microcracks on the inner and outer surfaces are greater than a certain threshold, severe safety accidents can occur, such as cracking and crushing. Therefore, it is necessary to carry out nondestructive testing of thin-walled tubes in the production process to eliminate the potential safety hazards.

For the nondestructive testing of thin-walled tubes, the commonly used testing methods include eddy current [1,2], magnetic powder [3], X-ray [4,5], liquid penetrant [6–8], and ultrasonic [9] methods. With eddy current detection, it is difficult to distinguish irregular surface defects and detect them quantitatively, which can easily lead to misjudgment [10]. Magnetic particle testing is more suitable for surface defect detection, and it is difficult to quantitatively detect the depth of a flaw [10]. The X-ray method is mainly used to detect the volume of the defect inside the object. For an area defect, there will be an overlap of the front and back, and the angle between the ray beam and the orientation of the crack cannot exceed 10°; otherwise, it cannot be detected [10]. Penetration detection is limited to the defects of the surface opening. The penetrating liquid pollutes the parts and the environment and is not suitable for on-line detection [10]. Ultrasonic testing is sensitive to the area of the defect inside the object and can locate the relative size and location of the defect. The detection depth is much greater than other detection methods, and thus, it has been widely used [10]. For thin-walled tubes, ultrasonic Lamb waves are usually used for testing [11].

The ultrasonic detection principle of thin-walled tubes is the same as that of pipe, and there are many studies on the principle, method, and application of Lamb wave detection of pipe for reference. Therefore, based on the analysis of Lamb wave detection of pipe, the ultrasonic detection method of thin-walled tubes was explored. Defect types in pipe include lamination [12], corrosion [13], and cracking [14–16], which can occur on both inner and outer surfaces. The modes of Lamb waves in the pipe are mainly divided into longitudinal [17,18], torsional [19–23], and flexural [24,25].

According to the relevant references, the following work must be completed to realize the Lamb wave detection of pipe defects. (i) Based on analyzing and studying the dispersion characteristics of Lamb waves, the transducer frequency and Lamb wave mode with the highest sensitivity to the detected defect type should be selected. (ii) To study the interaction between defect and Lamb waves in the pipe, simulation should be carried out under a three-dimensional (3D) model. (iii) A test platform should be built to compare the difference between simulation and test results by scanning artificial defects in the pipe with transducers. (iv) Signal processing is used to determine the location and size of defects in the pipe. The difficulties in the detection process mainly focus on the two aspects: mode selection and 3D model simulation. On the one hand, the dispersion characteristics of Lamb waves in pipe have three modes to be analyzed and studied. Compared with only symmetric and antisymmetric modes in a thin plate, the difficulty of modal analysis and selection is increased. On the other hand, a 3D model is needed to analyze the interaction between defects and Lamb waves in the pipe. Compared with the 2D model of a thin plate, there are great challenges in terms of computational complexity.

When the diameters of hollow cylindrical components are much larger than their thicknesses, the dispersion characteristics of Lamb waves in hollow cylindrical components can be approximated by studying the characteristics of Lamb waves in plates with equal thicknesses [26–28]. He et al. [29] found that when the diameter to thickness ratio of hollow cylindrical components is less than 0.05, the dispersion curves of the hollow cylindrical components were similar to those of components with the corresponding plate thicknesses. In this study, the diameters and thicknesses of the thin-walled tubes were 35 mm and 0.8–1.2 mm, respectively. Selecting the Lamb-wave dispersion characteristics of plates with equal thicknesses to approximate the propagation characteristics of Lamb waves in thin-walled tubes could not only simplify the model but also facilitate the application of practical engineering practice.

1.2. The State of the Art

1.2.1. Transducers for Lamb Waves

Qiu et al. [30] proposed a multi-physical simulation method for the Lamb wave propagation of piezoelectric transducers under load conditions. The results showed that the phase velocity and amplitude changes of Lamb waves obtained by simulation were in good agreement with experiment. He et al. [31] designed two electromagnetic acoustic transducers to generate and receive S_0 Lamb waves in ferromagnetic steel plates, which were kept within 5 and 19 mm from the sample surface, respectively. This allowed the electromagnetic acoustic transducer to interact with the sample with less force and facilitated easier scanning. Zhang et al. [32] established the finite element model and detection system of laser ultrasound and realized the simulation and test detection of multilayer bonded composite materials. The results showed that this method not only facilitated the visualization of debonding but also provided accurate bonding quality information.

1.2.2. Lamb Mode

Gupta et al. [33] studied the interaction of a basic symmetric Lamb wave mode S_0 with laminations in eight layers of glass-fiber-reinforced polymer (GFRP) laminated composite by numerical simulation and experimental verification. The results showed that the sensitivity of S_0 mode reflection depended on the thickness direction and layering direction of the composite laminate. Among all laminates tested, the S_0 mode was highly sensitive to DL23 and DL34 (Delamination is considered between the layers 2&3 and 3&4 locations, which are abbreviated as DL12 and DL34, respectively. The thickness is 0.42–1.26 mm). Mori et al. [34] experimentally studied the transmission characteristics of Lamb wave S_0 and A_0 modes at the contact edges of two plates. The results showed that the transmission coefficient of the S_0 mode increased monotonically with increasing contact pressure, but the transmission coefficient of the A_0 mode had a non-monotonic relationship with contact pressure and local minimum frequency. Cho et al. [35] calculated the reflection and propagation coefficients of the selected guided wave modes using simulation and compared the results with experimental data. The results showed that defect characterization was possible if a suitable criterion for modal selection was established. Ochoa et al. [36] used two zeroth-order Lamb wave modes to detect various types of invisible impact damage in composite materials. Because the digital shearography and ultrasonic C-scan technology could identify 5- and 10-J impact damage and the impact damage of 3-J could not be detected, the results of the Lamb wave detection were confirmed by comparison.

1.2.3. Characteristics of Lamb Waves

Pant et al. [37] analyzed the influence of material properties, such as E_{11} (axial elastic modulus), E_{22} (transversal elastic modulus), G_{12} (plane shear modulus), and ρ (density), on Lamb wave propagation/dispersion of three kinds of composite laminates. These composite laminates are unidirectional, cross-layered, and quasi-isotropic, which are all composed of 16 layers of unidirectional carbon fiber epoxy prepreg. The results showed that when Lamb waves are used for damage detection, the A_0 mode should be selected when fd (frequency times thickness) is small, and the S_0 mode should be selected when fd is large. Shen et al. [38] proposed an effective numerical method with which to study the nonlinear scattering and mode conversion of Lamb waves interacting with breathing cracks. The results showed that the scattering phenomenon was related to the incident wave mode. Symmetric and antisymmetric scattering modes exhibit alternate properties, which, in turn, carry most of the energy and dominate the harmonic scattering. Ai et al. [39] used the finite element method to study the propagation characteristics of Lamb waves in thin plates, and when compared with the frequency dispersion curve, the error of the simulation was 1.3%. Senyurek et al. [40] studied the notch and impact damage of slats, and the results showed that Lamb waves could provide time and cost efficiency in aircraft structure detection compared with other technologies. The finite element method was used by Alkassar et al. [41] to simulate the propagation of Lamb waves in a thin metal plate. The results showed that when there was damage in the plate, a new packet-wave formed by a mode conversion phenomenon could be observed.

1.2.4. Defects Identification Method

Yelve et al. [42] studied the interaction between the Lamb wave of the S_0 mode and the damage in an aluminum plate and used an artificial neural network (ANN) to locate the damage. The results showed that the ANN was an effective tool for damage location in a plate structure. The local wavenumber method of Lamb waves was used by Fan et al. [43] to characterize the flat bottom defects in isotropic thin plates. The simulation and experimental results showed that the average wavenumber could be used to not only identify the shape, position, and size of the damage but also quantify the depth of the damage.

1.3. The Contribution

In the development process of on-line Lamb wave detection equipment for microcracks in thin-walled tubes, the following work items have been accomplished.

(i) The dispersion characteristics of Lamb waves in thin plates with the same thickness were used to approximate the ultrasonic characteristics of thin-walled tubes, which reduced the dimensions of the model and laid a foundation for the engineering application of Lamb waves.
(ii) Based on the calculation and simulation of phase velocity, incident angle, and particle amplitude displacement, the Lamb wave mode with the highest sensitivity to the inner wall and outer wall cracks of a thin-walled cylinder was determined.
(iii) Multi-channel technology was adopted to solve the problem of rapid detection and improve the stability and reliability of the equipment.
(iv) The defect signal extraction circuit was designed, which greatly reduced the amount of data acquired and improved the data processing speed.
(v) Rapid on-line detection equipment for microcracks in thin-walled tubes was developed, and the engineering application of rapid detection of microcracks by Lamb wave was realized.
(vi) Qualitative and quantitative non-destructive testing of the minimum depth crack of $10 \times 0.15 \times 0.045$ (length × width × depth) was realized.

2. Theoretical Principle

2.1. Testing Object and Artificial Defect

The structural sketch of a thin-walled tube is shown in Figure 1. Each tube was a steel pressure vessel with a variable wall thickness formed by cold stamping and stretching. The wall thickness of the mouth was 0.8 mm, the wall thickness near the bottom was 1.2 mm, and the inner diameter of the mouth was 37 mm. The detection range was a thin-walled tube 255 mm away from the mouth.

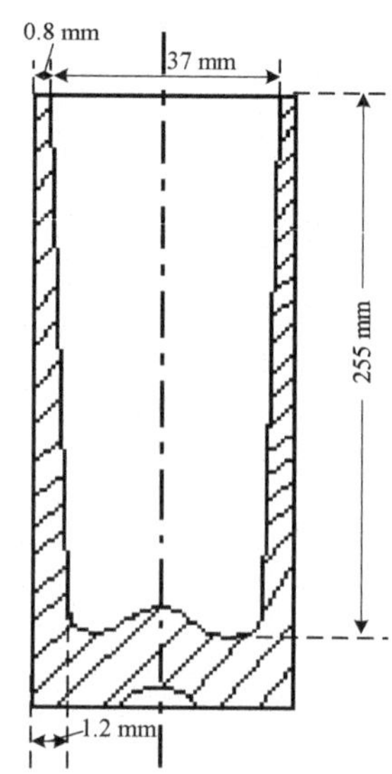

Figure 1. Structural sketch of a thin-walled tube.

Because thin-walled tubes are stamped and stretched in the production process, the inner and outer surfaces are prone to scratches with axial and circumferential distributions. In the actual production process, it was determined by manual inspection that the main defects are axially distributed. Therefore, to study the ultrasonic signal characteristics of these defects, groove defects with axial and circumferential distributions were fabricated on the inner and outer surfaces of thin-walled tube samples. Artificial defects were manufactured by the national measurement unit using electro-processing technology and measured by an optical image method. The lengths and widths of the defects were 10 ± 0.1 mm and 0.15 ± 0.01 mm, respectively. Three depths were used: 0.045 ± 0.005 mm, 0.073 ± 0.005 mm, and 0.10 ± 0.005 mm, and the detailed information is shown in Table 1.

Table 1. Artificial defects of different depths.

Defect Types	Defect Position (Away from the Mouth)/mm	Defect Depth/mm
Inner circumferential distribution	40	0.045 ± 0.005
Inner axial distribution	60	0.045 ± 0.005
Outer axial distribution	50	0.045 ± 0.005
Inner axial distribution	80	0.045 ± 0.005
Outer axial distribution	60	0.073 ± 0.005
Inner axial distribution	80	0.073 ± 0.005
Outer axial distribution	60	0.10 ± 0.005
Outer axial distribution	80	0.10 ± 0.005

2.2. Dispersion Characteristics of Lamb Waves

In the process of Lamb wave flaw detection, different modes of the Lamb wave will be excited when different incident angles are used, and thus, different sensitivities can be obtained for defects at different locations. Therefore, theoretical analysis is needed to determine which mode is most sensitive to the axial and circumferential distribution defects before testing. The Rayleigh–Lamb frequency equation is used to describe the wave characteristics of Lamb waves [11]. They are expressed mathematically as follows.

Symmetric mode:

$$4pq\tan\frac{\pi fd}{c_p}q+(p^2-1)^2\tan\frac{\pi fd}{c_p}p=0, \tag{1}$$

Antisymmetric mode:

$$(p^2-1)^2\tan\frac{\pi fd}{c_p}q+4pq\tan\frac{\pi fd}{c_p}p=0, \tag{2}$$

where $p=\sqrt{(\frac{c_p}{c_T})^2-1}$, $q=\sqrt{(\frac{c_p}{c_L})^2-1}$, c_p is the phase velocity of Lamb wave, c_T is the S-wave velocity, c_L is the P-wave velocity, f is the frequency of Lamb wave, and d is the thickness of the plate.

In calculating the dispersion characteristics of the Lamb waves, the P-wave velocity c_L and S-wave velocity c_T of the steel plate were 5970 and 3200 m/s, respectively. The Rayleigh–Lamb frequency equation was solved numerically using Matlab, and the phase velocity c_p vs. frequency-times-thickness fd dispersion curves of the Lamb wave was obtained, as shown in Figure 2.

Because the wall thickness of the thin-walled tubes varied from 0.8 to 1.2 mm, when a 5 MHz transducer was used for detection, the frequency-times-thickness was 4–6 MHz·mm, as shown by the red dotted line in Figure 2. As shown in the figure, the possible modes of the Lamb wave included the A_0, S_0, A_1, S_1, A_2, and S_2 modes. Because the phase velocities of the A_1, S_1, A_2, and S_2 modes vary greatly, the position of the defect echo signal is not fixed on the time axis, and it is difficult to identify the defect signal. Therefore, only the A_0 and S_0 modes were considered in the detection process.

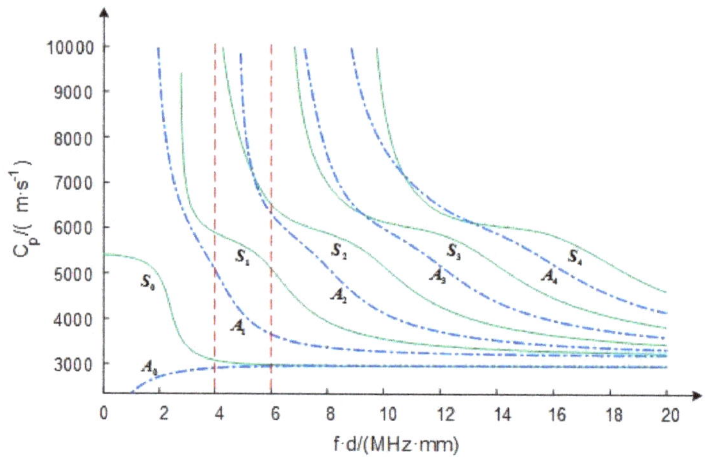

Figure 2. Relationship between the phase velocity c_p and frequency-times-thickness fd for Lamb waves in a steel plate (the S-wave velocity was 5970 m/s, the P-wave velocity was 3200 m/s, and Poisson's ratio was 0.28).

2.3. Mode Selection

Through the dispersion curves of the Lamb waves, we determined that the optional modes were A_0 or S_0. However, how to distinguish between them required further study of the direction of particle vibrations and the distributions of particle displacements for the A_0 and S_0 modes in plates. The particle displacement-amplitude curve of the Lamb waves reflects the variation and distribution of Lamb wave energy in the plate, which is one of the main bases for mode selection. The in-plane and out-of-plane displacement amplitudes of the particle vibrations are denoted as U and V, respectively. If the vertical distance between the particle and waveguide axis is x, their mathematical expressions are as follows [44]:

Symmetric mode:

$$U_s = \frac{(k_z^2 - s^2)\sin(0.5sd)}{2q'\sin(0.5q'd)}\cos(q'x)B + s\cos(sx)B, \tag{3}$$

$$V_s = \frac{(k_z^2 - s^2)\sin(0.5sd)}{2k_z\sin(0.5q'd)}\sin(q'x)B + k_z\sin(sx)B, \tag{4}$$

Antisymmetric mode:

$$U_a = \frac{(k_z^2 - s^2)\cos(0.5sd)}{2q'\cos(0.5q'd)}\sin(q'x)D - s\sin(sx)D,\ V_a = \frac{(k_z^2 - s^2)\cos(0.5sd)}{2k_z\cos(0.5q'd)}\cos(q'x)D + k_z\cos(sx)D, \tag{5}$$

where $k_z = \frac{2\pi f}{c_p}$, $s = k_z\sqrt{\left(\frac{c_p}{c_T}\right)^2 - 1}$, $q' = k_z\sqrt{\left(\frac{c_p}{c_L}\right)^2 - 1}$, and c_p is the phase velocity of Lamb wave.

In the symmetric mode, the vibration of a particle in a plate is symmetrical in the center of the plate, and its in-plane displacement is symmetric along the thickness of the plate, while the out-of-plane displacement is antisymmetric. Under the antisymmetric mode, the vibration phase of the particle in the plate is the same on the upper and lower surfaces, and the out-of-plane displacement is symmetric along the thickness of the plate, while the in-plane displacement is antisymmetric along the thickness of the plate. For the steel plate with 1-mm thickness, the particle displacement-amplitude curves of the A_0 and S_0 modes are shown in Figure 3.

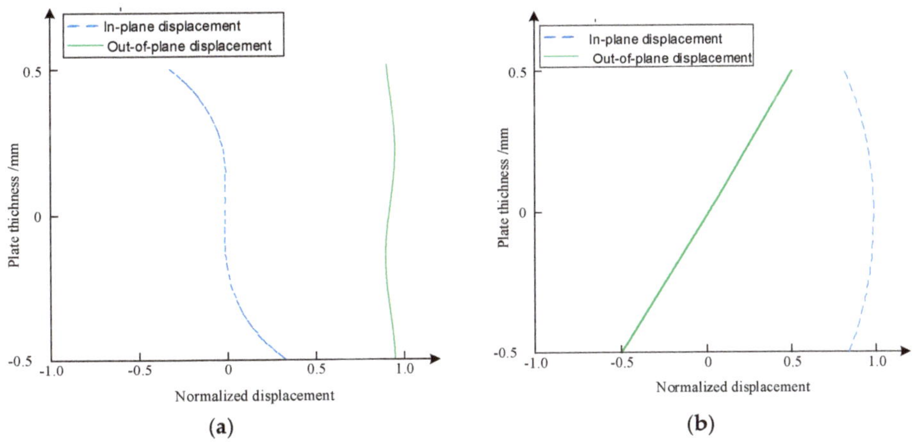

Figure 3. Particle displacement-amplitude curves of A_0 and S_0 modes in steel plate ((**a**) A_0 mode; (**b**) S_0 mode).

In Figure 3, the dashed line is the in-plane displacement along the thickness direction; that is, the vibration direction of the particle is parallel to the plate surface, while the real line is the out-of-plane displacement, that is, the vibration direction of the particle is perpendicular to the plate surface. The figure shows that the in-plane displacement of mode S_0 had a larger component in the plate, which is suitable for detecting the defects perpendicular to the plate surface, and the out-of-plane displacement of mode A_0 had a larger component in the plate, which is suitable for detecting defects on the inner and outer surfaces of the plate. Therefore, the A_0 mode, which is sensitive to the surface defects of the inner and outer walls of thin-walled tubes, was selected as the detection mode. This is consistent with the conclusion of Ref. [37].

2.4. Calculation of Incident Angle

The method of water immersion was adopted in the detection process, so it was necessary to calculate the incident angle of the A_0 mode Lamb wave produced by the ultrasonic transducer when the Lamb waves propagate from water into the steel. The mathematical formula is as follows:

$$\alpha = \arcsin\frac{c_L}{c_p}, \tag{6}$$

where α is the incident angle, c_L is the P-wave velocity in water, and c_p is the phase velocity of Lamb wave in steel. The relationship between the incident angle and frequency-times-thickness is shown in Figure 4.

As shown in Figure 4, when the frequency-times-thickness was 4–6 MHz mm, the incident angle of the mode A_0 corresponded to 30.5–29.99°. In practical applications, 30° was chosen as the incident angle to obtain the A_0 mode of the Lamb waves.

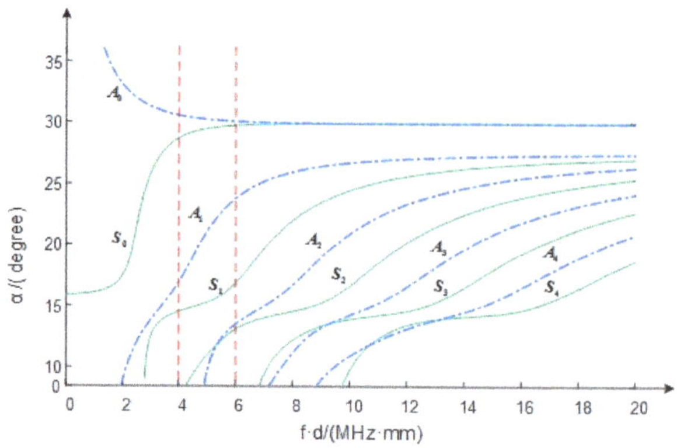

Figure 4. Relationship between the incident angle and frequency-times-thickness in steel (the P-wave velocity in water was 1480 m/s).

3. Testing Equipment

3.1. Appearance and Core Components

As shown in Figure 5, the testing equipment was mainly composed of three parts: a Programmable Logic Controller (PLC) control unit, a Personal Computer (PC) control unit, and a mechanical device. The control unit of the PLC was mainly responsible for the motion control of the mechanical device and communication with the PC through the RS232 serial module. The PC control unit mainly included a multi-channel pulse transmitter/receiver, the extraction device of the detection signal feature, the multi-channel data acquisition card, and the PC. The function of the multi-channel pulse transmitter/receiver was to detect the object using an excitation pulse to trigger the ultrasonic transducer and receive the detection signal from the detected object. The feature extraction device for the testing signal mainly ensured that only the defect echo signal was extracted through the hardware gate to reduce the amount of data acquisition and improve the speed of data processing. The mechanical device mainly performed the mechanical actions of grasping, rotating, and moving the thin-walled tube into place for testing.

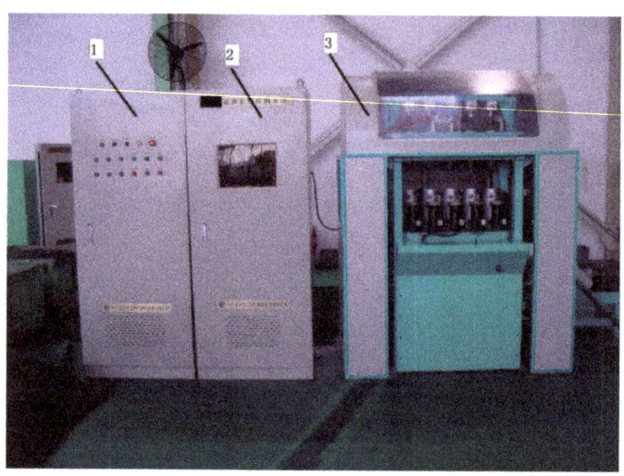

Figure 5. Appearance of testing equipment (1. PLC control unit; 2. PC control unit; 3. mechanical device).

The core components of the mechanical equipment are shown in Figure 6, which mainly include the material-preparation container, the testing container, the ultrasonic transducer, and the built-in manipulator. The material-preparation container was mainly used to place the thin-walled tube to be inspected. By moving up and down, the thin-walled tube was conveyed to a position that the built-in manipulator could grasp. There was an overflow outlet at the mouth of the testing container, which was used to discharge the overflow amount of water as the thin-walled tube slowly sank into the testing container. There was a water inlet at the bottom of the testing container, which was used to replace water when the thin-walled tube was removed from the testing container after the detection to ensure the coupling between the ultrasonic transducer and the thin-walled tube. A pair of ultrasonic probes were installed on the side of the testing container to scan the axial and circumferential distribution defects of the thin-walled tube. The movement of the testing container and the rotation of the built-in manipulator carrying the thin-walled tube could cooperate, which allowed the spiral scanning of the thin-walled tube by the ultrasonic probes. The built-in manipulator held the thin-walled tube from the inside. Meanwhile, an airbag at the top of the built-in manipulator sealed the opening of the thin-walled tube to prevent water from penetrating the thin-walled tube during the detection process. The built-in manipulator moved back and forth along the guide rail to grasp, rotate, and move the thin-walled tube into place.

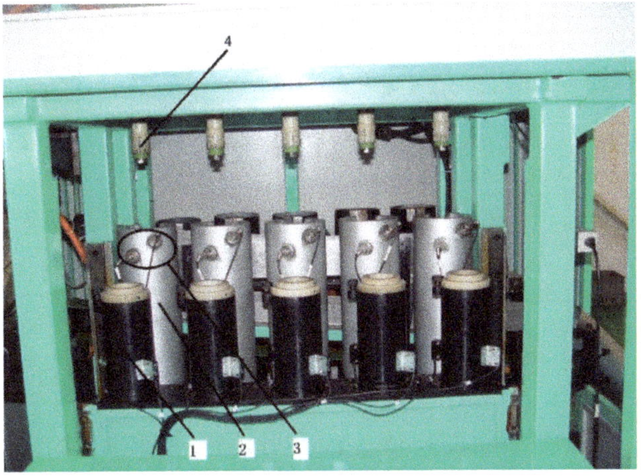

Figure 6. Core components of testing equipment (1. material-preparation container; 2. testing container; 3. ultrasonic transducer; 4. built-in manipulator).

The detection process was as follows. Firstly, when all five material-preparation containers were placed into the thin-walled tubes, the built-in manipulator moved backward to the central position of the material-preparation container. When the material-preparation container rose to a certain position, the built-in manipulator held the thin-walled tube from the inside and allowed the top airbag to seal the opening of the thin-walled tube. The material-preparation container quickly dropped to the original position. Secondly, the built-in manipulator carrying the thin-walled tube moved forward to the center of the testing container and began to rotate at 320 rpm. The testing container quickly rose to a certain position, and the PLC control unit notified the PC through the RS232 serial port to begin the detection. The PC started the data acquisition card and the pulse transmitter/receiver, and the testing container began to rise slowly and uniformly so that the ultrasonic transducer could achieve full coverage spiral scanning of the thin-walled tube to be detected. Thirdly, after the detection, the built-in manipulator stopped rotating, and the testing container quickly dropped to the original position. The PC stopped the data acquisition and transferred the data processing results to the PLC control unit through an RS232 serial port. Finally, the built-in manipulator continued to move forward and based on the data processing results, the thin-walled tube identified as the unqualified product was placed on the conveyor belt of rejected product. Then, the built-in manipulator moved on, and the remaining thin-walled tubes were placed on the conveyor belt of qualified product to complete the detection process.

3.2. Multichannel Detection Method

Because the detection speed required 6 s/piece, there were three high-speed requirements for single-piece detection: (i) the assistant time was zero; (ii) the rotational speed of thin-walled tube was over 3000 rpm; (iii) the sampling frequency was over 100 MHz. These three requirements directly affected the stability, reliability, and service life of the system, and the design risk was very high. If five groups of 10-channel parallel structures were adopted, the detection time could be extended to 30 s (18 s for a detection time, 12 s for auxiliary time; the rotational speed could be reduced to 300 rpm, and the sampling frequency could be reduced to 100 kHz). Thus, not only was the impact of mechanical transmission alleviated but also sufficient processing time was provided for data acquisition and defect signal processing. The problem of detection time, which is difficult to solve, was transferred to a multi-channel problem that was relatively easy to solve. Furthermore, this laid a foundation for the stable and reliable design of the system.

4. Results and Analysis

In the detection process, the ultrasonic transducer was a point focus probe with a frequency of 2.5 MHz, a diameter of 12 mm, a focal length of 20 mm, and a focal spot of 2 mm. The diameter of the thin-walled tube was 37 mm, the length of the detection area was 255 mm, and the rotation speed was 320 rpm. Thus, 0.1875 s was required for the tube to rotate once. Based on the single-channel sampling frequency of 0.75 kHz, the number of points sampled during the rotation of the tube was calculated to be 140. The total sampling length was 20,800 points. For the longitudinal defect with a 10-mm length (i.e., the defect distributed along the axis), the theoretical detection period was 10/1.72 = 5.81 (where 10 is the length of longitudinal defect, in-unit of mm; 1.72 is the pitch; that is, the axial interval of the helix formed by ultrasonic transducer scanning along the circumferential direction of the thin-walled tube, in-unit of mm; 5.81 is the number of defect waves obtained by the ultrasonic transducer scanning the longitudinal artificial defect with 10 mm length, in-unit of cycle). For the transverse defect with a 10-mm length (i.e., the defect distributed along the circumference), the theoretical detection period was 12/1.72 = 6.98 (where 12 is the diameter of the transducer, in-unit of mm; the transverse defect is distributed along the circumference of the thin-walled tube, and the transducer is also spiral-scanned along the circumference of the thin-walled tube, so the number of cycles scanned by the transducer for the transverse defect depends on the ratio of the diameter of the transducer to the pitch; 1.72 is the pitch, in-unit of mm; 6.98 is the number of defect waves obtained by the ultrasonic transducer scanning the transverse artificial defect with a 10-mm length, in-unit of cycle). The actual number of defect detection cycles is shown in Figure 7a–h. The number of transverse defect detection cycles was seven, and the number of longitudinal defect detection cycles was 8–10. The actual detection period of the longitudinal defect was larger than the theoretical value, while that of the transverse defect was equal to the theoretical value. This was due to the different scanning and defect distribution directions.

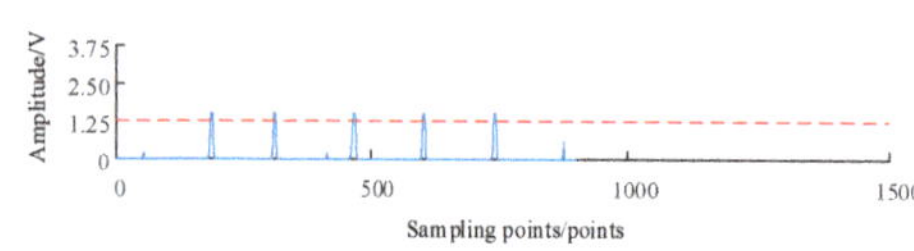

(a)

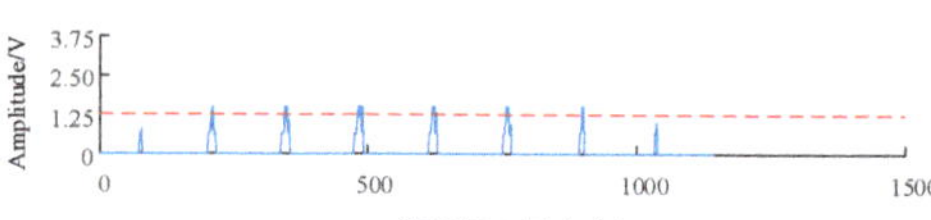

(b)

Figure 7. *Cont.*

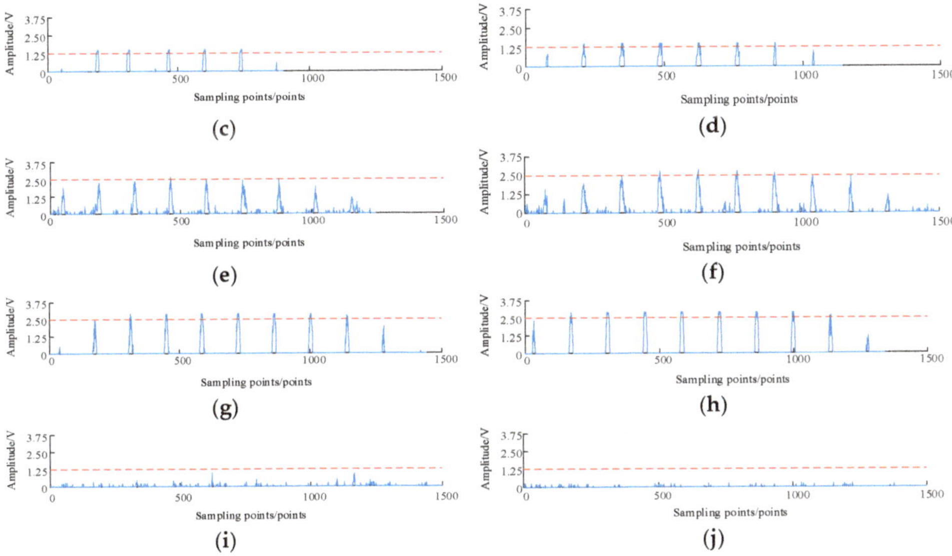

Figure 7. Signal with and without flaw ((**a**) Depth of inner transverse flaw was 0.045 mm; (**b**) Depth of inner longitudinal flaw was 0.045 mm; (**c**) Depth of outer transverse flaw was 0.045 mm; (**d**) Depth of outer longitudinal flaw was 0.045 mm; (**e**) Depth of inner longitudinal flaw was 0.073 mm; (**f**) Depth of outer longitudinal flaw was 0.073 mm; (**g**) Depth of inner longitudinal flaw was 0.10 mm; (**h**) Depth of outer longitudinal flaw was 0.10 mm; (**i**) Signal without flaw and with strong noise; (**j**) Signal without flaw and with weak noise).

4.1. Analysis of Qualitative Detection

The signals of the ultrasonic testing with and without flaws are shown in Figure 7. For (a), (c) and (b), (d) in Figure 7, the defects with same depth, same distribution direction, and different location, from the amplitude observation for defect wave, seemed to indicate that (a), (c) and (b), (d) in Figure 7 overlap; in fact, they were ultrasonic echo signals of defects at different locations, but the difference was very small. For (a), (b) and (c), (d) in Figure 7, the defects with the same depth, same location, and different distribution directions, the difference between them was obvious. Among them, Figure 7a–h showed the echo signals of defects with different depths, and Figure 7i,j showed the echo signal without defects. Therefore, regardless of the signal's periodicity or amplitude, the differences in the echo signals, when flaws were present or absent, were very significant, which showed that the method adopted in this study exhibited a good detection performance for qualitative detection.

4.2. Analysis of Quantitative Detection

To better analyze the quantitative relationship between the defect depth and signal amplitude, all the defect signals were smoothed first to eliminate the influence of single-peak extrema. Second, the peak values of the echo signals at different depths were extracted and plotted. Finally, the first and last peaks of the echo signals with different depths were eliminated, and the mean of residual peaks in the signals was calculated to make the mean curve approach the peak value of the defect signals as stably as possible.

As shown in Figure 8, for longitudinal internal defects, when the defect depth increased from 0.045 to 0.073 mm, the echo signal peak value increased from 1.36 to 2.05 V. When the defect depth increased from 0.073 to 0.10 mm, the echo signal peak value increased from 2.05 to 2.76 V. For the longitudinal defect, when the defect depth increased from 0.045 to 0.073 mm, the peak value of the echo signal increased from 1.36 to 2.39 V. When the defect depth increased from 0.073 to 0.10 mm, the peak value of the echo signal increased from 2.39 to 2.86 V.

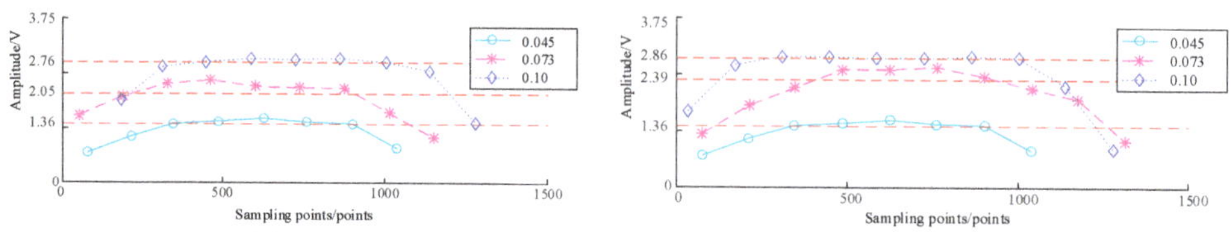

(**a**) (**b**)

Figure 8. The peak of flaw signal with different depths ((**a**) The peak of inner flaw signal with different depths; (**b**) The peak of outer flaw signal with different depths).

The results showed that the quantitative relationship of the internal longitudinal defects was slightly better than that of external longitudinal defects. There was a linear quantitative relationship between the depth of the longitudinal defect and the peak value of the detection signal. The depth of the detected defect could be represented by the peak value of the defect detection signal.

5. Issues and Conclusions

5.1. Issues

5.1.1. Misjudgment

Based on the minimum defect standard of 10 × 0.15 × 0.045 (length × width × depth) (taking a thin-walled tube with a 0.045-mm depth artificial defect as a standard test block, when the signal amplitude of the Lamb wave for the defect in the detected thin-walled tube is greater than or equal to the signal amplitude of the Lamb wave for the artificial defect in the standard test block, and the signal periods of the Lamb wave for the defect is greater than or equal to the signal periods of Lamb wave for the artificial defect in the standard test block, the tested thin-walled tube will be judged as unqualified product. Otherwise, the tested product will be judged as a qualified product. The missed detection rate is the ratio of unqualified products that are identified as qualified products to the total detected quantity, while the reject rate is the ratio of unqualified products to the total detected quantity), 100,000 thin-walled tubes were tested. The test results showed that the leakage rate was 0%, the rejection rate was 0.3%, and the average detection time of each product was 5.8 s, which fully meets the actual application requirements of a production line. Owing to the airbag located at the top of the built-in manipulator being cut, a small amount of water penetrated the thin-walled tubes during detection, resulting in misjudgment. This problem was solved by modifying the sawing technology of the mouth of the thin-walled tube.

5.1.2. Signal-to-Noise Ratio

To solve the problem of the rapid detection of thin-walled tubes, a multi-channel parallel detection method was adopted. Owing to the differences in the components between channels, some channels had high signal-to-noise ratios and no noise, as shown in Figure 7a–d,g,h. Other channels had echo noise with small amplitudes, as shown in Figure 7e,f,i,j. In the follow-up equipment development process, it is necessary to further optimize the design of related hardware circuits to further improve the consistency of multi-channel detection signals.

5.2. Conclusions

In the development process of on-line Lamb wave detection equipment for microcracks in thin-walled tubes, the following conclusions were drawn.

(i) Based on the calculation and simulation of phase velocity, incident angle, and particle amplitude displacement, the Lamb wave A_0 mode with the highest sensitivity to the inner wall and outer wall cracks of thin-walled tubes was determined. This is consistent with the results of Ref. [37].
(ii) A defect signal extraction circuit was designed, which greatly reduced the amount of data acquired and improved the data processing speed.
(iii) Rapid on-line detection equipment for microcracks in thin-walled tubes was developed, and the engineering application of rapid detection of microcracks by Lamb waves was realized.
(iv) For the 0.045-mm depth of the artificial defect, whether the inner wall, or outer wall of the longitudinal defect, or the transverse defect, there was almost no difference in the shape and amplitude of the defect waveform. However, for the inner wall of the longitudinal defect and the transverse defect, or the outer wall of the longitudinal defect and the transverse defect, waveform amplitude and contour had certain differences.
(v) For the artificial defects with depths of 0.073 and 0.10 mm, the longitudinal defects of the inner and outer walls were quite different in terms of waveform profile and amplitude.
(vi) This study not only solved the practical issue of the rapid detection of microcracks in thin-walled tubes but also provided a reference for the application of ultrasonic technology to detect other components.

Author Contributions: Conceptualization, S.Y. and M.W.; Methodology, S.Y.; Software, S.Y. and L.Y.; Validation, L.Y., and S.Y.; Formal Analysis, M.W.; Investigation, L.Y.; Resources, M.W.; Data Curation, L.Y.; Writing—Original Draft Preparation, S.Y.; Writing—Review and Editing, S.Y. and M.W.; Visualization, L.Y.; Supervision, M.W.; Project Administration, S.Y. and M.W.; Funding Acquisition, M.W.

Funding: This work was supported by the National Key Scientific Instrument and Equipment Development Projects of China (Grant No. 2013YQ240803) and the International Science and Technology Cooperation Program of Shanxi, China (Grant No. 201803D421032).

Conflicts of Interest: The authors declare no conflict of interest.

References

1. Shim, H.; Choi, M.S.; Lee, D.H.; Hur, D.H. A prediction method for the general corrosion behavior of Alloy 690 steam generator tube using eddy current testing. *Nucl. Eng. Des.* **2016**, *297*, 26–31. [CrossRef]
2. Buck, J.A.; Underhill, P.R.; Morelli, J.; Krause, T.W. Intranets: Evaluation of machine learning tools for inspection of steam generator tube structures using pulsed eddy. In Proceedings of the 43th Annual Review of Progress in Quantitative Nondestructive Evaluation, Atlanta, GA, USA, 17–22 July 2016; AIP: Atlanta, GA, USA, 2016; Volume 36. [CrossRef]
3. Liu, Y.; Lian, Z.; Lin, T.; Shen, Y.; Zhang, Q. A study on axial cracking failure of drill pipe body. *Eng. Fail. Anal.* **2015**, *59*, 434–443. [CrossRef]
4. Dong, B.; Fang, G.; Liu, Y.; Dong, P.; Zhang, J.; Xing, F.; Hong, S. Monitoring reinforcement corrosion and corrosion-induced cracking by X-ray microcomputed tomography method. *Cem. Concr. Res.* **2017**, *100*, 311–321. [CrossRef]
5. Lauridsen, T.; Willner, M.; Bech, M.; Pfeiffer, F.; Feidenhans, R. Detection of sub-pixel fractures in X-ray dark-field tomography. *Appl. Phys. A* **2015**, *121*, 1243–1250. [CrossRef]
6. Irek, P.; Slania, J. Material factors in relation to development time in liquid-penetrant inspection. Part 1. Material factors. *Arch. Metall. Mater.* **2016**, *61*, 509–514. [CrossRef]
7. Irek, P.; Slania, J. Material factors in relation to development time in liquid-penetrant inspection. Part 2. InvestIgatIon programme and preliminary tests. *Arch. Metall. Mater.* **2016**, *61*, 1697–1708. [CrossRef]
8. Irek, P.; Slania, J. Material factors in relation to development time in liquid-penetrant inspection. Part 3. Testing of model plates. *Arch. Metall. Mater.* **2017**, *62*, 41–49. [CrossRef]
9. Rose, J.L. *Ultrasonic Waves in Solid Media*; Cambridge University Press: New York, NY, USA, 2004; pp. 73–143.
10. Hellier, C. *Handbook of Nondestructive Evaluation*; McGraw-Hill: New York, NY, USA, 2003; pp. 1–7.
11. Rose, J.L. *Ultrasonic Guided Waves in Solid Media*; Cambridge University Press: New York, NY, USA, 2014; pp. 48–137.

12. Li, Z.; He, C.; Liu, Z.; Wu, B. Quantitative detection of lamination defect in thin-walled metallic pipe by using circumferential Lamb waves based on wavenumber analysis method. *NDT E Int.* **2019**, *102*, 56–67. [CrossRef]
13. Howard, R.; Cegla, F. Detectability of corrosion damage with circumferential guided waves in reflection and transmission. *NDT E Int.* **2017**, *91*, 108–119. [CrossRef]
14. Liu, K.; Wu, Z.; Jiang, Y.; Wang, Y.; Zhou, K.; Chen, Y. Guided waves based diagnostic imaging of circumferential cracks in small-diameter pipe. *Ultrasonics* **2016**, *65*, 34–42. [CrossRef]
15. Yeung, C.; Ng, C.T. Time-domain spectral finite element method for analysis of torsional guided waves scattering and mode conversion by cracks in pipes. *Mech. Syst. Signal Process.* **2019**, *128*, 305–317. [CrossRef]
16. Zhanga, X.; Tangb, Z.; Lva, F.; Yang, K. Scattering of torsional flexural guided waves from circular holes and crack-like defects in hollow cylinders. *NDT E Int.* **2017**, *89*, 56–66. [CrossRef]
17. Lowea, P.S.; Sandersonb, R.; Pedrama, S.K.; Boulgouris, N.V.; Mudge, P. Inspection of pipelines using the first longitudinal guided wave mode. *Phys. Procedia* **2015**, *70*, 338–342. [CrossRef]
18. Liu, Z.; Hu, Y.; Fan, J.; Yin, W.; Liu, X.; He, C.; Wu, B. Longitudinal mode magnetostrictive patch transducer array employing a multi-splitting meander coil for pipe inspection. *NDT E Int.* **2016**, *79*, 30–37. [CrossRef]
19. Heinlein, S.; Cawley, P.; Vogt, T.K. Reflection of torsional T (0,1) guided waves from defects in pipe bends. *NDT E Int.* **2018**, *93*, 57–63. [CrossRef]
20. Nakamura, N.; Ogi, H.; Hirao, M. EMAT pipe inspection technique using higher mode torsional guided wave T (0,2). *NDT E Int.* **2017**, *87*, 78–84.
21. Niu, X.; Duan, W.; Chen, H.P.; Marques, H.R. Excitation and propagation of torsional T (0,1) mode for guided wave testing of pipeline integrity. *Measurement* **2019**, *131*, 341–348. [CrossRef]
22. Kim, Y.; Park, K. Characterization of axial and oblique defects in pipes using fundamental torsional guided modes. *NDT E Int.* **2017**, *92*, 149–158. [CrossRef]
23. Muggleton, M.; Kalkowski, M.; Gao, Y.; Rustighi, E. A theoretical study of the fundamental torsional wave in buried pipes for pipeline condition assessment and monitoring. *J. Sound Vib.* **2016**, *374*, 155–171. [CrossRef]
24. Zhang, X.; Tang, Z.; Lv, F.; Pan, X. Helical comb magnetostrictive patch transducers for inspecting spiral welded pipes using flexural guided waves. *Ultrasonics* **2017**, *74*, 1–10. [CrossRef]
25. Zhang, X.; Tang, Z.; Lü, F.; Pan, X. Excitation of dominant flexural guided waves in elastic hollow cylinders using time delay circular array transducers. *Wave Motion* **2016**, *62*, 41–54. [CrossRef]
26. Li, J.; Rose, J.L. Natural beam focusing of non-axisymmetric guided waves in large-diameter pipes. *Ultrasonics* **2005**, *44*, 35–45. [CrossRef] [PubMed]
27. Nishino, H.; Takashina, S.; Uchida, F.; Takemoto, M.; Ono, K. Modal analysis of hollow cylindrical guided waves and applications. *Jpn. J. Appl. Phys.* **2001**, *40*, 364–370. [CrossRef]
28. Velichko, A.; Wilcox, P.D. Excitation and scattering of guided waves: Relationship between solutions for plates and pPipes. *J. Acoust. Soc. Am.* **2009**, *125*, 3623–3631. [CrossRef] [PubMed]
29. He, C.; Li, L.; Wu, B. Guided circumferential waves in hollow cylinders. *Chin. J. Mech. Eng.* **2004**, *40*, 7–12. [CrossRef]
30. Qi, L.; Ya, X.X.; Li, X.D.; Yu, S.F. Multiphysics simulation method of lamb wave propagation with piezoelectric transducers under load condition. *Chin. J. Aeronaut.* **2019**, *32*, 1071–1086. [CrossRef]
31. He, J.; Dixon, S.; Hill, S.; Xu, K. A new electromagnetic acoustic transducer design for generating and receiving S0 lamb waves in ferromagnetic steel plate. *Sensors* **2017**, *17*, 1023. [CrossRef]
32. Zhang, K.; Zhou, Z. Quantitative characterization of disbonds in multilayered bonded composites using laser ultrasonic guided waves. *NDT E Int.* **2018**, *97*, 42–50. [CrossRef]
33. Gupta, S.; Rajagopal, P. Effect of ply orientation and through-thickness position of delamination on the reflection of fundamental symmetric S_0 Lamb mode in GFRP. *Ultrasonics* **2018**, *90*, 109–119. [CrossRef]
34. Mori, N.; Biwa, S. Transmission characteristics of the S_0 and A_0 Lamb waves at contacting edges of plates. *Ultrasonics* **2017**, *81*, 93–99. [CrossRef]
35. Cho, Y.; Hongerholt, D.D.; Rose, J.L. Lamb wave scattering analysis for reflector characterization. *IEEE Trans. Ultrason. Ferroelectr. Freq. Control* **1997**, *44*, 44.
36. Ochôa, P.; Infante, V.; Silva, J.M.; Groves, R.M. Detection of multiple low-energy impact damage in composite plates using Lamb wave techniques. *Compos. Part B Eng.* **2015**, *80*, 291–298. [CrossRef]
37. Pant, S.; Laliberte, J.; Martinez, M.; Rocha, B.; Ancrum, D. Effects of composite lamina properties on fundamental Lamb wave mode dispersion characteristics. *Compos. Struct.* **2015**, *124*, 236–252. [CrossRef]

38. Shen, Y.; Cesnik, C.E.S. Nonlinear scattering and mode conversion of Lamb waves at breathing cracks: An efficient numerical approach. *Ultrasonics* **2019**, *94*, 202–217. [CrossRef] [PubMed]
39. Ai, C.; Cai, X.; Li, J.; Liu, K. Intranets: Finite element simulation of Lamb wave detection in thin plate. In Proceedings of the International Conference on Mechatronics Engineering & Information Technology, Xi'an, China, 27–28 August 2016; Atlantis Press: Xi'an, China, 2016. [CrossRef]
40. Senyurek, V.Y. Detection of cuts and impact damage at the aircraft wing slat by using Lamb wave method. *Measurement* **2015**, *67*, 10–23. [CrossRef]
41. Alkassar, Y.; Agarwal, V.K.; Alshrihi, E. Simulation of Lamb wave modes conversions in a thin plate for damage detection. *Procedia Eng.* **2017**, *173*, 948–955. [CrossRef]
42. Yelve, N.P.; Mulla, I.B. Locating damage in thin metallic plates using Lamb waves and Artificial Neural Network. In Proceedings of the International Conference on Nascent Technologies in Engineering, Navi Mumbai, India, 27–28 January 2017.
43. Fan, G.; Zhang, H.; Zhang, H.; Zhu, W.; Chai, X. Lamb wave local wavenumber approach for characterizing flat bottom defects in an isotropic thin plate. *Appl. Sci.* **2018**, *8*, 1600. [CrossRef]
44. Viktorov, I.A. *Rayleigh and Lamb Waves: Physical Theory and Applications*; Plenum Press: New York, NY, USA, 1967; pp. 75–82.

applied sciences

MDPI

Article

A Deep Learning Based Method for the Non-Destructive Measuring of Rock Strength through Hammering Sound

Shuai Han [1], Heng Li [2], Mingchao Li [1,*] and Timothy Rose [3,*]

[1] State Key Laboratory of Hydraulic Engineering Simulation and Safety, Tianjin University, Tianjin 300354, China

[2] Department of Building and Real Estate, The Hong Kong Polytechnic University, Hong Kong, China

[3] Science and Engineering Faculty, Queensland University of Technology, Queensland, QLD 4000, Australia

* Correspondence: lmc@tju.edu.cn (M.L.); tm.rose@qut.edu.au (T.R.)

Received: 14 July 2019; Accepted: 20 August 2019; Published: 23 August 2019

Abstract: Hammering rocks of different strengths can make different sounds. Geological engineers often use this method to approximate the strengths of rocks in geology surveys. This method is quick and convenient but subjective. Inspired by this problem, we present a new, non-destructive method for measuring the surface strengths of rocks based on deep neural network (DNN) and spectrogram analysis. All the hammering sounds are transformed into spectrograms firstly, and a clustering algorithm is presented to filter out the outliers of the spectrograms automatically. One of the most advanced image classification DNN, the Inception-ResNet-v2, is then re-trained with the spectrograms. The results show that the training accurate is up to 94.5%. Following this, three regression algorithms, including Support Vector Machine (SVM), K-Nearest Neighbor (KNN), and Random Forest (RF) are adopted to fit the relationship between the outputs of the DNN and the strength values. The tests show that KNN has the highest fitting accuracy, and SVM has the strongest generalization ability. The strengths (represented by rebound values) of almost all the samples can be predicted within an error of [−5, 5]. Overall, the proposed method has great potential in supporting the implementation of efficient rock strength measurement methods in the field.

Keywords: transfer learning; spectrogram analysis; selecting samples; non-destructive testing; hammering sound; regression algorithm

---

1. Introduction

Rock strength is an important parameter for analyzing geological problems. In the field, geological engineers are always concerned with how to estimate the strength of rocks rapidly, accurately, and conveniently. There are many methods for estimating rock strength. The most widely used is the unconfined compressive strength (UCS) test [1]. However, this method damages rock mass, and is time-consuming and laborious in many cases. Moreover, it is sometimes impossible to damage the rocks to determine their strength, especially in the field. For these reasons, many non-destructive methods have been presented to predict the UCS indirectly, such as the rebound method and sonic technique. The rebound method establishes the relationship between rock strength and rebound value produced by rebound hammers. However, variations in operational procedures may lead to large deviations, and there are too many equations used in this method, making it difficult to select a suitable one. The sonic technique method establishes the relationship between rock strength and sonic information. However, there are operational challenges in using sonic instruments, which are often costly and impractical for use in the field.

Due to the lack of facile methods for determining rock strength during geology surveys, experienced geologists usually prefer to determine rock strength from hearing the sounds produced by hammering a rock with a geological hammer. The rationale is that the hammering sound can reflect the surface strength (the outer surface at a depth of 30–50 mm) of the rock to some extent [2], which is directly related to its UCS. This method is convenient but subjective. The estimation heavily depends on the experience of geological engineers and is easily affected by many factors, such as noise, dust, and even the health condition of the geological engineer.

In recent years, deep learning, as one of the emerging fields of machine learning, has gained increasing interest in various fields. More and more researchers find that many complex problems can be solved well based on deep neural networks (DNNs). Generally, the two most active application fields of deep learning are speech recognition and image processing [3,4]. Transfer learning is also an active field of machine learning and has gained more attention because it can help build effective deep learning networks without vast amounts of data [5,6].

In this research, we aimed to develop a rapid, accurate, and convenient method for measuring the surface strength of rocks in the field. Inspired by the success of deep learning and the practice of estimating rock strength from hammering sounds, a retrained DNN was developed as the core of the proposed method. In the following of this paper, a literature review is firstly presented, and the development of non-destructive methods of rock strength and the usage of DNN in geology are introduced. Then the methodology of the research is described in detail, including the overall process, the DNN technique, the short-time Fourier transformation, the data filtering algorithm, and the regression algorithms. Finally, an experiment is carried out to test the proposed method.

2. Literature Review

2.1. Non-Destructive Methods for Measuring Rock Surface Strengths

Non-destructive testing is always the research hotspot of engineering [7]. For the non-destructive measuring of rock strength, the most commonly used non-destructive methods are the rebound method [8] and the sonic technique [9]. Many scholars have completed successful research with the rebound method [10]. For example, Yaşar and Erdoğan [11] carried out a series of experiments to investigate the relationship between Schmidt hammering rebound value and physicomechanical properties, including UCS, porosity, and unit volume weight. Aoki and Matsukura [12] developed a portable and simple equipment to test rock strength in the field, and validated the effectiveness of the equipment using the rebound method. Lai et al. [13] estimated the mean quantification of rock mass of five distinct locations through Schmidt hammer rebound tests. Despite these positive results, it remains difficult to find out a general formula to describe the relationship between rebound values and the surface strengths of rocks. Moreover, the measurements are of low accuracy, especially when the operation is not standardized.

In contrast, sonic technology is a relatively accurate method. For example, Sharma and Singh [14] established an empirical equation to predict rock strength based on P-wave velocity. Tziallas et al. [15] fitted the relationship between Young's modulus, UCS, and the velocity of an ultrasonic wave. Liu et al. [16] proposed a machine learning based method to determine the UCS of rocks with P-wave and some other indexes such as mineral composition and specific density. Son and Kim [17] used the sound signal obtained by hammering a rock to calculate the total energy of the sound, and then used the total energy calculation to calculate the strength of the rock. Azimian [18] developed a model for predicting UCS with P-wave and Schmidt hammer rebound. However, the equipment used in this experiment was laboratory-based and had limited application in the field.

2.2. Deep Learning in Geological Engineering

Deep learning has been used in many fields, and geological engineering is no exception. For example, Palafox et al. [19] adopted a deep convolutional neural network (CNN) to automatically recognize volcanic rootless cones and impact craters from images of Martian terrain. Xu et al. [20] established a DNN to classify different kinds of land covers automatically. Sidahmed et al. [21] trained a DNN to recognize reservoir rock types to help in identifying hydrocarbon resources. Furthermore, Yu et al. [22] combined the deep CNN and region growing algorithm to recognize landslides.

Training an effective DNN needs a large number of training data; however, in many situations, there are not enough samples for researchers, such as in geology tasks. Fortunately, transfer learning offers a good solution to such problems [23,24]. For instance, Li et al. [25] proposed a classification method for recognizing the features of microscopic sandstone pictures based on transfer learning. Zhang et al. [26] used transfer learning to identify different geological structures from images.

Generally, there are two methods for transfer learning: (1) continue to train the pre-trained network model to adjust the structure or weights of the network [27], or (2) remove the last layer of the pre-trained network model, then use a new dataset to train a new output layer [25]. Considering that the former is not effective when the sample types and sizes are different, the latter was adopted in this research.

3. Methodology

3.1. Overall Process

The overall process can be presented as Figure 1. Firstly, several rocks were selected as subjects. For each rock, the strength data was measured with a rebounder and the hammering sound was collected using a geological hammer and a voice recorder. Next, the hammering sounds were transformed into spectrograms, and then the spectrograms were filtered with a clustering algorithm. On the other hand, the measured strength data was pre-processed. After that, some spectrograms and their corresponding strength data (Batch A) were used to retrain the Inception-ResNet-v2 model, and the other spectrograms and their corresponding strength data (Batch B) were used for the final test of the method.

In the next sub-sections, the core technologies of this research are described in details, including the Inception-ResNet-v2 model, the short-time Fourier transformation (STFT), the clustering algorithm, and the regression algorithms.

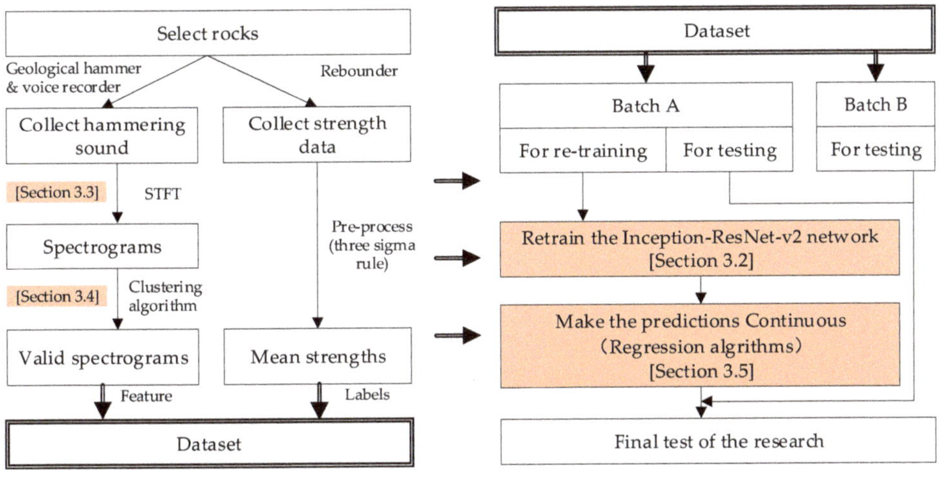

Figure 1. Flow chart of this research. The left scheme shows how to collect and process the data, the right scheme shows how to train the deep neural network (DNN) model and how to test our method.

3.2. Inception-ResNet-v2 Model

Inception-ResNet-v2 was developed from the Inception Net model by Google [28]. Different from Inception v1 to v3, the Inception-ResNet-v2 model takes advantage of the residual networks, successfully improving the accuracy and convergence speed of the original model. In a residual network, as shown in Figure 2, the output of the previous layer is inputted into the middle layer and the next layer together with the output of the middle layer [29]. Therefore, when adjusting the weights using the back propagation method, the gradient from the upper layer can skip over the middle layer up to the lowest layer, ensuring that all the weights can be adjusted effectively. More details of the Inception-ResNet-v2 model can be found in the publications [28,29].

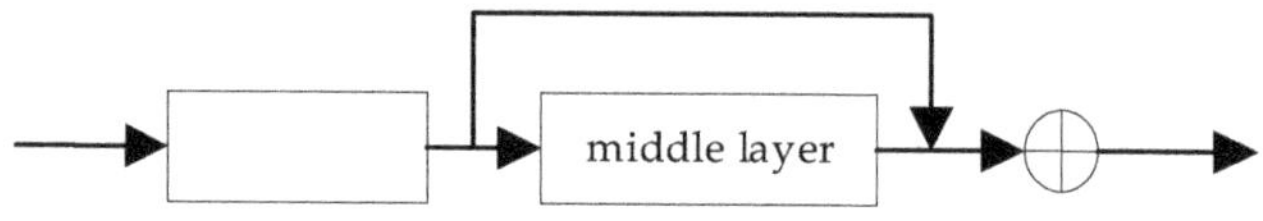

Figure 2. Diagram of a residual network unit.

Generally, the fine-tuning process includes four key steps: (1) train a deep learning network based on the source data. In this research, the Inception-ResNet-v2 model has been well-trained. (2) remove the output layer of the network, and reform the size of the output layer according to the target data, (3) initialize the weights of output layer and keep the weights of other layers unchanged, and (4)'re-train the network, as illustrated in Figure 3.

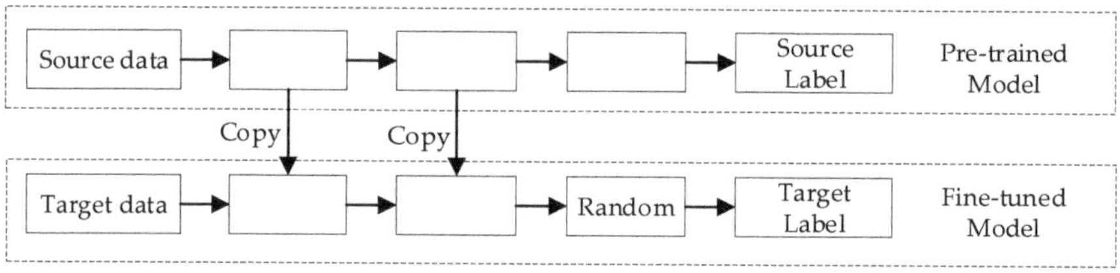

Figure 3. Process of fine-tuning a deep learning network.

3.3. Short-Time Fourier Transformation (STFT) and Spectrogram

In this research, hammering sounds were firstly transformed into spectrograms because of the Inception-ResNet-v2's strength in image identification. Hence, the STFT was utilized to process the sounds, as follows:

$$S_x(n,\omega) = \sum_{-\infty}^{\infty} x(m)w(n-m)e^{-j\omega m} \tag{1}$$

where $w(n)$ represents the window function, and ω is the frequency in radians [30]. In an STFT process, it is important to determine the size of the window. A large window leads to poor time resolution, while a small window leads to a poor frequency resolution. Either window size will degrade the quality of a spectrogram. Moreover, during programming, the sampling frequency fs, the time lapse L, and the frequency discretization N also affect the resolution of the spectrograms. However, it is difficult to determine the exact parameters in different cases. In this research, we found out a set of parameters through repeated experiments to make the spectrograms look clear.

Spectrograms generated by the STFT are presented in Figure 4. In a spectrogram, there are three dimensions that correspond to frequency (vertical axis), time (horizontal axis), and sound pressure (color gradient) or power spectral density (PSD). In the spectrograms example presented in Figure 4, yellow represents high sound pressure, and blue represents low sound pressure. It is easy to distinguish how sound pressures are distributed along with the frequencies in a certain period, how the sound pressures change over time in a certain frequency range, and how the principal frequency changes over time. Specifically, Figure 4a corresponds to the hammering sound of a rock with low strength, Figure 4b corresponds to a rock with moderate strength, and Figure 4c corresponds to a rock with high strength. It is obvious that in Figure 4c the proportion of high frequency is the largest, the attenuating speed is the slowest, and the differentiation of different frequency bands is the clearest.

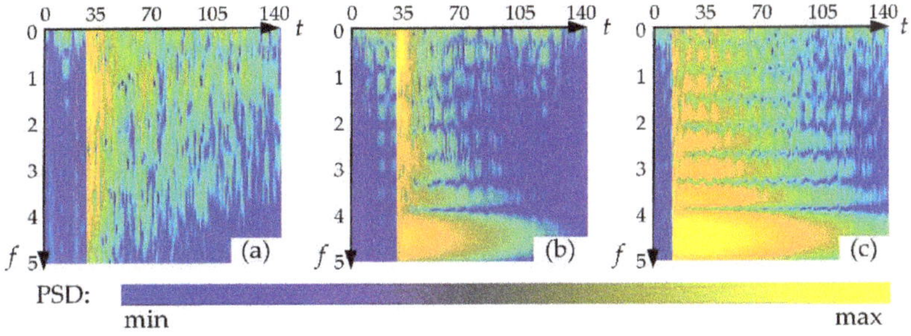

Figure 4. Examples of spectrograms. The rebound values of (**a**), (**b**), and (**c**) are 20, 54, and 73 respectively. The horizontal axes represent time (/ms), and the vertical axes are frequencies (/kHz).

Additionally, we found that there were no differences in parts of the spectrograms that had frequencies greater than 5 kHz, no matter for high-strength rocks or for low-strength rocks; thus, the parts with high frequency do not contribute to the analysis. Moreover, research has shown that human's most sensitive sound frequency is 2 kHz–5 kHz [31]. Therefore, only the 0–5 kHz part of the spectrograms was used for identifying strengths.

3.4. Selection of Training Samples Based on Clustering

3.4.1. Script Program for Cutting Hammering Sound

In this study, thousands of hammering sounds were transformed into spectrograms. Therefore, a simple script was programmed to split the hammering sounds automatically. First, for a sound file that contained hundreds of hammering sounds, the time series was iterated to determine a series of mutational points, $[t_1, t_2, \ldots, t_n]$, according to the amplitude, as shown in Figure 5. Next, it was assumed that the hammerings occurred at 10 milliseconds before the mutational points and each hammering sound lasted 150 ms, as shown in the right part of Figure 6. The two durations, 10 milliseconds and 150 ms, were determined by the statistical data in this research. The individual hammering sounds were then extracted, and the STFT was used to transform all the sound fragments into spectrograms.

However, these spectrograms could not be directly used for training because some sound fragments may not contain the right hammering sound. For example, some sound fragments may contain loud talking voices, or some may be produced by hammering at a wrong point by mistake. Therefore, invalid spectrograms should be eliminated to maximize the effectiveness of training.

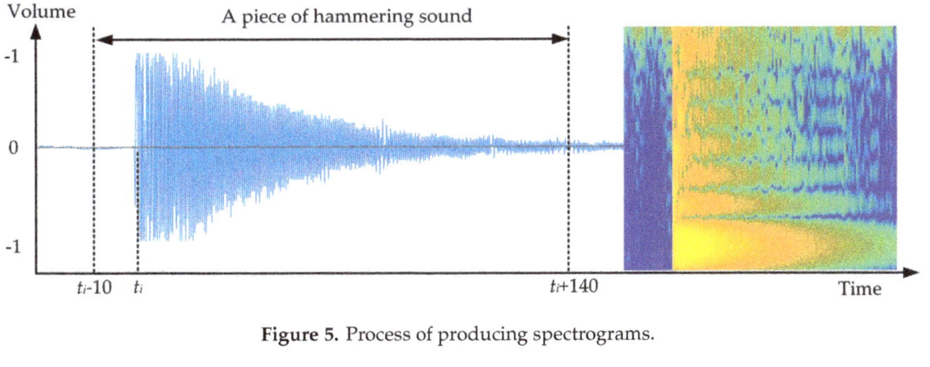

Figure 5. Process of producing spectrograms.

3.4.2. Binarization and Feature Extraction

For a set of spectrograms that obtained from the same rock, the biggest difference between the valid and invalid spectrograms is the distribution of frequencies. According to this principle, we suggest binarizing the spectrograms firstly to emphasize their features, as shown in Figure 6. The binarization threshold of each spectrogram can be determined using Otsu's method [32]. Otsu's method entails dividing an image into two sections (background and object) according to the gray-scale gradient. The formula of Otsu's method is as follows:

$$g = w_1 \times w_2 \times (\mu_1 - \mu_2)^2 \tag{2}$$

where w_1 is the proportion of the pixels of the object, w_2 is the proportion of the pixels of the background, μ_1 is the mean gray value of the object, and μ_2 is the mean gray value of the background. The optimal gray threshold is determined by finding out the maximum of g.

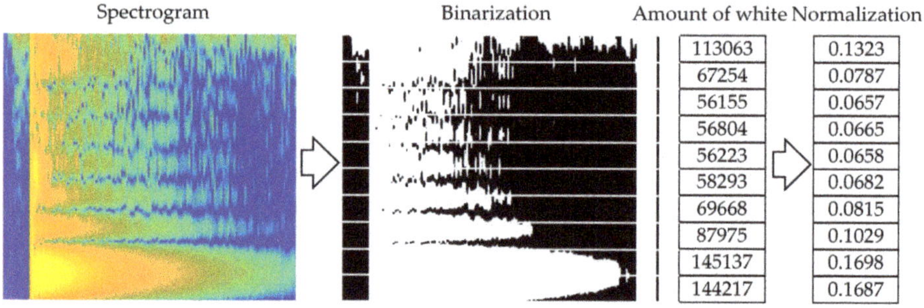

Figure 6. Progress of binarization and feature extraction.

Following this, each binarization was divided into 10 equal parts along the vertical axis. The number of white pixels in every part was calculated, then normalized by dividing them by the total amount of the white pixels, as shown in Figure 6. In this way, each spectrogram could be represented by 10 features. For example, the spectrogram shown in Figure 6 can be represented by [0.1323, 0.07887, 0.0657, 0.0665, 0.0682, 0.0815, 0.1029, 0.1698, 0.1687].

For a set of spectrograms that obtained from the same rock, the features of the valid spectrograms are similar to each other, while the invalid spectrograms are various. In this study, we assume that the spectrograms can be divided into three categories with a clustering method, and the categories that contain the most spectrograms are regarded as valid spectrograms.

3.4.3. Clustering Based on the Modified K-means Algorithm

The K-means algorithm, as a classical clustering algorithm, is famous for its simplicity and strong clustering ability [33]. As mentioned in Section 3.4.2, we set the value of K as three to divide the binarization spectrograms into three clusters, meaning that there is a major cluster that contained the most binarization spectrograms, and the other two clusters represented two extremes that differed from the major cluster.

However, the K-means algorithm is highly random because the initial seeds are selected randomly, making it easily plunge into local optima. To solve this problem, the initial seed selection is modified. The modified K-means algorithm is as follows:

Algorithm Modified K-means Algorithm

Input: Dataset $X = \{x_1, x_2, \ldots, x_n\}$, numbers of clusters $N = 3$
Output: Clustering result *LabelX*

```
For i = 1 to n−1
    For j = i to n
        Distance(i, j) = the Euclidean distance between x_i and x_j
    End
End
MaxDistance = the maximum of Distance
If number(MaxDistance) > 1
    Select one of the MaxDistances randomly and find out its corresponding x_max_i, x_max_j
End
Seeds(1) = x_max_i, Seeds(2) = x_max_j, Seeds(3) = Mean(x_max_i, x_max_j)
LabelX = Kmeans(X, Seeds)
```

The modified seed selection method identifies three seeds that have the longest distance from each other, and can ensure that the differences between the three categories are maximized.

The next step is to determine which categories are valid. The category that contains the most spectrograms is regarded as a valid category firstly. By experience, the amount of valid spectrograms occupies approximately more than 85% of the total. Therefore, for the other two categories, if the sum of them is less than 15%, then the spectrograms of the two categories are determined to be invalid; if the sum of them is larger than 15%, then only the smaller category is determined to be invalid.

3.5. Prediction Using Machine Learning

Using the re-trained Inception-ResNet-v2 model to determine the surface strength of rocks is virtually an image classification process. In training the DNN, the inputs are the spectrograms, and the labels are the values of surface strengths. After training, for a new spectrogram, the DNN can give the probabilities that the spectrogram belongs to each of the strengths, and classify it into a class according to the maximum probability, as shown in Table 1.

Table 1. Prediction result of a spectrogram.

Strength	10	20	30	40	50	60	70	80	90	100
Probability	0.0002	0.0000	0.0098	0.0005	0.0000	0.8806	0.1017	0.0004	0.0006	0.0051

It can be seen that the prediction results are discrete. However, surface strength is a continuous variable. To resolve this problem, the probabilities are regarded as the spectrograms' features extracted by the DNN model. Therefore, every spectrogram has 10 features and one label (strength value). Then, the relationship between the features and the strength values can be fitted with regression algorithms. The regression algorithms used in this research included the K-Nearest Neighbor [34], the Support Vector Machine [35,36], and the Random Forest [37].

4. Experiment and Analysis

4.1. Data Collection

Data for re-training the DNN model comprised the surface strengths and hammering sounds. In this experiment, two batches of data, Batch A and Batch B, were collected, and the rocks in Batch B were different from the rocks in Batch A. Batch A was used for re-training and conducting the preliminary tests. Batch B was used for the final test of the method.

The surface strengths of rocks were measured by an N-type rebound device. An N-type rebound device can measure the rebound values by hitting a rock. As mentioned in Section 2.1, current research has demonstrated that there is a strong link between the rebound value and the UCS. Therefore, in this study, the surface strengths of the rocks were represented by their rebound values.

The objects used in the experiment were rocks that existed in the natural environment. Every rock was intact and at least 0.05 m^3 in volume. Rocks were not limited to a particular type. It was not possible to measure the strength of the whole rock mass due to the rock anisotropy, and what was measured was just the strength of one point on the rock. To avoid damage to the surface of the rock caused by measuring one point too many times (especially when the rock was weak), three measuring points close to each other were set on one rock. Each point was hit five times by the rebound hammer. In total, there were 15 rebound values for each rock. Then, the surface strength of a rock could be calculated by filtering the values with the three sigma rule and calculating the mean of the remaining rebound values. Moreover, before measuring, the weathered layers of the rocks were removed. Table 2 shows the measurements of 15 rock samples A1–A10 (strengths of Batch A) and B1–B5 (strengths of Batch B). In addition, the rocks included granites, basalts, killas, and andesites. However, the types and mineral compositions of rocks were not regarded as the influential factors.

The reason why we divided the whole dataset into two batches was that an algorithm (even DNN) may be sensitive to the data that similar to the training data. The rocks in Batch B were different from the rocks in Batch A, and the strengths of Batch-B rocks were also different from the Batch-A rocks. A part of Batch A was for the primary test of our method, and the whole data of Batch B was for the further verification of the generalization of the method.

Table 2. Rebound values of A and B.

No.	Mean	Variance	No.	Mean	Variance	No.	Mean	Variance
A1	20	1.12	A6	54	1.56	B1	61	1.88
A2	73	1.13	A7	67	1.18	B2	48	2.37
A3	48	1.50	A8	58	1.72	B3	64	2.39
A4	39	1.73	A9	26	1.00	B4	72	2.43
A5	62	1.15	A10	44	1.88	B5	66	1.92

After measuring the strengths, each rock was hammered 200–260 times with a geological hammer around the three measuring points. A voice recorder was used to record the hammering sounds. The hammering rate was approximately one-two times per second, and the hammering force was slightly varied every 20–30 times to ensure the variability of the hammering force. The voice recorder had two channels, and its sampling frequency was 24 kHz.

It should be noted that in fact the hammering force mainly affects the amplitude of the hammering sound (sound level). Therefore, we did not hammer the rocks too hard, but put the voice recorder very near the hammering points to obtain clear sounds, and in this way prevented (or reduced) damage of the rock surface. Despite that, some rocks were still damaged in our experiment, and in these cases, new hammering points near the old ones were selected to continue the experiment. However, for those rocks of which nearby areas of the measuring points were all damaged, the measuring processes were terminated immediately.

4.2. Producing Spectrograms

With the script described in Section 3.4.1, 2410 hammering sounds were extracted from the sound files. The parameters used in STFT were as follows: the size of the window function $R = 64$, the time lapse $L = 32$, the sampling frequency $fs = 24$ kHz, and the frequency discretization $N = 8192$.

As shown in Table 2, there are 10 different rebound values in total, each of which corresponded to 200–260 spectrograms. According to the rebound values, all the spectrograms of Batch A were assigned to 10 different file folders, and the folders were named by their rebound values. After that, the modified K-means clustering algorithm described in Section 3.4.3 was used to remove the invalid spectrograms in every folder. Table 3 shows the filtering result of Batch A.

Table 3. Clustering results of spectrograms.

No.	Total	Cluster 1	Cluster 2	Cluster 3	Removed	Remained
1	201	24	52	125	Cluster 1	177
2	217	85	103	29	Cluster 3	188
3	258	73	16	169	Cluster 2	242
4	239	136	67	36	-	239
5	260	96	18	146	Cluster 2	242
6	232	105	99	28	Cluster 3	204
7	252	10	81	161	Cluster 1	242
8	253	139	20	94	Cluster 2	233
9	259	154	90	15	Cluster 3	250
10	239	2	74	163	Cluster 1	237
Total	2410	-	-	-	-	2254

4.3. Re-Training the DNN

In Batch A, there were 2254 spectrograms. Figure 7 shows some of the spectrograms with different strengths.

These spectrograms were then used to re-train the Inception-ResNet-v2 model. The parameters set for training comprised an initial learning rate of 0.002, a learning rate decay of 0.7, an epoch number of 100, and a batch size of 10. About 80% of the data were used to train the network, and the remaining 20% were used for validation. The evaluation indicators included the accuracy and the loss. After training, the structure and weights of the network were determined and were not changed any further. The training process is illustrated in Figure 8.

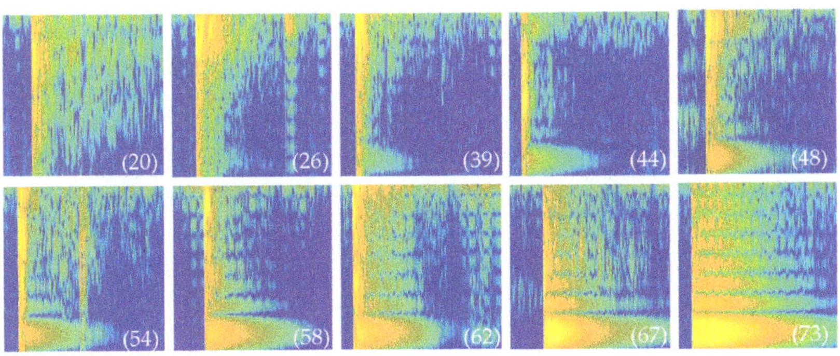

Figure 7. Spectrograms with different strengths.

As presented in Figure 8, in the first 1500 iterations, the accuracy grew rapidly. After about 5000 iterations, the accuracy reached 0.9, and then grew slowly. At the end of the training, the accuracy was 0.945. In the first 3000 iterations, the loss dropped quickly, then held steady between 0.50 and 1.00. We stopped the training at the 18000th step, because the loss showed that it would not be further decreased, and more training would lead to an over-fitting.

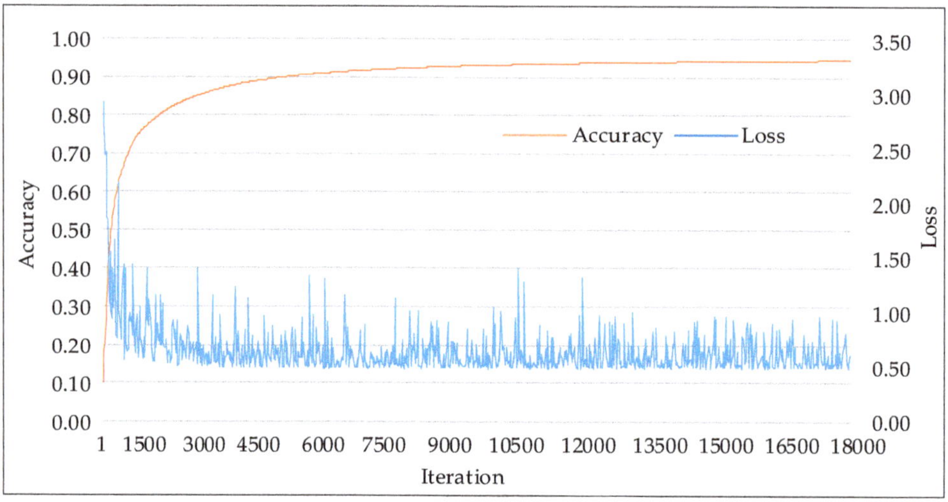

Figure 8. Re-training of the Inception-ResNet-v2 model.

4.4. Predicting Rock Surface Strength

The training results demonstrate that the fine-tuned DNN can classify the spectrograms with high accuracy. In the next step, the outputs of the network were regarded as the features of the spectrograms. Three regression algorithms, including the KNN, the SVM, and the RF, were then tried to fit the relationship between the features and the strengths.

4.4.1. Predictions of Batch A

First, Batch A was used as the input to test the method. The regression algorithms were used as follow: (1) in the SVM, the Gaussian kernel function was adopted; (2) in the KNN, the number of N was set to 20; (3) in the RF, the number of trees was 800, and the maximum depth was nine. Figure 9 displays the regression results.

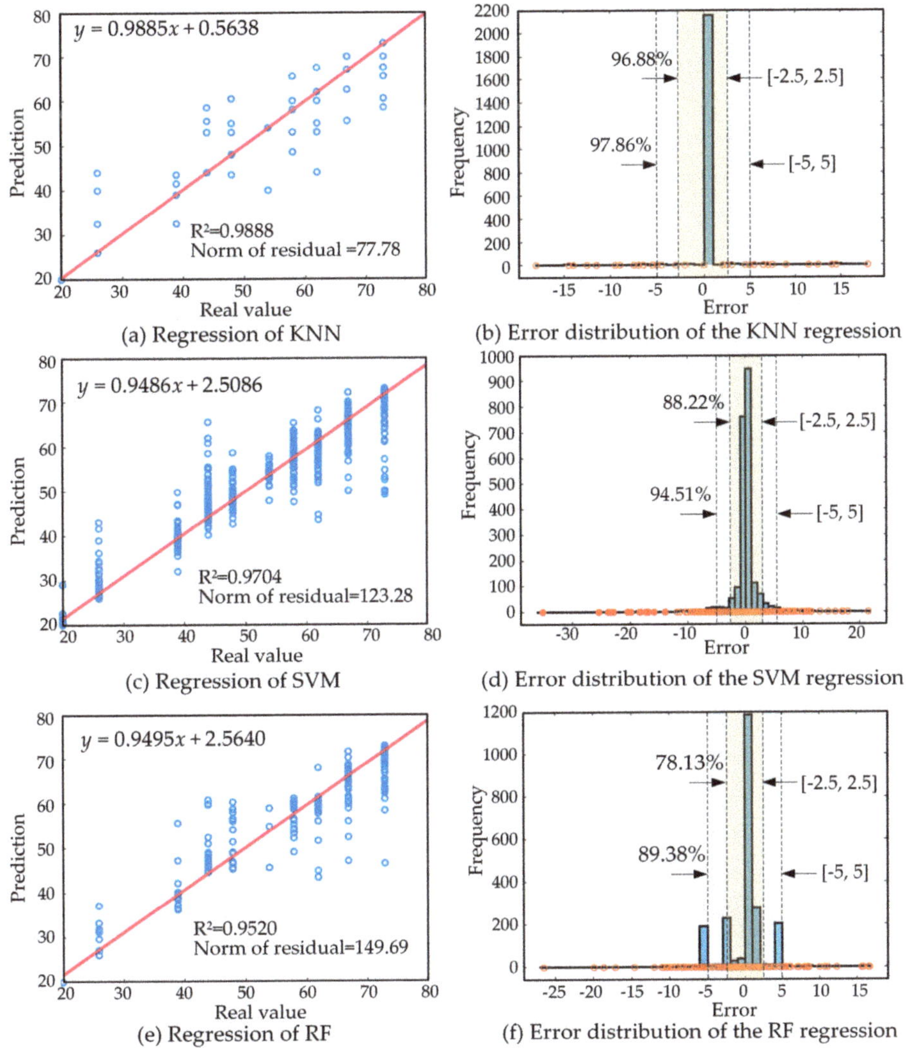

Figure 9. Predictions of Batch A.

The R^2 is used to measure the goodness of fit (GOF). Figure 9a,c, and e show that all the three algorithms can reach a high R^2 of more than 0.95. The R^2 of the KNN algorithm is the largest and is more than 0.98. The norm of residual is another measure of GOF, and a lower norm signifies a better fit. Among the three algorithms, KNN has the smallest norm of residual. Figure 9b,d,f are the distributions of the errors of the three algorithms. It also indicates that KNN is more accurate than the other two algorithms: the range of the errors is within [−20, 20], the errors that larger than −2.5 and smaller than 2.5 occupy 96.88%, and the errors in the range of [−5, 5] occupy 97.86%. Comprehensively, KNN gets the best performance, and the followings are the SVM and the RF.

4.4.2. Predictions of Batch B

The samples in Batch B had neither participated in the deep network re-training nor the regression process. Moreover, four out of five of the labels (rebound values) of Batch B were out of the range of

Batch A. The configurations of the regression algorithms were the same as that in Section 4.4.1. Table 4 shows the mean errors and variances between the predictions and the real strength values.

As presented in Table 4, the predictions of the SVM have the minimum mean errors except for the fourth and the fifth sample sets, and have the minimum variance except for the second set. The minimum mean error of the 5th samples sets is predicted by KNN. The minimum variance of the second set is also predicted by KNN. Overall, SVM predictions are the best, the following is KNN, then RF.

Table 4. Regression results of Batch B.

Strength	Mean Errors			Variance		
	KNN	SVM	RF	KNN	SVM	RF
48.00	−1.75	−1.11	−1.96	2.19	1.23	2.33
61.00	1.68	1.17	1.49	2.14	2.31	2.41
64.00	−1.75	−0.58	−1.79	6.30	4.66	5.60
66.00	2.44	1.62	1.18	3.15	1.81	3.29
72.00	5.19	6.95	5.73	5.65	4.34	6.47

4.5. Discussion

Based on the results presented in Section 4.4, it can be seen that the predictions of Batch A are significantly better than the predictions of Batch B. For one thing, the data of Batch A and Batch B were collected by two groups of researchers, and there might be some non-standard operations when collecting Batch B—for example, the hitting directions and the hitting speeds when using the rebound device. Further, the re-trained deep learning network was used for image classification, and there were only 10 clusters in the training sample. By enlarging the size of training data, the prediction of Batch B can be improved. For Batch A, the KNN algorithm made the best predictions, meaning that the fitting accuracy of KNN was the highest. For Batch B, the SVM achieved the best results, meaning that it had the strongest generalization ability in this regression.

5. Conclusions

In this paper, a new non-destructive measuring method for rock surface strength is presented based on the DNN technique and spectrogram analysis. The process comprises four steps: (1) collect hammering sounds and strength data, (2) produce spectrograms of the hammering sounds and remove invalid spectrograms using a modified K-means algorithm, (3) re-train the Inception-ResNet-v2 model by taking the spectrograms and strength data as the inputs and labels respectively, and (4) use regression algorithms to make the prediction results continuous. The validation shows that the strengths (represented by rebound values) of almost all the samples can be predicted within an error of [−5, 5].

Moreover, the combination of the re-trained DNN and KNN has the highest fitting accuracy, and the combination of the re-trained DNN and SVM has the strongest generalization ability. Therefore, if the size and the number of clusters of training samples are large, we recommend KNN for regression. Otherwise, SVM is recommended.

The proposed method was accomplished using Python and Tensorflow programming; however, it should be noted that the contribution of this research is not about a new deep neural network, and both the Inception-ResNet-v2 model and transfer learning are well-established techniques. This research is an application of them, and is aimed at presenting an effective and simple method for field survey. Overall, the proposed method offers great potential in supporting the implementation for efficient rock strength measurement methods in the field.

A noted limitation of this research is related to the use of a rebound device to capture the raw rock strength data, under experimental conditions. The precision of the rebound method is low, and the measurement is easily affected by hitting directions and hitting speed. In the subsequent work, higher precision techniques, such as acoustic emission techniques, will be considered to optimize the performance of the method further.

Author Contributions: S.H. collected and analyzed the data, wrote the code and the paper; H.L. gave professional geological guidance; M.L. provided the idea and edited the manuscript; T.R. gave professional geological guidance.

Funding: This research was funded by the National Natural Science Foundation for Excellent Young Scientists of China (Grant no. 51622904), the Tianjin Science Foundation for Distinguished Young Scientists of China (Grant no. 17JCJQJC44000), and the Research Grants Council of Hong Kong (PolyU 152093/14E).

Conflicts of Interest: The authors declare no conflict of interest.

References

1. Quinones, J.; Arzúa, J.; Alejano, L.R.; García-Bastante, F.; Ivars, D.M.; Walton, G. Analysis of size effects on the geomechanical parameters of intact granite samples under unconfined conditions. *Acta Geotech.* **2017**, *12*, 1229–1242. [CrossRef]
2. Selçuk, L.; Yabalak, E. Evaluation of the ratio between uniaxial compressive strength and Schmidt hammer rebound number and its effectiveness in predicting rock strength. *Nondestruct. Test. Eval.* **2015**, *30*, 1–12. [CrossRef]
3. Hannun, A.; Case, C.; Casper, J.; Catanzaro, B.; Diamos, G.; Elsen, E.; Prenger, R.; Satheesh, S.; Sengupta, S.; Coates, A. Deep speech: Scaling up end-to-end speech recognition. *arXiv* **2014**, arXiv:1412.5567.
4. Zhou, B.; Lapedriza, A.; Xiao, J.; Torralba, A.; Oliva, A. Learning deep features for scene recognition using places database. In Proceedings of the Advances in Neural Information Processing Systems (NIPS), Montreal, Canada, 8–13 December 2014; MIT Press: Cambridge, MA, USA; pp. 487–495.
5. Cireşan, D.C.; Meier, U.; Schmidhuber, J. Transfer learning for Latin and Chinese characters with deep neural networks. In Proceedings of the 2012 International Joint Conference on Neural Networks (IJCNN), Brisbane, Australia, 10–15 June 2012; pp. 1–6.
6. Zhou, J.T.; Pan, S.J.; Tsang, I.W.; Yan, Y. Hybrid Heterogeneous Transfer Learning through Deep Learning. In Proceedings of the 28th AAAI Conference on Artificial Intelligence, Québec City, Québec, Canada, 27–31 July 2014; pp. 2213–2220.
7. Noori Hoshyar, A.; Rashidi, M.; Liyanapathirana, R.; Samali, B. Algorithm Development for the non-destructive testing of structural damage. *Appl. Sci.* **2019**, *9*, 2810. [CrossRef]
8. Haramy, K.; DeMarco, M. Use of the Schmidt hammer for rock and coal testing. In Proceedings of the 26th US Symposium on Rock Mechanics (USRMS), American Rock Mechanics Association, Rapid City, SD, USA, 26–28 June 1985; pp. 549–555.
9. McNally, G. Estimation of coal measures rock strength using sonic and neutron logs. *Geoexploration* **1987**, *24*, 381–395. [CrossRef]
10. Goudie, A.S. The Schmidt Hammer in geomorphological research. *Prog. Phys. Geogr.* **2006**, *30*, 703–718. [CrossRef]
11. Yaşar, E.; Erdoğan, Y. Estimation of rock physicomechanical properties using hardness methods. *Eng. Geol.* **2004**, *71*, 281–288. [CrossRef]
12. Aoki, H.; Matsukura, Y. A new technique for non-destructive field measurement of rock-surface strength: An application of the Equotip hardness tester to weathering studies. *Earth Surf. Process. Landf.* **2007**, *32*, 1759–1769. [CrossRef]
13. Lai, G.T.; Mazlan, N.A.; Nadzir, M.S.M.; Rafek, A.G.; Serasa, A.S.; Hussin, A.; Ern, L.K.; Yeok, F.S. Uniaxial Compressive Strength of Antarctic Peninsula Rocks: Schmidt Hammer Rebound Test. *Sains Malays.* **2017**, *46*, 677–684.
14. Sharma, P.; Singh, T. A correlation between P-wave velocity, impact strength index, slake durability index and uniaxial compressive strength. *Bull. Eng. Geol. Environ.* **2008**, *67*, 17–22. [CrossRef]
15. Tziallas, G.P.; Tsiambaos, G.; Saroglou, H. Determination of rock strength and deformability of intact rocks. *Electron. J. Geotech. Eng.* **2009**, *14*, e12.
16. Liu, Z.; Shao, J.; Xu, W.; Wu, Q. Indirect estimation of unconfined compressive strength of carbonate rocks using extreme learning machine. *Acta Geotech.* **2015**, *10*, 651–663. [CrossRef]

17. Son, M.; Kim, M. Estimation of the Compressive Strength of Intact Rock Using Non-Destructive Testing Method Based on Total Sound-Signal Energy. *Geotech. Test. J.* **2017**, *40*, 643–657. [CrossRef]
18. Azimian, A. Application of statistical methods for predicting uniaxial compressive strength of limestone rocks using nondestructive tests. *Acta Geotech.* **2017**, *12*, 321–333. [CrossRef]
19. Palafox, L.F.; Hamilton, C.W.; Scheidt, S.P.; Alvarez, A.M. Automated detection of geological landforms on Mars using Convolutional Neural Networks. *Comput Geosci.* **2017**, *101*, 48–56. [CrossRef]
20. Xu, G.; Zhu, X.; Fu, D.; Dong, J.; Xiao, X. Automatic land cover classification of geo-tagged field photos by deep learning. *Environ. Model. Softw.* **2017**, *91*, 127–134. [CrossRef]
21. Sidahmed, M.; Roy, A.; Sayed, A. Streamline Rock Facies Classification with Deep Learning Cognitive Process. In Proceedings of the SPE Annual Technical Conference and Exhibition, Society of Petroleum Engineers, San Antonio, TX, USA, 9–11 October 2017; pp. 1–11.
22. Yu, H.; Ma, Y.; Wang, L.; Zhai, Y.; Wang, X. A landslide intelligent detection method based on CNN and RSG_R. In Proceedings of the Mechatronics and Automation (ICMA), Takamatsu, Japan, 6–9 August 2017; pp. 40–44.
23. Carneiro, G.; Nascimento, J.; Bradley, A.P. Unregistered multiview mammogram analysis with pre-trained deep learning models. In Proceedings of the Medical Image Computing and Computer-Assisted Intervention, Munich, Germany, 5–9 October 2015; pp. 652–660.
24. Schwarz, M.; Schulz, H.; Behnke, S. RGB-D object recognition and pose estimation based on pre-trained convolutional neural network features. In Proceedings of the Robotics and Automation, Seattle, WA, USA, 26–30 May 2015; pp. 1329–1335.
25. Li, N.; Hao, H.; Gu, Q.; Wang, D.; Hu, X. A transfer learning method for automatic identification of sandstone microscopic images. *Comput. Geosci.* **2017**, *103*, 111–121. [CrossRef]
26. Zhang, Y.; Li, M.; Han, S. Automatic identification and classification in lithology based on deep learning in rock images. *Acta Petrol. Sin.* **2018**, *34*, 333–342.
27. Chi, J.; Walia, E.; Babyn, P.; Wang, J.; Groot, G.; Eramian, M. Thyroid nodule classification in ultrasound images by fine-tuning deep convolutional neural network. *J. Digit. Imagin.* **2017**, *30*, 477–486. [CrossRef]
28. Szegedy, C.; Vanhoucke, V.; Ioffe, S.; Shlens, J.; Wojna, Z. Rethinking the inception architecture for computer vision. In Proceedings of the IEEE Conference on Computer Vision and Pattern Recognition, Seattle, WA, USA, 27–30 November 2017; pp. 2818–2826.
29. He, K.; Zhang, X.; Ren, S.; Sun, J. Deep residual learning for image recognition. In Proceedings of the IEEE Conference on Computer Vision and Pattern Recognition, Las Vegas, NV, USA, 26 June–1 July 2016; pp. 770–778.
30. Allen, J. Short term spectral analysis, synthesis, and modification by discrete Fourier transform. *IEEE Trans. Acoust. Speech Signal Process.* **1977**, *25*, 235–238. [CrossRef]
31. Gelfand, S.A. *Essentials of Audiology*, 2nd ed.; Thieme: New York, NY, USA, 2001.
32. Otsu, N. A threshold selection method from gray-level histograms. *IEEE Trans. Syst. Man. Cybern.* **1979**, *9*, 62–66. [CrossRef]
33. Li, M.; Han, S.; Shi, J. An enhanced ISODATA algorithm for recognizing multiple electric appliances from the aggregated power consumption dataset. *Energ. Build.* **2017**, *140*, 305–316. [CrossRef]
34. Li, M.; Miao, L.; Shi, J. Analyzing heating equipment's operations based on measured data. *Energ. Build* **2014**, *82*, 47–56. [CrossRef]
35. Hearst, M.A.; Dumais, S.T.; Osuna, E.; Platt, J.; Scholkopf, B. Support vector machines. *IEEE Intell. Syst. Their Applic.* **1998**, *13*, 18–28. [CrossRef]
36. Ballabio, C.; Sterlacchini, S. Support vector machines for landslide susceptibility mapping: The Staffora River Basin case study, Italy. *Math. Geosci.* **2012**, *44*, 47–70. [CrossRef]
37. Breiman, L. Random forests. *Mach. Learn.* **2001**, *45*, 5–32. [CrossRef]

applied sciences

MDPI

Article

Detecting the Void behind the Tunnel Lining by Impact-Echo Methods with Different Signal Analysis Approaches

Rongning Cao [1,2], Meng Ma [1,2,*], Ruihua Liang [1,2] and Chao Niu [3]

1 Key Laboratory of Urban Underground Engineering of Ministry of Education, Beijing Jiaotong University, Beijing 100044, China
2 School of Civil Engineering, Beijing Jiaotong University, Beijing 100044, China
3 CCCC Highway Consultants Co., Ltd., Beijing 100088, China
* Correspondence: mameng_02231250@163.com

Received: 4 July 2019; Accepted: 8 August 2019; Published: 10 August 2019

check for updates

Abstract: A void behind the lining in a tunnel is considered to be a critical condition as it can significantly impair the tunnel service life. In this study, we adopted the impact-echo (IE) method to detect the voids. We designed two test conditions (tunnel lining with and without a void) for our experiments performed in a laboratory environment. The influences of void size and impact-void position were analysed using numerical simulations. The vibration response signals were analysed in the time, frequency, and time–frequency domains using various signal analysis approaches. The results were comparatively analysed to determine the best approach for void detection. The study helped establish that a tunnel void can be evaluated through the vibration energy (amplitude and duration) in the time domain, the resonance frequency and dynamic stiffness in the frequency domain, and the energy distribution in time–frequency domain. The wavelet transform analysis is the most appropriate method to observe the energy flow during the state changing and the dynamic stiffness method can determine the void position precisely.

Keywords: non-destructive test; impact-echo method; tunnel void; experimental model; numerical simulation; signal analysis

1. Introduction

With the rapid development of railway and highway constructions, the number of tunnels increases, nevertheless, because of the complicated geological conditions and inappropriate constructions, the tunnel diseases such as the lining cracks, tunnel leakage, vault roof slab, voids behind the lining, etc. are also increasing. Particularly, a void located behind the lining in a tunnel may cause leakage and deformation of the surrounding rocks that can significantly impair the tunnel service life.

The non-destructive testing (NDT) methods commonly adopted for the investigation of a tunnel void includes ground-penetrating radar (GPR) [1,2], ultrasonic resilience method [3], infrared photographing technology of temperature field [4], etc. Similarly, impact-echo (IE) is one another widely used NDT method to detect the piles [5] and other engineering structures [6], and it is also employed to detect the voids behind a tunnel lining. This method is especially known for locating the voids in the very early stage of evaluation whereby the stress waves get reflected between the interfaces; accordingly, the IE method is based on the analysis of the vibration response signal. Ni et al. [7] assessed the quality of long drilled piles, such as the pile length, by analysing the test results in time domain and time–frequency domain. Cheng et al. [8] used the transfer function simulated by the IE method to analyse the bond condition of the concrete and substrate layers. The effect of the acoustic impedance and thickness of the substrate layer on the simulated transfer function was analysed in the frequency domain. Aggelis et al. [9] reported the effectiveness of grouting through a study that combined the time domain characteristics, spectral content, and wavelet transform. Song and Cho [10] analysed the response signals in the time–frequency domains through short-time Fourier transform (STFT) to evaluate the bonding state of tunnel shotcrete. They also studied the effects of the ground types, thickness of the shotcrete, undulating surfaces, and impact sources in both frequency and time–frequency domains [11]. Davis et al. [12] calculated the dynamic stiffness, mobility and damping, and peak/mean mobility ratio of the tested concrete elements to evaluate the tunnel lining grouting. Ryden et al. [13] evaluated the quality of backfill material in the segmental lining of a tunnel based on the IE amplification factor (Q-factor).

In a previous study, the IE method was used to study the grouting quality, the tunnel shotcrete bonding state, the material quality, etc. The response signals derived from IE method were analysed in the time, frequency, time–frequency domains using various methods such as Q-factor, dynamic stiffness, STFT, etc. The methods of analysis and the aspects of the response signal were selected based on the quality or bonding state of the material under study. Various analysis methods were used to analyse the properties of the data obtained by the IE method. They are usually compared in the situation of detecting the bonding state or the material quality. However, the tunnel structure that we studied in this article is in a complex loaded situation compared with the structure mentioned above such as the concrete slab. Accordingly, in order to study the specific aspects of the tunnel lining and to select an appropriate method of analysis to detect the void behind tunnel lining, it is necessary to apply various methods and analyse the test results. The properties and applicability of these methods should be compared.

In this paper, to compare different types of signal analysis method for detection of tunnel voids, an experimental study was first performed to compare two conditions, tunnel lining with and without voids. The responses were then analysed in the time, frequency and time–frequency domains using several methods that include fractal box-counting dimension for time history analysis, dynamic stiffness, STFT, wavelet analysis, and Hilbert Huang transform (HHT). Then, the influences of void size and impact-void position were calculated using numerical simulations. It was ensured that the best methods were used for numerical analysis.

2. Experimental Study

2.1. Experimental Outline

Impact echo method is based on the stress wave that is excited by an impact force. During the impact echo test, an impact force is applied on the surface of the structure by a hammer. The generated stress waves which include the body waves (P wave and S wave) and the surface wave (Rayleigh wave) propagate in the structure. Reflections and refractions happen at the interface of two media with different wave impedance. The reflectional signals are then collected by the sensor adjacent to the impact point. The P wave propagates faster than other types of waves, and if the transducer is placed close to the impact point, the response is dominated by P-wave echoes [14]. The signals are recorded and restored by the data acquisition instrument and are analysed in the computer. The response signals are analysed in the time domain, the frequency domain and the time–frequency domain. Finally, the property of the structure, such as the location of void and the thickness of the structure [6], can be detected from the patterns of waveform versus time and spectra versus frequency. The test flew is shown in Figure 1.

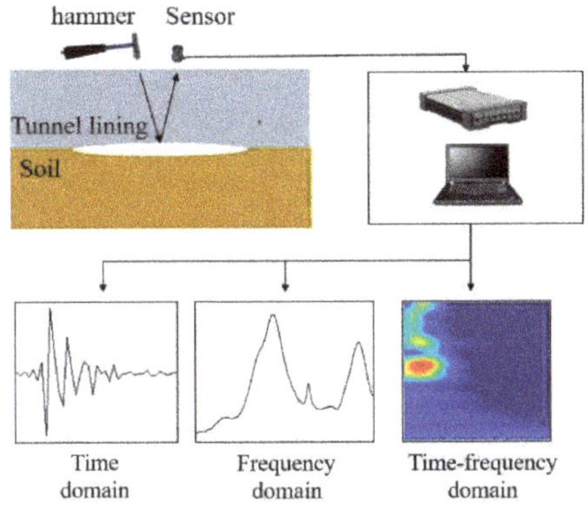

Figure 1. Test flow chart.

When the IE method is used to detect the thickness of a slab structure, the excited P wave reflects in the slab structure repetitively, which forms a periodic modal related to the thickness of the slab. In turn, the resonance frequency corresponding to the thickness occurs. Therefore, the thickness of a slab structure can be detected by the thickness frequency according to the IE method. When there is a void in the slab, the excited frequency needs to be higher than the thickness frequency. For example, if the velocity of the P wave is 4000 m/s and the depth of the void is 0.2 m, the resonance frequency is calculated as about 10 kHz, which should be contained in the excited frequency range. Therefore, when a void is inside the slab structure, another resonance frequency corresponding to the void can be seen except for the resonance frequency related to the thickness. Then the void can be detected by the IE method by observing the resonance frequencies.

In the situation of detecting voids behind the tunnel lining, the resonance frequencies of the conditions with and without void will be the same, which are both related to the thickness of the tunnel lining. Thus, the void cannot be detected by the difference in the resonance frequency. However, a void can be recognised by IE method due to different responses caused by different reflective ratios. The reflective ratios are caused by the distinct difference in the wave impedance among tunnel lining, soil and air. The wave impedance Z is expressed as:

$$Z = \rho V_{\mathrm{P}}, \tag{1}$$

where ρ is density and V_{P} is P wave velocity. For the P waves propagate from structure A to structure B, the reflective ratio is expressed as:

$$R = \frac{A_{\mathrm{r}}}{A_{\mathrm{i}}} = \frac{Z_{\mathrm{B}} - Z_{\mathrm{A}}}{Z_{\mathrm{B}} + Z_{\mathrm{A}}}, \tag{2}$$

where A_{r} is the reflected amplitude and A_{i} is the incident amplitude. When Z_{A} is much larger than Z_{B}, the reflected ratio R is closed to -1, which means the amplitude of the reflected wave remains the same, while the phase of the reflected changes. This situation happens at the interface of tunnel lining and air, where almost all of the energy is reflected. When Z_{A} is approximately equal to Z_{B}, the reflected ratio R is closed to 0, which means most of the energy propagates through the interface. This situation is similar to the condition that the tunnel lining is tightly surrounded by the soil.

This section presents the design of a laboratory principle experiment (Figure 2), whereby different signal analysis approaches were then employed to compare the vibration responses for the conditions A and B.

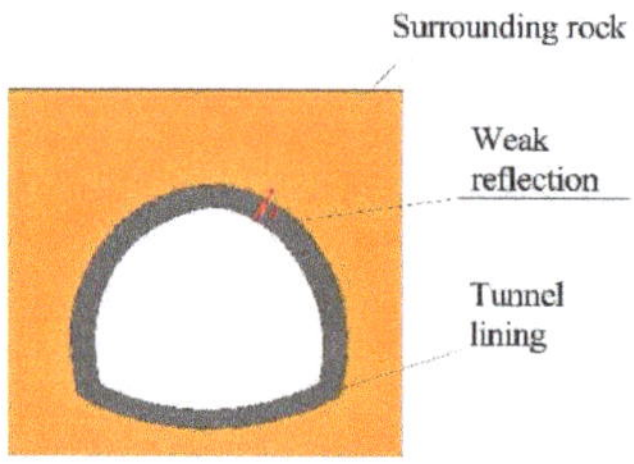

(a) Condition A

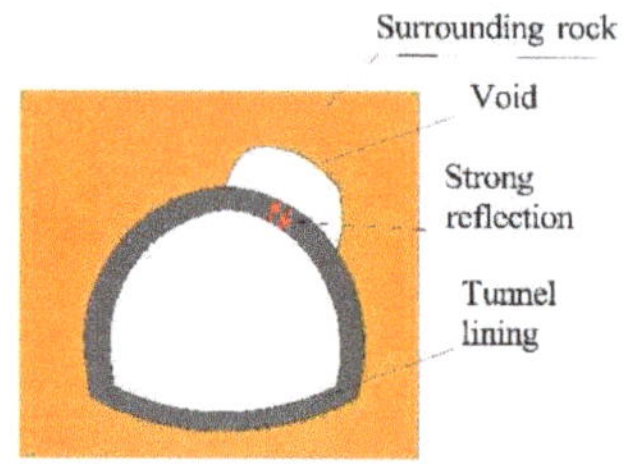

(b) Condition B

Figure 2. The basic principle of detecting the void behind tunnel lining by impact-echo (IE) method.

The experiment was carried out in a brick box of size 3.6 × 0.94 × 1.44 m^3 (Figure 3a). A rockwool material with a thickness of 0.1 m was placed inside the bunker to weaken the wave reflection at the boundaries. The soil was filled into the bunker. The horseshoe tunnel models were buried in the soil. The cross-sectional size of the tunnel model was 1:16 of a real railway tunnel and the thickness of the tunnel model was 0.025 m (Figure 3b). The tunnel models were made of polyvinyl chloride (PVC) material instead of the concrete. The material coefficients of the soil, the concrete, the PVC, and the air are shown in Table 1. The wave impedance of the PVC and the concrete is within one order of magnitude, which is much larger than the wave impedance of the air. This illustrates that the concrete tunnel can be instead by the PVC, a cheaper material, in this principle experiment. Two different conditions (A and B) were designed. Condition A referred to a tight contact between the tunnel lining and the surrounding soils. The reflected ratio R_A at the interface is −0.31. For Condition B, a void behind the tunnel wall was modelled using a pre-embedded empty box of 285 × 145 × 155 mm^3 placed between the lining and the soil. The reflected ratio R_B at the interface is −1.

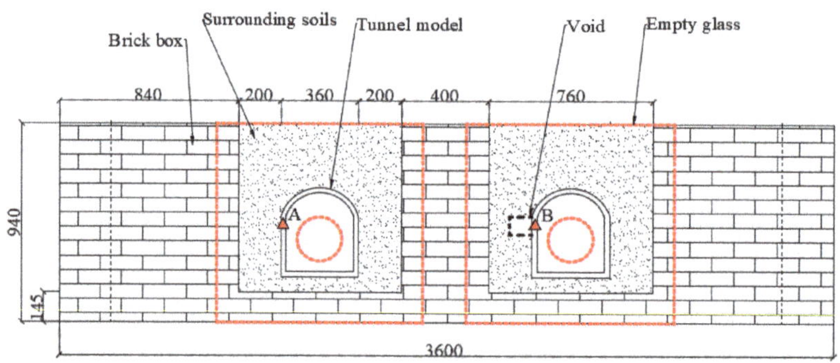

(a) Layout of the experiment table (unit: mm)

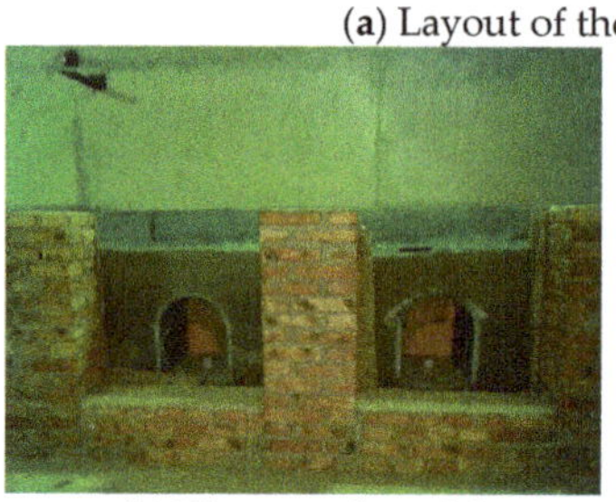

(b) The tunnel model

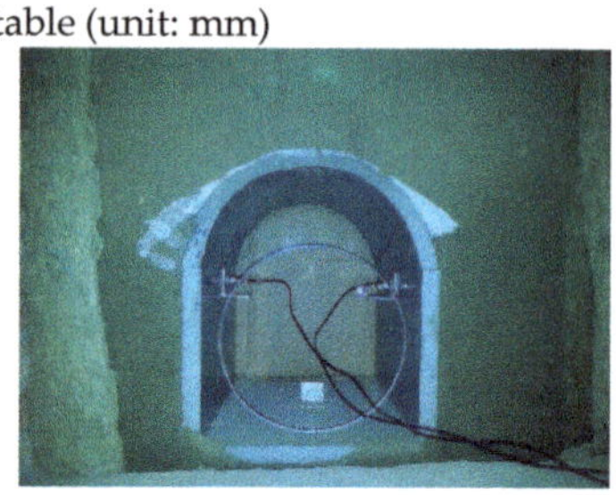

(c) Accelerometer arrangement

Figure 3. The test system.

Table 1. Material parameters used in the principle experiment.

Material	Density ρ/(kg/m^3)	P Wave Velocity V_p/(m/s)	Wave Impedance Z/(kg/m^{2}s)
Soil	2000	576	1.1×10^6
Concrete (C30)	2400	3691	8.8×10^6
PVC	1380	1500	2.1×10^6
Air	1.2	343	411.6

The test system consisted of a hand hammer, two unidirectional accelerometers, a data acquisition instrument with 16 channels, and a computer. The hand hammer, consisted of a hammer head, a force sensor and a hammer body, was used to apply an impact force on the tunnel model. The type of the hammer is Dytran Dytranpulse™5800B4T, developed by the Dytran Instruments company in California, USA. The head of the hammer is made by aluminium with the weight of 100 g and Young's modulus of 70 GPa. There is a special acceleration compensated piezoelectric force sensor in the hammer head at the striking face, ensuring a smooth frequency spectrum, which is used to measure the value of the impact force. The hammer has a sensitivity of 10 mV/lbf and a maximum force of 1000 lbf. The upper limit of the frequency range is 75 kHz. The type of unidirectional accelerometers is LC0104, developed by LANCE in Ohio, USA. They are used to collect the accelerations perpendicular to the tunnel lining, which has a sample frequency range of 0.5 Hz-9000 Hz, a sensitivity of 100 mV/g, a resolution of 0.0002 g, and an acceleration range of 50g. The type of data acquisition instrument is INV3018C, developed by the China Orient Institute of Noise & Vibration (COINV).

The testing point B was arranged at the tunnel lining behind the empty box on the right tunnel model, while testing point A was arranged at the same location on the left tunnel model, as shown in Figure 3a. According to the thickness and the P wave velocity of the tunnel lining in the test, the thickness resonance frequency is obtained approximately at 28,800 Hz. Since the thickness frequency of the two conditions is the same, this experiment of detecting the void behind tunnel lining does not intend to focus on the amplitude of resonance frequency, but to analyse the dynamic responses of these two conditions in a lower frequency range. The sampling frequency of the accelerometers is set at 2560 Hz. The impact points are 2 cm nearer to the testing points. For each condition, the impact was repeated 10 times. For each time, the peak of impact force is controlled to be close to 400 N, and the sampling frequency is 2560 Hz, as shown in Figure 4. It should be noted that the hammer should be controlled to be perpendicular to the surface of the tested object. To obtain this goal, the coherence function of the response and the force was calculated in each test. If the coherence function was close to 1, the impulse force would be qualified. Otherwise, the data would be removed.

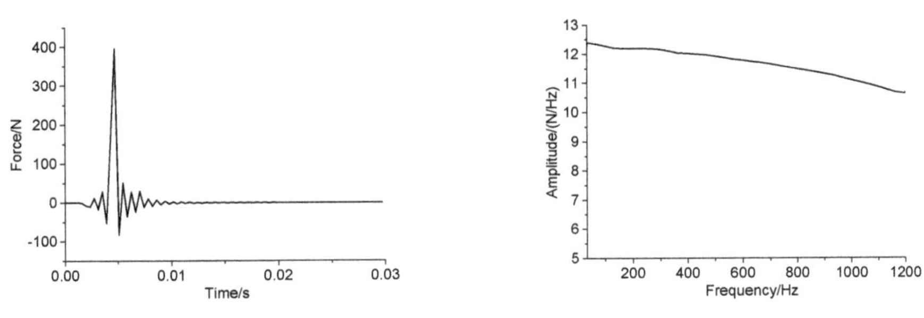

Figure 4. (**a**) Time history and (**b**) spectrum of the impact force.

2.2. Time Domain Analysis

Although the impact forces generated by the hand hammer are controlled within a range, the signals still need to be normalized in order to compare the responses of different conditions. Each impact force and its related response need to multiply by a coefficient C_i, which is obtained as:

$$C_i = \frac{400}{F_i^{\text{peak}}}, \tag{3}$$

where F_i^{peak} is the peak value of the i-th test. Figure 5 shows the time-domain waveforms of the two conditions, with an intercepted time duration of 0–50 ms. For Condition A, the waveform attenuates smoothly with the amplitude almost decaying to zero in 10 ms. For condition B, an obvious secondary peak occurs at 7 ms, while the minor peaks continuing until 15 ms. This phenomenon of a time-domain waveform with a stronger reflection energy and a prolonged response is suggestive of the existence of a void. It indicates that the stress waves reflect at the interface between the tunnel and the void, and the reflective ratio of concrete–air interface is higher than that of concrete–soil interface.

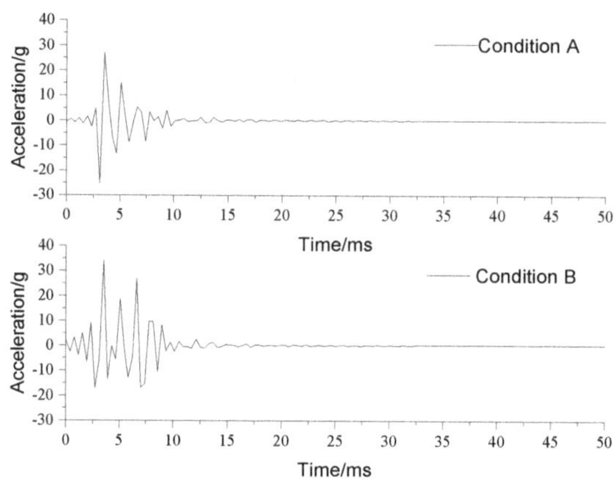

Figure 5. Time histories of the two conditions (A and B).

The vibration response can be analysed quantitatively using the fractal box-counting dimension method. The concept of fractal dimension was proposed by Mandelbrot [15]. The vibration signal is irregular but self-similar. It occupies a larger area than the one-dimensional line, smaller area than the two-dimensional plane. Its fractal dimension is measured by creating multiscale covers around the signal graph. Square boxes with side length (δ) are used to cover the signal waves on the time-amplitude plane. Then, the number of covered boxes (N) can be obtained (Figure 6). The fractal dimension is defined as:

$$D = -\lim_{\delta \to 0} \frac{\lg N}{\lg \delta}, \tag{4}$$

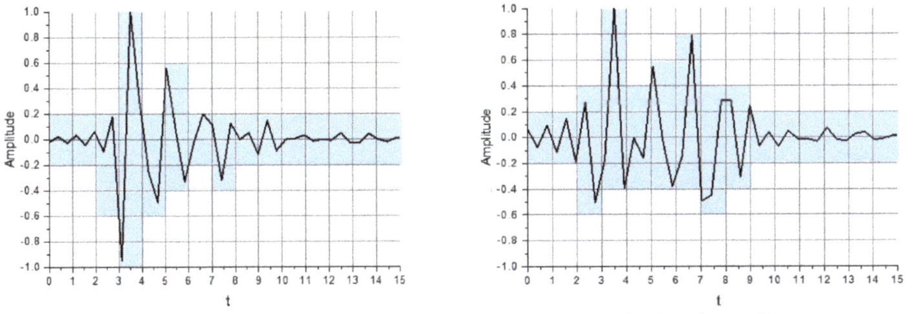

(**a**) Condition A (**b**) Condition B

Figure 6. Number of boxes with side length δ.

For the measured response signals in this paper, square boxes of a series of side lengths ($\delta_1, \delta_2, \cdots, \delta_n$) were used to cover the signal waves on the time-amplitude plane, and then a series of covered box quantities ($N_1, N_2, \cdots, N_n$) would be obtained. The minimum δ should be larger than the sampling interval Δt, because the signal wave was a straight line on every sampling interval. After calculating the fitting line of ($-\lg\delta - \lg N$), the fractal dimension would be obtained as the slope of the fitting line (Figure 7).

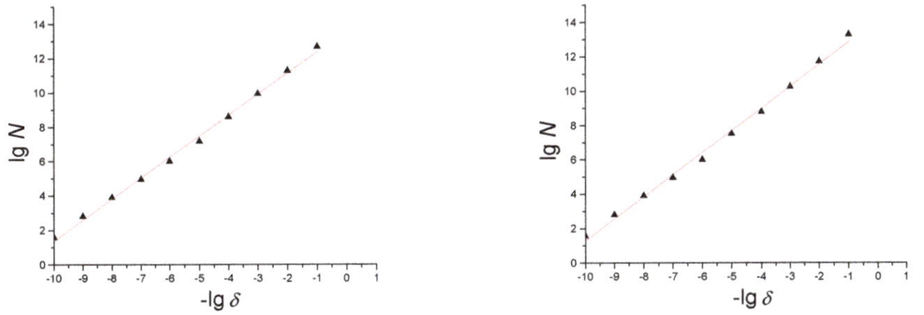

(**a**) Condition A (**b**) Condition B

Figure 7. The fitting lines of ($-\lg\delta - \lg N$).

The results show that the fractal dimensions of the conditions A and B are 1.221 and 1.288, respectively. This illustrates that when the waveforms of the two conditions are in the coordinate system, with the same time scale and the same amplitude scale, the waveform B is more complicated. Therefore, when the void exists, the fractal dimension is larger, and the waveform in the time domain is more complicated.

2.3. Frequency Domain Analysis

Figure 8 shows the frequency spectra of ten test runs conducted on each of the two conditions. In condition A, all the spectra are similar, and a dominant frequency of approximately 690 Hz is observed. The amplitudes increase steadily between 0 and 500 Hz. In condition B, larger differences among the tested spectra are observed. The dominant frequency remains the same (~690 Hz) as that of condition A. Besides, the amplitude fluctuates between 0 and 500 Hz with two additional peaks between 300 and 500 Hz. For each condition, the average frequency spectra of these test runs are calculated, except for the incoherence test runs (test 1 of condition A and test 4 for condition B), as shown in Figure 8c. The amplitude of condition B is higher than that of condition A in the frequency range of 800 Hz–1100 Hz. There are more dominant frequencies in condition B.

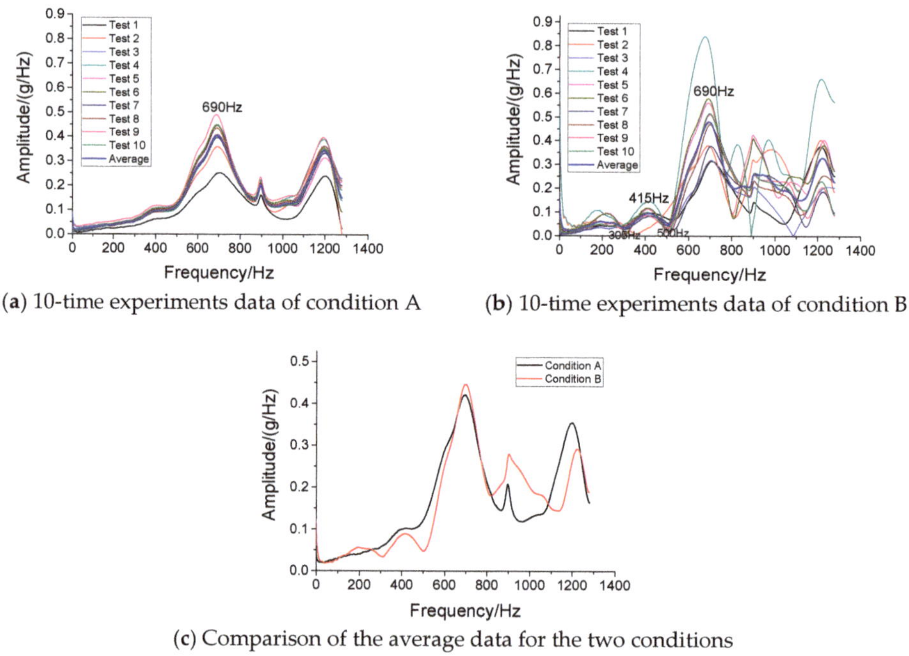

Figure 8. Fourier spectra for two conditions.

The dynamic stiffness was also used to study the influential differences between the two conditions. It expresses the resistance to deformation under the impact force [16]. In this test, dynamic stiffness reflects the supporting performance of the tunnel-soil structure. The impact force $F(t)$ and the obtained acceleration response $a(t)$ can be transformed into the frequency domain, using Fourier transform, as $F(f)$ and $a(f)$, respectively. The acceleration response is expressed as the superposition of a series of simple harmonic vibrations, so the amplitude of velocity can be represented as:

$$V(f) = \frac{a(f)}{2\pi f}, \tag{5}$$

The velocity admittance is defined as:

$$G_v = \frac{V(f)}{F(f)}, \tag{6}$$

Then, the dynamic stiffness can be calculated by:

$$K_d = \frac{2\pi f}{G_v} = \frac{2\pi f \cdot F(f)}{V(f)}, \tag{7}$$

Figure 9 compares the K_d of the two conditions. The K_d in condition B is lower than condition A. The reason is that the wave propagation in condition B is approximated as guided by a free-free boundary condition while the wave in condition A is guided by a free-constrained boundary condition. This phenomenon shows that the void existence can significantly reduce the dynamic stiffness in most frequencies, which means it reduces the ability of the tunnel structure to resist the load.

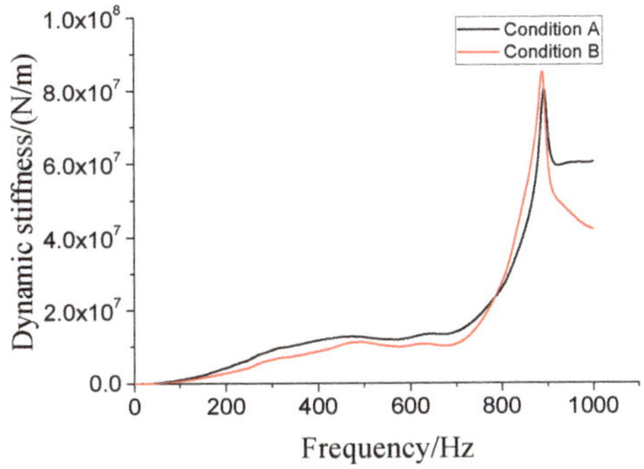

Figure 9. Dynamic stiffness analysis.

2.4. Time–Frequency Analysis

The frequency-domain analysis is widely used to research the frequency distribution of the signals. However, it is only suitable to the stationary process, not for the non-stationary process. In other words, the frequency domain analysis illustrates the frequency components of the signals but fails to figure out the moment that a certain frequency component appears.

The time–frequency analysis expresses the signal as a function of both time and frequency using the local transform. It helps understand changes in energy with both frequency and time upon the occurrence of a certain frequency. The methods usually adopted for time–frequency analysis are short-time Fourier transform, wavelet transform, Hilbert Huang transform, and so on.

2.4.1. Short-Time Fourier Transform (STFT)

The STFT is used to describe the frequency spectra of local signal sections changing over time. The original signal is first divided into several sections with equal time length by an appropriate sliding time window function $\eta(t)$. These sections are approximated as stationary processes. Then, by applying the Fourier transform from t to f to each section, the frequency spectra vary with time are obtained. The STFT can be expressed as [17]:

$$STFT_z(t,f) = \int_{-\infty}^{\infty} z(t')\eta^*(t'-t)e^{-j2\pi f t'}\,\mathrm{d}t', \tag{8}$$

where $z(t')$ is the original signal, $\eta(t)$ is the window function centered around zero.

The width of the time window should be properly selected. For example, if a wide time window is selected, the frequency resolution will increase, but the time resolution will decrease, which means these frequency components can only be determined to appear in this wide time band, not a certain instantaneous moment.

Figure 10 shows 3-dimensional (3D) graphs of the two conditions calculated by the STFT. For condition A, three energy peaks can be observed at approximately 700, 900, and 1200 Hz. For condition B, the number of energy peaks increases, and these peaks are all higher than those of condition A. Among them, the energy peak at approximately 700 Hz is the highest. On the low-frequency band between 0 and 100 Hz, the energy distributes widely along the time axis.

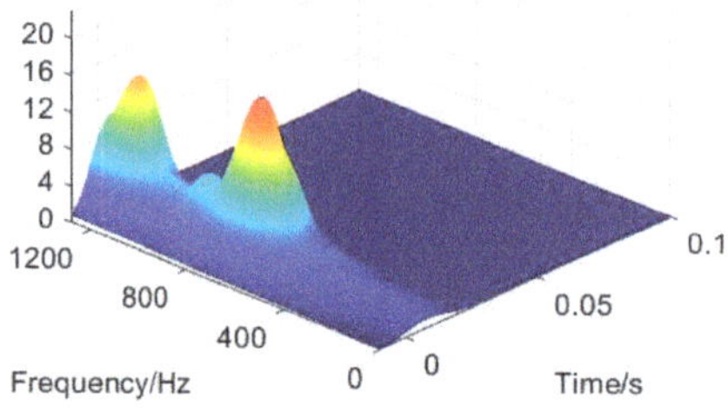

(a) Condition A

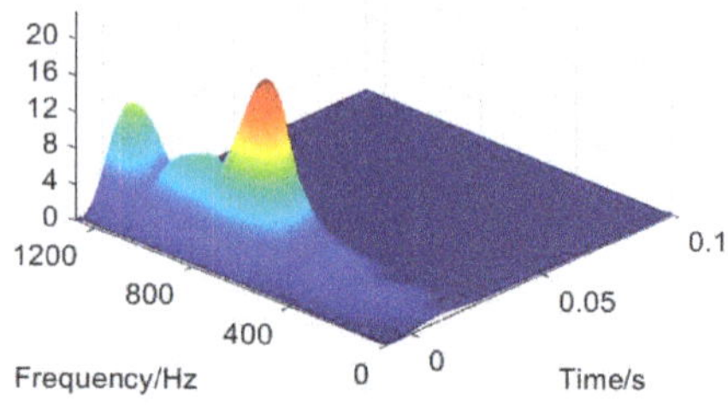

(b) Condition B

Figure 10. Time–frequency spectra by STFT.

In summary, when the void exits, the vibration energy becomes stronger, which is indicative of increased reflections of the stress waves from the tunnel-air interface. In addition, more energy moves to the frequency of approximately 700 Hz, which in turn becomes the dominant frequency. It can also be seen that the low-frequency energy attenuates slowly when the void exists.

2.4.2. Wavelet Transform

Wavelet transform uses the mother wavelet function to express the original signal, which can be expressed as:

$$\mathrm{WT}_z(a,b) = \frac{1}{\sqrt{a}}\int_{\mathrm{R}} z(t)\varphi(\frac{t-b}{a})\mathrm{d}t, \quad (9)$$

where $z(t)$ is the original signal. The $\varphi(\frac{t-b}{a})$ is the mother wavelet function. The a and b are the scaling and translating parameters, respectively. The scaling parameter represents the dilation or compression of the mother wavelet, and the translating parameter determines the mother wavelet moving along the time domain. The dilative mother wavelet corresponds to low-frequency resolution. The compressive mother wavelet corresponds to high-frequency resolution [18]. By calculating the cross-correlation of the original signal $z(t)$ and the mother wavelet function under different scales, the frequency components are obtained. The moment that a certain frequency appears is obtained by moving the mother wavelet along the time domain.

For the STFT, when the width of the time window is determined, the resolution of time and frequency on the whole time–frequency range is constant. However, in signal analysing, the high-frequency resolution is needed on the low-frequency range, while the high-time resolution is needed on the high-frequency range. On this aspect, the wavelet transform meets the requirements more closely.

Figure 11 shows the contour maps of the two conditions analysed by Cmor wavelet transform. Except for the characters that are shown in Section 2.4.1 by STFT, new phenomena are found in Figure 11. When the void exists, the frequency components between 1000 Hz and 1200 Hz appears from 0 s to 0.005 s, which is the period of the primary peak in Figure 5. In addition, the frequency components above 1200 Hz occurs from 0.005 s to 0.01 s, which is the period of the secondary peak in Figure 5.

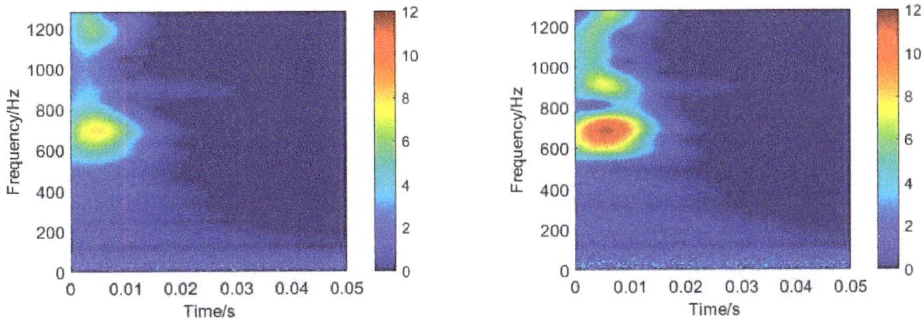

(a) Condition A (b) Condition B

Figure 11. Time–frequency spectra by wavelet transform.

For the STFT, it could only be roughly determined when the frequency components appear because of the low time resolution. However, for the wavelet transform, the moment that a certain frequency component appears could be clearly figured out. This means that the wavelet transform can tell more detail about the characters.

2.4.3. Hilbert Huang Transform (HHT)

The signal can also be presented in the time–frequency domain using the Hilbert Huang Transform (HHT). The HHT decomposes a signal into intrinsic mode functions (IMFs) along with a trend, and then obtains instantaneous frequency data. Compared to other transforms, HHT is more like an algorithm (an empirical approach) that can be applied to a data set, rather than a theoretical tool. The calculation of the HHT includes two steps: empirical mode decomposition (EMD) and Hilbert transform. By means of EMD, the acceleration data can be decomposed into different simple non-sinusoidal signals, called IMFs [19]. Each IMF is transformed into a time–frequency domain by Hilbert transform, and then the time–frequency spectrum can be obtained after the superposition.

Figure 12 shows the time–frequency spectra of the signals analysed by HHT. The original signals were separated into several IMFs, and the first step of IMF was in the range of 600–1200 Hz. According to the previous analysis, the study would focus on the first-step IMF (600–1200 Hz). For condition A, the vibration energy peak is at 700 Hz and 5 ms, and the energy attenuates significantly after 5 ms, almost approaching zero at 10 ms. For condition B, there are two energy peaks, of which one appears at 1100 Hz and 3.5 ms, while the other one is at 700 Hz and 6 ms. The energy remains high until 10 ms.

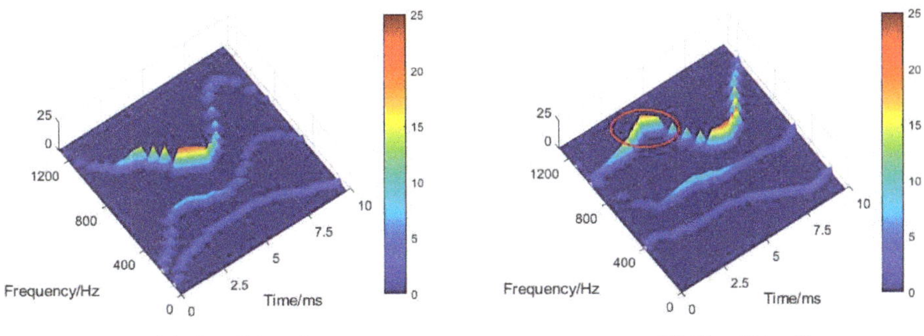

(a) Condition A (b) Condition B

Figure 12. The time–frequency spectrum.

In summary, when the void exists, another vibration energy peak of 1100 Hz appears at 3.5 ms that is indicative of the reflection of the stress waves at the tunnel-air interface. This phenomenon also explains that the stress waves take longer time to propagate to the soil, which causes the energy peak of 700 Hz to appear 1 ms later in condition B. In addition, the vibration lasts longer when a void exists.

By means of HHT, a signal was separated into several sub-signals dominant in different frequency ranges. As the key frequency band for the sub-signals could be studied individually, the stress wave propagation and the frequency component distribution could be seen clearly, however, it is less holistic because of the separation of the original signal.

The STFT and wavelet transform analysis to focus on the overall vibration energy in a time–frequency plane that could be observed clearly. Similarly, the energy flow between different conditions could also be studied clearly. In principle, unlike the STFT, the wavelet transform can change the scale while analysing different frequencies of the signal. As a result, both high-time and high-frequency resolution can be achieved using wavelet a transform. Based on these aspects, the wavelet analysis method would be chosen for the following numerical study to analyse the signals in time–frequency domains.

3. Numerical Study

The laboratory experiment compared the vibration responses of the two conditions (with and without voids) using IE method. The approach proved to be more satisfactory in the experimental study was employed in the signal analysis using the numerical model, whereby the size and the position of the void were considered.

3.1. Numerical Model

A finite element (FE) model considering the soil, tunnel, and void on the soil-structure interface was built in Midas GTS, shown as Figure 13. The material parameters are shown in Table 2. The coordinate system based on the right-hand rule was shown, and the dimensions of the model were 84 × 84 × 40 m^3. The external diameter of the tunnel was 6 m, and the thickness of the tunnel lining was 0.3 m. In order to simulate the wave shape, the size of the mesh needs to be divided smaller than 1/8 of the smallest wavelength, which is expressed as:

$$\Delta l \leq \frac{V_s}{8 f_{\max}}, \tag{10}$$

where V_s is the shear wave velocity and f_{max} is the highest analysis frequency, which is 150 Hz in this model. According to the equation, the element of soil should be smaller than 0.25 m and the element of tunnel lining should be smaller than 1.88 m. In this model, the tunnel elements were set at 0.3 m. The soil elements near to the top of the tunnel were set as 0.2 m, and the size was getting bigger as the elements spread from the centre to the boundary of the model. The viscoelastic artificial boundaries were applied to the model to simulate the infinite soil. The viscoelastic artificial boundary is a continuous distributed parallel spring-damping system with calculated stiffness and damping factors. Once these coefficients are input into the Midas GTS software, the equivalent spring-damping elements will be applied on the boundary nodes.

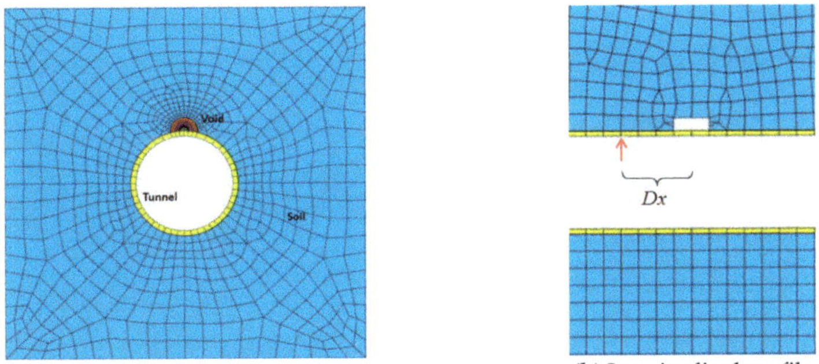

(**a**) Cross section (**b**) Longitudinal profile

Figure 13. Tunnel-soil FE model.

Table 2. Material parameters used in the numerical models.

Material	Young's Modulus E/(kPa)	Poisson's Ratio v	Density ρ/(kg/m^3)	Compressive Wave Velocity V_p/(m/s)	Shear Wave Velocity V_s/(m/s)
Soil	475,000	0.32	2000	576	297
Tunnel lining	30,000,000	0.20	2400	3691	2260

In the experimental study, the effect of void size on the response was not considered. In contrast, for the numerical study, four conditions with varying void sizes were designed as listed in Table 3. The cross-sections of the voids were fan-shaped with different radius, and their longitudinal length was kept the same as 4 m. An impact force signal was applied on the inner surface of the tunnel lining in the model. The duration of the force was 0.0032 s, and the peak value was 88.26 kN. The frequency range influencing the impact force was between 1 and 150 Hz (Figure 14). The vibration detecting point was 0.1 m close to the impact point. To analyse the effect of the impact-void position, the impact and response points were varied with reference to the centre of void along the outer surface of the tunnel lining. The distance between the impact point and the centre of the void, represented by Dx (Figure 13b), was varied from 1 to 6 m.

Table 3. Conditions with different void sizes.

Conditions	Void Size/m^3
Condition S0	0 (without void)
Condition S1	0.39
Condition S2	3.14
Condition S3	7.07

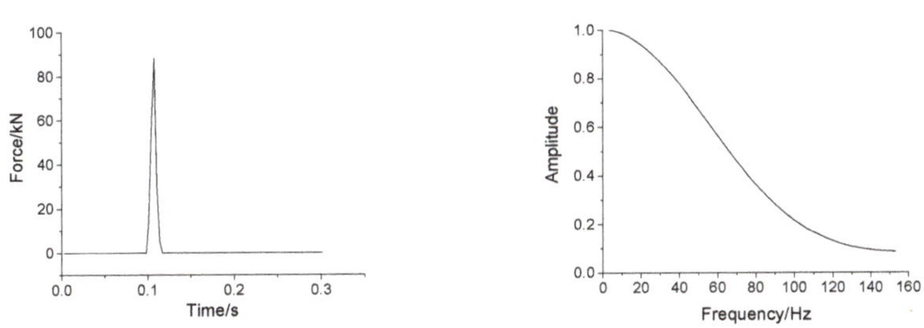

(**a**) (**b**)

Figure 14. (**a**) Time history and (**b**) spectrum of the impact force.

3.2. Effect of Void Size

Figure 15 illustrates the time histories of the four void size conditions. For condition S0, the acceleration is relatively smaller with a peak value of 6.49 m/s^2. For condition S1, the acceleration is larger with a peak value of 7.02 m/s^2. This is because, part of the vibration energy is absorbed by the soil beside the void, and part of the energy reflects at the tunnel-void interface. For conditions S2 and S3, with the increase of the void size, the peak value is higher, which means more energy reflects at the interface rather than transmitting into the soil. In short, as the void is larger, the peak value of the waveform is higher, which means more energy is reflected back.

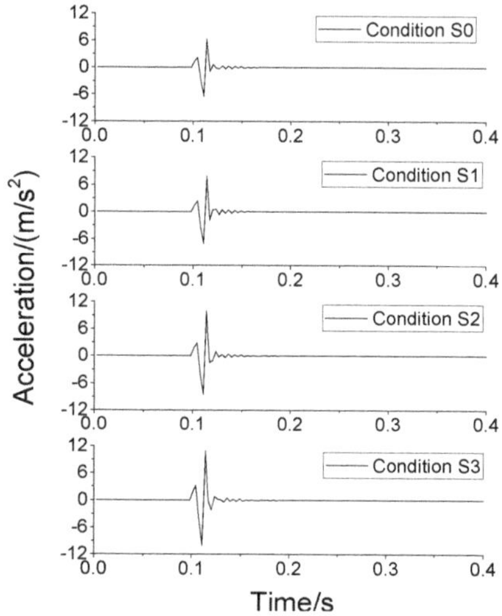

Figure 15. Time histories of void conditions.

Figure 16 shows normalised Fourier spectra of the four void size conditions. For condition S0, the vibration amplitude distributes in a wide frequency range, and the dominant frequency is not obvious. The normalised peak amplitude is approximately 0.50. With the increase of void size, the amplitude increases, and the dominant frequency can be clearly observed. In general, as the void is bigger, the frequency distributes a narrower range, and the dominant frequency at 90 Hz is more obvious.

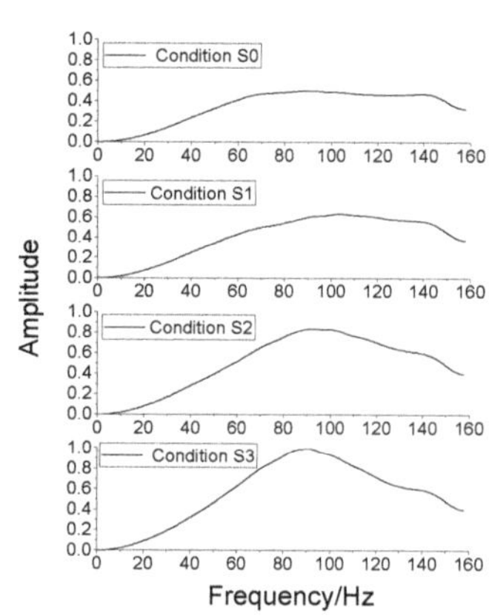

Figure 16. Normalised Fourier spectra of void conditions.

Figure 17 illustrates that the dynamic stiffness varies with frequency for the four conditions. With the increase of void size, the dynamic stiffness decreases in almost all the frequencies.

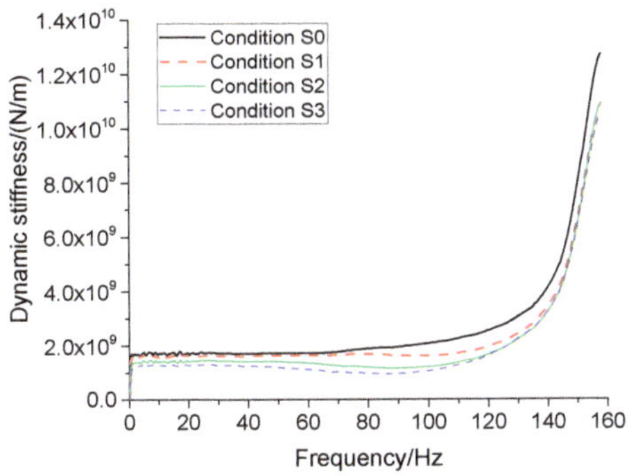

Figure 17. Dynamic stiffness analysis.

Figure 18 shows the colour contours of the wavelet spectra for the four void size conditions. With the increase of the void size, the spectrum becomes larger, and an increasing energy concentration can be obviously found. More energy tends to concentrate over 80–120 Hz, which in turn becomes the dominant frequency band. The vibration duration is longer especially in the range of 60–120 Hz. While the frequency is lower, the vibration energy lasts longer.

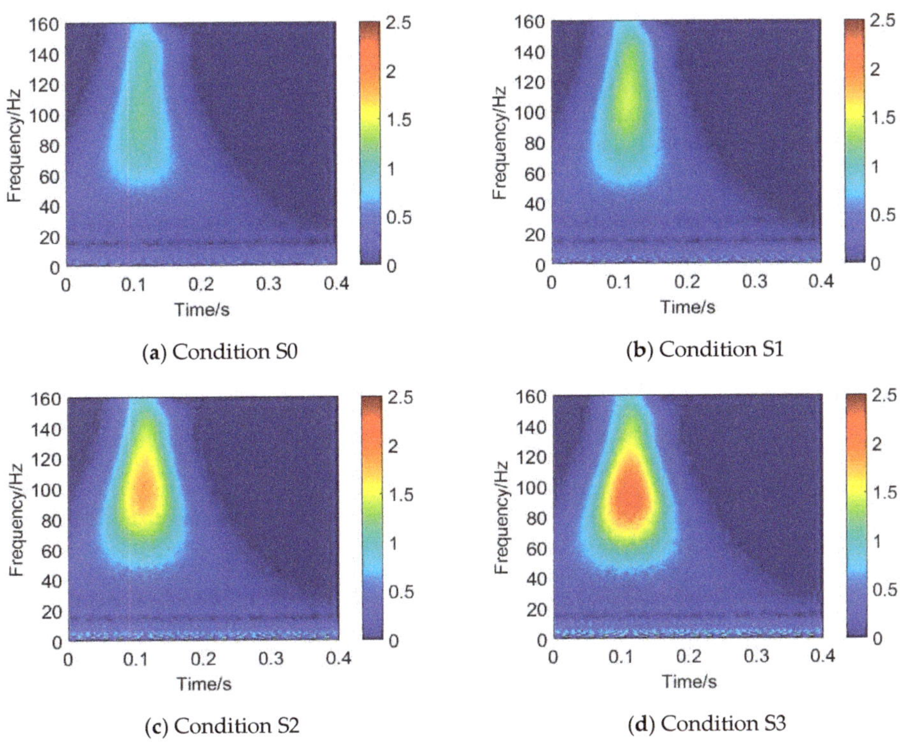

(**a**) Condition S0 (**b**) Condition S1

(**c**) Condition S2 (**d**) Condition S3

Figure 18. Wavelet spectra for different void size conditions.

When the impulse force is applied to the tunnel, the P wave propagates into the structure as a semi-spherical wave. Figure 19 shows a sectional draw of the reflection of the stress wave with different void sizes. According to the wave impendence introduced in Section 2.1, there is a strong reflection (red part) in the tunnel-void interface and a weak reflection (blue part) in the tunnel-soil interface. While the void becomes larger, the angle of the red part is larger, which means more energy reflects back and the attenuation lasts longer. This may cause the spectrum becomes larger in proportion to the void size.

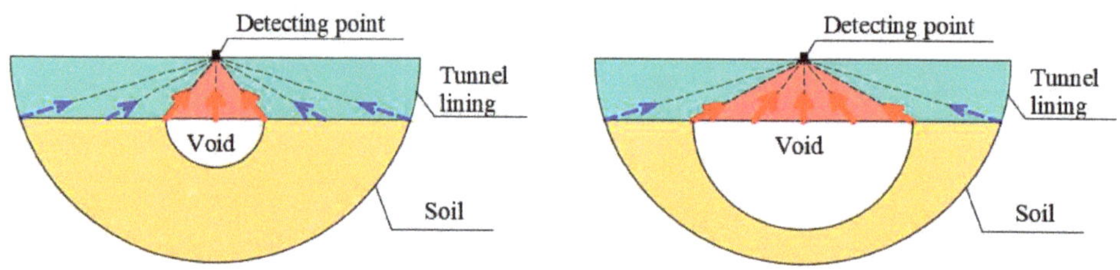

Figure 19. The reflections of P wave in the conditions with (**a**) small void and (**b**) big void.

3.3. *Effect of Distance between Impact and Void Position*

When the impact location is far from the void, the void existence is not distinctly recognised. Accordingly, in this subsection, different impact positions were analysed. The distance between the impact location and the void centre was defined as Dx, which was varied from 0 to 6 m for the different void size conditions. Figure 20 shows the dynamic stiffness K_d of the responses. The average values of the dynamic stiffness were calculated in the range of 20–100 Hz due to the steady amplitudes.

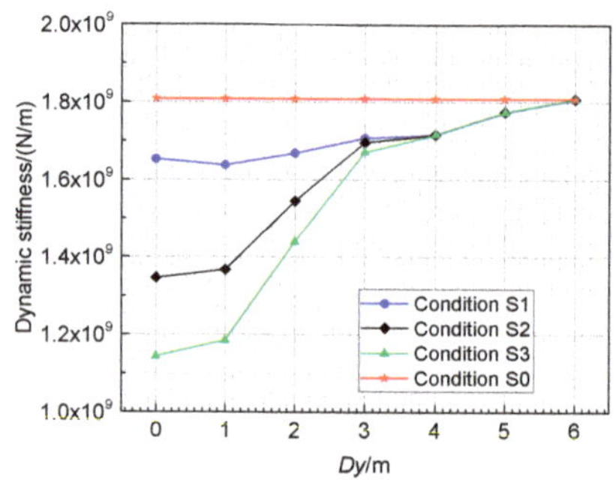

Figure 20. Dynamic stiffness of different void positions.

For condition S0, K_d is not affected by Dx. When Dx changes from 0 to 1 m, in the range of void size, K_d for condition S1 decreases slightly, while for conditions S2 and S3 increase slightly.

For conditions S1, S2, and S3, when Dx changes from 1 to 3 m, K_d increases significantly. That is, the dynamic stiffness changes quickly when the impact point moves through the void edge. When Dx changes from 3 to 6 m, K_d increases slowly. The dynamic stiffness for void conditions becomes almost the same as that of the non-void condition when Dx reaches to 6 m.

4. Conclusions

To investigate the effect of a void behind a tunnel lining using the IE method, both experimental and numerical studies were performed. The different signal analysis approaches were compared, and the influence of the void size and impact-void position were analysed. The following conclusions can be drawn:

- In the analysis of the vibration response in the time, frequency, and time–frequency domains, the fractal box-counting dimensions and dynamic stiffness can be used for void recognition quantitatively, and the wavelet analysis can provide a relatively better visual energy distribution among the different time–frequency analysis approaches.
- When the void is larger, the dynamic stiffness decreases. When the impact location is beyond the range of void size, the dynamic stiffness changes quickly.

By comparison of different signal analysis approaches, wavelet transform is believed to be the most appropriate one to show the energy flow during the void state changing. Besides, the dynamic stiffness could be used to detect the void position accurately.

Author Contributions: Data curation, R.C. and R.L.; formal analysis, R.C. and C.N.; funding acquisition, M.M.; investigation, R.L. and C.N.; methodology, M.M.; project administration, M.M.; software, R.C.; supervision, M.M.; writing—original draft, R.C.; writing—review & editing, R.C. and M.M.

Funding: This research was funded by fundamental research funds for the central universities, grant number 2018JBM036 and the research funding of China academy of railway sciences, grant number 2018YJ164.

Conflicts of Interest: The authors declare no conflict of interest. The funders played role in the design of the study, the collection and the decision to publish the results.

References

1. Lalagüe, A.; Lebens, M.A.; Hoff, I.; Grøv, E. Detection of Rockfall on a Tunnel Concrete Lining with Ground-Penetrating Radar (GPR). *Rock Mech. Rock Eng.* **2016**, *49*, 2811–2823. [CrossRef]
2. Cardarelli, E.; Marrone, C.; Orlando, L. Evaluation of tunnel stability using integrated geophysical methods. *J. Appl. Geophys.* **2003**, *52*, 93–102. [CrossRef]
3. Huang, J.; Qi, H.; Diao, Z. Application of ultrasonic-rebound combined method in concrete strength test for tunnel secondary lining. *Water Resour. Hydropower Northeast* **2011**, *23*. [CrossRef]
4. Wang, W.; Xiang, Y. Non-destructive testing and its application in tunnels. *J. Railw. Eng. Soc.* **2001**, *3*, 85–88.
5. Sansalone, M.; Carino, N. Detecting Voids in Metal Tendon Ducts using the Impact-Echo Method. *Mater. J. Am. Concr. Inst.* **1992**, *89*, 8.
6. Sansalone, M.J.; Streett, W.B. *Impact-Echo: Nondestructive Evaluation of Concrete and Masonry*; Bullbrier Press: Jersey Shore, PA, USA, 1997; ISBN 0961261064.
7. Ni, S.-H.; Lehmann, L.; Charng, J.-J.; Lo, K.-F. Low-strain integrity testing of drilled piles with high slenderness ratio. *Comput. Geotech.* **2006**, *33*, 283–293. [CrossRef]
8. Cheng, C.-C.; Yu, C.-P.; Liou, T. Evaluation of interfacial bond condition between concrete plate-like structure and substrate using the simulated transfer function derived by IE. *NDT E Int.* **2009**, *42*, 678–689. [CrossRef]
9. Aggelis, D.; Shiotani, T.; Kasai, K. Evaluation of grouting in tunnel lining using impact-echo. *Tunn. Undergr. Space Technol.* **2008**, *23*, 629–637. [CrossRef]
10. Song, K.-I.; Cho, G.-C. Bonding state evaluation of tunnel shotcrete applied onto hard rocks using the impact-echo method. *NDT E Int.* **2009**, *42*, 487–500. [CrossRef]
11. Song, K.-I.; Cho, G.-C. Numerical study on the evaluation of tunnel shotcrete using the Impact-Echo method coupled with Fourier transform and short-time Fourier transform. *Int. J. Rock Mech. Min. Sci.* **2010**, *47*, 1274–1288. [CrossRef]
12. Davis, A.G.; Lim, M.K.; Petersen, C.G. Rapid and economical evaluation of concrete tunnel linings with impulse response and impulse radar non-destructive methods. *NDT E Int.* **2005**, *38*, 181–186. [CrossRef]
13. Ryden, N.; Aurell, O.; Nilsson, P.; Hartlén, J. Impact echo Q-factor measurements towards non-destructive quality control of the backfill in segmental lined tunnels. In *Nondestructive Testing of Materials and Structures*; Springer: Berlin/Heidelberg, Germany, 2013; pp. 915–919. [CrossRef]
14. Sansalone, M.; Carino, N. *Impact-Echo: A Method for Flaw Detection in Concrete Using Transient Stress Waves*; National Bureau of Standards: Gaithersburg, MD, USA, 1986. [CrossRef]
15. Mandelbrot, B.B. *The Fractal Geometry of Nature*; WH Freeman: New York, NY, USA, 1983; Volume 173, ISBN 0716711869.

16. Ma, M.; Liu, J.; Ke, Z.; Gao, Y. Bearing Capacity Estimation of Bridge Piles Using the Impulse Transient Response Method. *Shock Vib.* **2016**, *2016*, 1–8. [CrossRef]
17. Ge, Z.; Chen, Z. *Matlab Guide to Time-Frequency Analysis*; Posts & Telecom Press: Beijing, China, 2006; ISBN 9787115141132.
18. Lee, I.-M.; Han, S.-I.; Kim, H.-J.; Yu, J.-D.; Min, B.-K.; Lee, J.-S. Evaluation of rock bolt integrity using Fourier and wavelet transforms. *Tunn. Undergr. Space Technol.* **2012**, *28*, 304–314. [CrossRef]
19. Shi, Z.; Liu, L.; Peng, M.; Liu, C.; Tao, F.; Liu, C. Non-destructive testing of full-length bonded rock bolts based on HHT signal analysis. *J. Appl. Geophys.* **2018**, *151*, 47–65. [CrossRef]

applied sciences

MDPI

Article

A PEC Thrice Subtraction Method for Obtaining Permeability Invariance Feature in Conductivity Measurement of Ferromagnetic Samples

Dongdong Wen [1,*], Mengbao Fan [1,*], Binghua Cao [2], Zhian Xue [1] and Ping Wang [3,4]

1 School of Mechatronic Engineering, China University of Mining and Technology, Xuzhou 221116, China
2 School of Information and Control Engineering, China University of Mining and Technology, Xuzhou 221116, China
3 School of Automation Engineering, Nanjing University of Aeronautics and Astronautics, Nanjing 210016, China
4 Nondestructive Detection and Monitoring Technology for High Speed Transportation Facilities, Key Laboratory of Ministry of Industry and Information, Nanjing University of Aeronautics and Astronautics, Nanjing 210016, China
* Correspondence: wdd@cumt.edu.cn (D.W.); wuzhi3495@cumt.edu.cn (M.F.); Tel.: +86-0516-83885829 (M.F.)

Received: 17 May 2019; Accepted: 5 July 2019; Published: 7 July 2019

check for updates

Featured Application: The work is potential to obtain a permeability invariance (PI) feature in pulsed eddy current (PEC) signals for reducing the permeability and environment magnetic field influence in the conductivity measurement of ferromagnetic samples.

Abstract: Conductivity, as an important index of structural health monitoring, can be used to evaluate heat treatment condition, and sort different materials or measure the stress of mechanical parts. However, the permeability of a measured sample has significant impact on the detected signal in pulsed eddy current (PEC) testing, which is prone to measurement errors due to the effect of permeability change. In this paper, a thrice subtraction method is investigated and utilized to obtain a permeability invariance (PI) feature for reducing permeability effect in conductivity measurement of ferromagnetic samples. The thrice subtraction method is based on the PEC signals of sample and air, the difference signal between the difference PEC signal and its normalization signal, and the difference signal between the difference normalization signal and its standard deviation. In the thrice subtraction signals, the behavior of the obtained PI feature is analyzed by experiments and simulations. The results demonstrate that the thrice subtraction method is a practicable program and the PI feature is potential to measure the conductivity of ferromagnetic samples. The work reported in this paper provides an effective approach to obtain a PI feature for estimating the conductivity of ferromagnetic samples without a permeability effect.

Keywords: pulsed eddy current; ferromagnetic sample; conductivity; permeability

1. Introduction

Conductivity, as a parameter of ferromagnetic materials, can serve as an index to sort different materials, and evaluate heat treatment condition or measure the stress of mechanical parts [1–3]. Due to the advances in electromagnetic nondestructive evaluation, pulsed eddy current (PEC) testing has become a feasible and preferred technique for measuring conductivity of materials in recent years [4,5]. However, for ferromagnetic samples, the measurement of conductivity is affected by the inhomogeneity of tested samples and the confounding cross-sensitivity to conductivity and permeability, which is undesirable in precision PEC testing [6,7]. Especially, the influence of permeability is complex and irregular. On the other hand, correspondingly, the permeability of ferromagnetic material will change when the environmental magnetic field changes. Therefore, reducing the permeability effect or the influence of environmental magnetic field in the measurement of electrical conductivity of ferromagnetic samples is an important problem to be solved in PEC testing. It is important and meaningful to improve the detection accuracy of conductivity of ferromagnetic samples.

In PEC testing, the measurements of conductivity and permeability have been studied by many scholars in recent years. Adewale et al. [8] decoupled the influence of permeability and conductivity in PEC measurements and the contributions of the electromagnetic properties (permeability and conductivity) were analyzed with a view to isolate the influence of these two properties. It was found that the conductivity effects are prominent in the rising edge of the transient response and the permeability effects are prominent in the stable phase of the transient response. Chen et al. [4] studied the electrical conductivity measurement of ferromagnetic metallic materials using the PEC technique. In the measurement of conductivity, an inverse problem method based on a PEC analytical model was proposed to determine the conductivity and permeability of ferromagnetic plates. Dziczkowski et al. [9] described the method of determining the equivalent parameters of the coil, a scaling method for a conductivity meter based on the elimination of liftoff on conductivity results, and a proposition for an effective method to measure the conductivity of rough elements. Lu et al. [1,10] found a conductivity lift-off invariance phenomenon and proposed an inverse problem method to determinate the conductivity and permeability of ferrite metallic plates. Wang et al. [2] proposed an eddy current measurement method to determinate electrical conductivity of samples, which indicated that the logarithm of phase signature of impedance change basically varies linearly with electrical conductivity. Yu et al. [11] observed a conductivity invariance phenomenon of eddy current non-destructive testing (NDT), and used it to estimate the magnetic permeability of metals without the influence of its conductivity. Ye et al. [12] proposed a novel approach based on decay time (DT) using PEC and a highly sensitive magnetoresistive (MR) sensor in conductivity measurement of materials. The results indicate that the DT parameter is linearly correlated with material conductivity allowing it to be used for estimating the conductivity from the PEC measurement. According to the above studies, it is found that the conductivity measurements of ferromagnetic samples with a permeability effect are rarely studied by using PEC testing when permeability changes. These works also revealed that the change of PEC signals caused by conductivity and permeability is indistinguishable in the measurement of ferromagnetic samples. In other words, when the conductivity is measured, the influence of permeability can not be reduced or eliminated in PEC signals. Thus, it is important and significant to reduce the influence of permeability on PEC signals, and accurately measure the conductivity of ferromagnetic samples when permeability changes.

It is known that liftoff point of intersection (LOI), as a desirable signal feature, is independent of liftoff distance [13]. The LOI is a common intersection point of PEC signals when liftoff distance changes. For the permeability effect on PEC signals, a similar method can be considered in the conductivity measurement to reduce the influence of permeability. Thus, similarly, a common intersection invariance point feature is needed in PEC signals of ferromagnetic samples for reducing the permeability effect when sample permeability changes or the environmental magnetic field changes. Therefore, in the measurement of conductivity, it is important and significant to propose a method to obtain a permeability invariance (PI) feature for reducing permeability effect or environmental magnetic field influence.

The main contribution of this paper is that a thrice subtraction method is proposed to obtain a PI point feature in PEC signals of ferromagnetic samples for reducing the permeability effect or environmental magnetic field influence. Correspondingly, the availability of the thrice subtraction method and the obtained PI feature is analyzed and discussed. The rest of this paper is organized as follows: in Section 2 the PEC analytical model is formulated from a harmonic eddy current model, the thrice subtraction method is introduced in Section 3, the simulation results are analyzed in Section 4, in Section 5 the experiments are performed for verifying the simulation results and finally, Section 6 presents the concluding remarks.

2. Formulation of the PEC Analytical Model

The PEC analytical model, as an effective tool, has been used to analyze the change of PEC signals caused by thickness, conductivity and permeability [4,14]. According to previous studies, it can be found that it is very difficult to directly construct a PEC analytical model. However, on the other hand, it is simpler to construct a PEC analytical model based on the harmonic eddy current analytical model. It is known that a periodic pulse signal can be expressed as a Fourier series of trigonometric functions. Therefore, in this paper, the PEC analytical model is reconstructed through using the harmonic eddy current analytical model. The main process is shown in Figure 1.

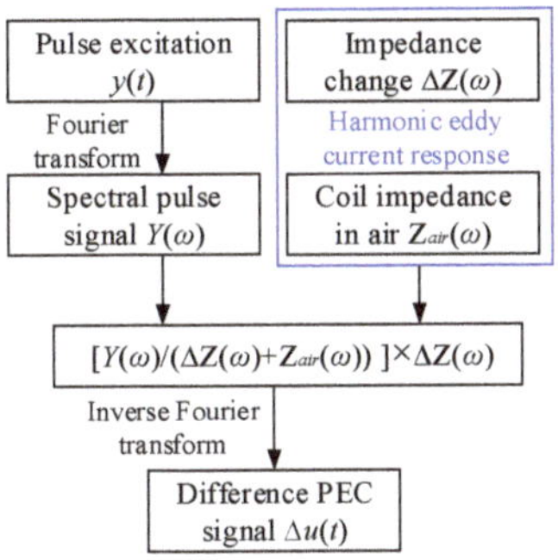

Figure 1. Flow chart of pulsed eddy current (PEC) model.

In the process of constructing the PEC analytical model, a pulse voltage signal $y(t)$ serves as an excitation signal of the coil sensor. The difference PEC signal $\Delta u(t)$ is the voltage change of the coil sensor when a plate is tested. The harmonic eddy current responses $\Delta Z(\omega)$ and $Z_{air}(\omega)$ in the frequency domain are the basis of the PEC model construction, which have been built in previous studies [15,16]. The case for the coil sensor above a conductive plate was considered, as illustrated in Figure 2. The PEC probe was an air-core cylindrical coil. The plate under research was assumed to be linear, homogeneous and isotropic. The electrical conductivity, permeability and thickness of the measured metallic plate are denoted as σ, μ and d, respectively, where $\mu = \mu_0\mu_r$ (μ_0 is the permeability of the vacuum, μ_r is the relative permeability of the metallic plate). An artificial magnetic insulation boundary was placed at h.

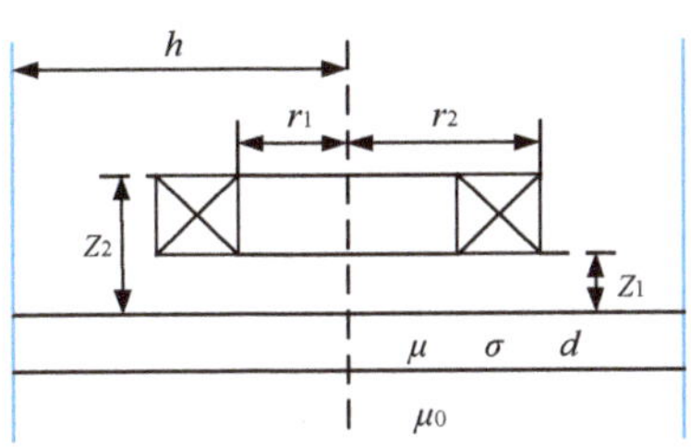

Figure 2. Coil sensor above a plate.

Based on the harmonic eddy current response and spectral pulse signals, the difference PEC response can be formulated by performing Kirchhoff's Law in the frequency domain. It signifies that the algebraic sum of the voltages of the coil sensor is always equal to zero in a closed loop, which can be expressed as $\Delta U = |U_{plate} - U_{air}| = |I \times (Z_{plate} - Z_{air})|$. Correspondingly, the difference PEC response in the frequency domain can be obtained. Where U_{plate} and Z_{plate} are the voltage and impedance signal of coil sensor when a plate is tested, U_{air} and Z_{air} are the voltage and impedance signal of the coil sensor when the coil sensor is in air, the voltage signal U_{air} and the impedance signal Z_{air} serve as a reference signal, ΔU is the difference PEC response between the detection signal U_{plate} and reference signal U_{air}, I is the current of coil sensor, $\Delta Z = |Z_{plate} - Z_{air}|$ is the impedance change response when a plate is tested. Correspondingly, the difference PEC response in the frequency domain can be expressed as

$$\Delta U(\omega) = \frac{Y(\omega)}{\Delta Z(\omega) + Z_{air}(\omega)} \times \Delta Z(\omega) \tag{1}$$

where $Y(\omega)$ denotes the Fourier transform of pulse voltage signal $y(t)$, $Z_{plate}(\omega) = \Delta Z(\omega) + Z_{air}(\omega)$, $Y(\omega)/(\Delta Z(\omega) + Z_{air}(\omega)) = I$.

By using inverse Fourier transform (IFT), the difference PEC signal in the time domain can be expressed as

$$\Delta u(t) = \text{IFT}[\Delta U(\omega)]. \tag{2}$$

Based on the results, a numerical calculation program can be written in MATLAB software (MTLAB R2014a, Math Works, Inc, Natick, Massachusetts, United States, 2014). Correspondingly, some scientific explanations can be verified by numerical method.

In Equation (1), when the air-core coil is excited by harmonic excitation [15], the impedance change of air-core coil can be written as

$$\Delta Z(\omega) = j2\pi\omega\mu_0 n_c^2 \sum_{i=1}^{\infty} \frac{\chi^2(\lambda_{0i}r_1, \lambda_{0i}r_2)}{\lambda_{0i}^7 h^2 [J_0(\lambda_{0i}h)]^2} (e^{-\lambda_{0i}z_1} - e^{-\lambda_{0i}z_2})^2 R_{0i,1i}. \tag{3}$$

On the other hand, for the case of the air-core coil in air [15], the impedance can be expressed as

$$Z_{air}(\omega) = j4\pi\omega\mu_0 n_c^2 \sum_{i=1}^{\infty} \frac{\chi^2(\lambda_{0i}r_1, \lambda_{0i}r_2)}{\lambda_{0i}^7 h^2 [J_0(\lambda_{0i}h)]^2} [\lambda_{0i}(z_2 - z_1) + e^{-\lambda_{0i}(z_2 - z_1)} - 1]. \tag{4}$$

In Equation (3), the $R_{0i,1i}$ can be expressed as

$$R_{0i,1i} = \frac{(\lambda_{0i}\mu_r)^2 - \lambda_{1i}^2 + e^{-2\lambda_{1i}d}[\lambda_{1i}^2 - (\lambda_{0i}\mu_r)^2]}{(\lambda_{1i} + \lambda_{0i}\mu_r)^2 - e^{-2\lambda_{1i}d}(\lambda_{1i} - \lambda_{0i}\mu_r)^2} \tag{5}$$

where $n_c = N_c/(r_2 - r_1)(z_2 - z_1)$, N_c denotes the number of the coil turns, r_2, r_1 are the outer and inner radius of the coil, $(z_2 - z_1)$ denotes the height of the coil, λ_{0i} is the $i-th$ positive root of the Bessel function $J_1(\lambda_0 h)$, J_m is the $m-th$ order Bessel function of the first kind, $R_{0i,1i}$ is a constant related to the conductivity, permeability and thickness of measured metallic plates, $\lambda_{1i} = \sqrt{\lambda_{0i}^2 + j\omega\sigma\mu}$.

In order to obtain the difference PEC response for simulating transient induced voltage, the periodic pulse signal is applied. Combined with the superposition principle and inverse Fourier transform, the PEC transient response can be reconstructed from the harmonic eddy current response [17,18]. For the T-periodic pulse, the formula can be expressed as

$$y(t) = A[(1 - e^{-\frac{t}{a}})step(t) + (e^{\frac{\tau - t}{a}} - 1)step(t - \tau)]\tau < T \tag{6}$$

where, a is the time constant, $step(t)$ is the step function, A is the amplitude value of excitation current, τ is the excitation time, T is the excitation period.

3. Thrice Subtraction Method

By analyzing previous research ideas [13,19], a thrice subtraction method was proposed to obtain a PI feature for reducing the effect of permeability change. The thrice subtraction method included three subtraction steps and the operating process is shown in Figure 3. The first subtraction was implemented to eliminate the effect of PEC signal from air. A difference PEC signal was obtained after the subtraction process. Then, the normalization signal of the difference PEC signal was obtained and the second subtraction was performed between the difference PEC signal and its normalization signal, which was used to amplify the variation of the difference PEC signal. Next, the standard deviation of the difference normalization signal was calculated and the third subtraction was implemented, which was used to reduce the effect of permeability. Finally, the PI feature could be extracted from thrice subtraction signals when permeability changed. The detailed steps are described below.

First, the subtraction between PEC signals of sample and air is implemented to obtain the difference PEC signal. The subtraction process can be shown as

$$\Delta U(t) = \left| U(t)_{plate} - U(t)_{air} \right| \tag{7}$$

where, $U(t)_{plate}$ is the PEC signal of sample, $U(t)_{air}$ is the PEC signal of air.

Second, the subtraction between the difference PEC signal and its normalization signal is performed to obtain the difference normalization signal. The difference normalization signal can be formulated as

$$U_d^{norm} = \Delta U(t) - \frac{\Delta U(t)}{\max(\Delta U(t))} \tag{8}$$

where, $\frac{\Delta U(t)}{\max(\Delta U(t))}$ denotes the normalization signal of the difference PEC signal, $\max(\Delta U(t))$ denotes the maximum value of the difference PEC signal.

Then, the third subtraction between the difference normalization signal and its standard deviation is calculated to obtain the thrice subtraction signal. The thrice subtraction signal can be expressed as

$$U_{std} = U_d^{norm} - std(U_d^{norm}) \tag{9}$$

where, $std(U_d^{norm})$ denotes the standard deviation of the difference normalization signal.

By changing the permeability of samples, the different thrice subtraction signals can be obtained and the PI feature can be extracted by mutual subtraction between them as

$$U_{std-1} - U_{std-2} = 0. \tag{10}$$

Finally, by solving Equation (10), the amplitude and time of the PI feature can be obtained. It is easy to think that the common point of both different signals, called the PI feature point, is unaffected by the change in permeability. In other words, the PI feature is immune to the permeability effect.

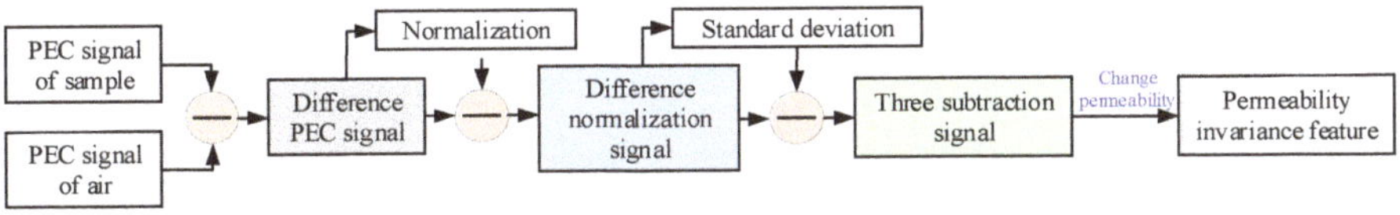

Figure 3. Flow diagram of thrice subtraction method.

4. Simulation Results

4.1. PI Feature Acquisition

According to the PEC analytical model, the numerical simulation code was written in MATLAB software to simulate the PEC signal with different relative permeability and verify the thrice subtraction method. In simulations, the height, inner and outer radius of the probe coil were 6 mm, 4 mm and 6 mm, respectively. There were 300 turns of coil. The boundary was set to $h = 20r_2 = 120$ mm. The pulse excitation had a 10 V amplitude, 200 Hz frequency and 50% duty cycle. The time constant of pulse excitation was set to 33 µs. The excitation signal of pulse voltage is shown in Figure 4. In Figure 5, the difference PEC signals are shown when permeability changes. It can be observed that the difference

PEC signals integrally decreased when relative permeability decreased. Especially, in the partially enlarged drawing shown in Figure 6, the variation of difference PEC signal is obvious. The results indicated that the difference PEC signals did not show intersection phenomenon and the permeability effect was obvious when permeability changes.

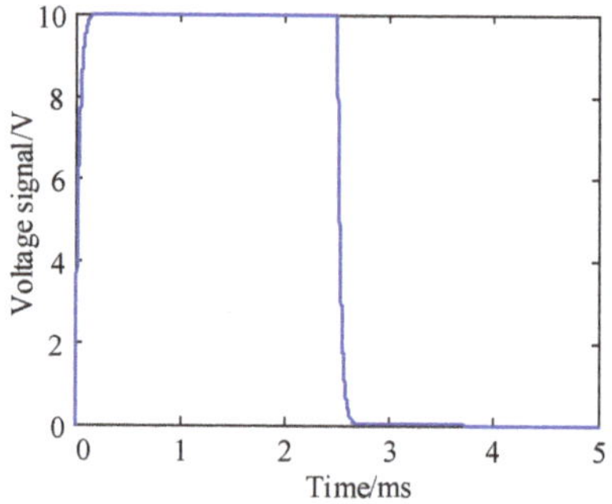

Figure 4. Pulse excitation voltage signal.

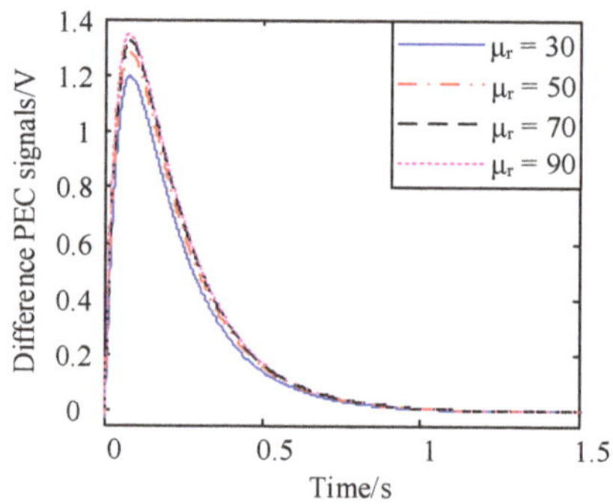

Figure 5. The difference in PEC signals.

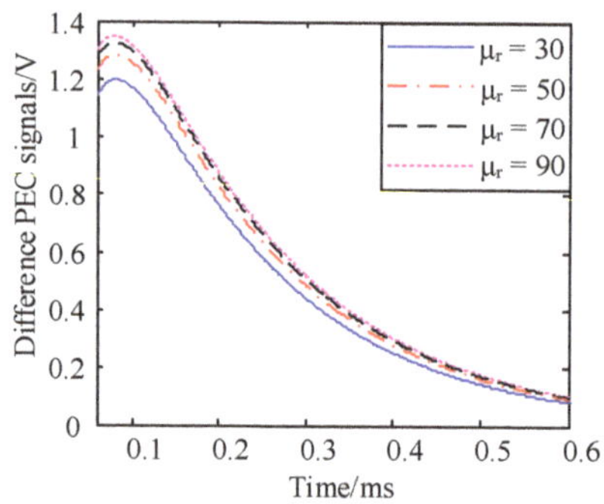

Figure 6. Partially enlarged drawing.

Under the condition that the rise time of pulse excitation is constant, when excitation pulse frequency increased, the difference PEC signals changed as shown in Figure 7. It was observed that the change of excitation pulse frequency led to the change of excitation pulse period. In experiments, the difference PEC signals were recorded when it was triggered by the rising or falling of pulse excitation. It can be seen that the difference PEC signal did not change in the first half part of the difference PEC signal, as shown in Figure 7a, when the excitation pulse frequency changes at lower excitation pulse frequency. When the pulse frequency was higher, the recorded difference PEC signal in first half part of periodic pulse was incomplete and there was no stationary part in difference PEC signal, as shown in Figure 7b. When the rising time of pulse excitation changed, as shown in Figure 8, it can be observed that the response time of the difference PEC signals to stationary section was shorter.

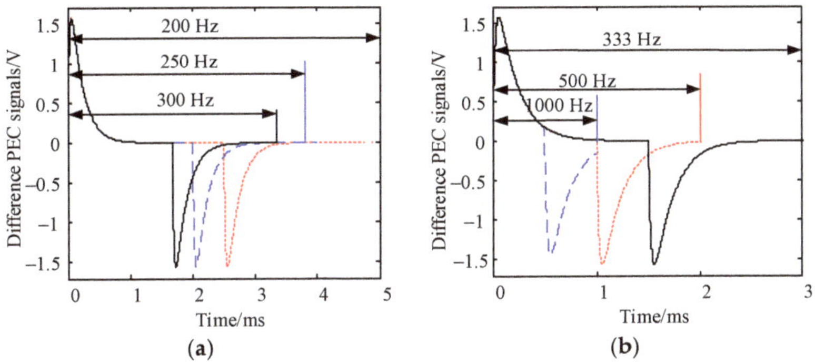

Figure 7. The difference PEC signals with different frequencies. (**a**) lower frequency; (**b**) higher frequency.

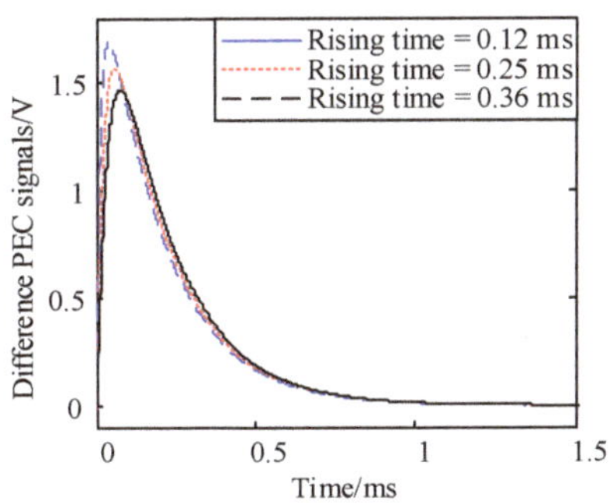

Figure 8. The difference in PEC signals with different rising times.

By using the thrice subtraction method based on the above difference PEC signals, the thrice subtraction signals could be obtained from the difference PEC signals, as shown in Figure 9. It was observed that the amplitude of thrice subtraction signals was smaller than that in the difference PEC signals. An intersection point phenomenon was exhibited in the thrice subtraction signals, which is called a PI signal feature. It was observed that at the intersection point of thrice subtraction signals, the thrice subtraction signals were independent of permeability effect. The PI feature was immune to permeability change. According to above analyses, it is known that the thrice subtraction signal was affected by excitation pulse frequency and rising time of pulse excitation. Thus, some limits are needed in the application of PI phenomenon. In order to keep the method practical, the excitation pulse frequency should be kept lower when rising time of pulse excitation is longer.

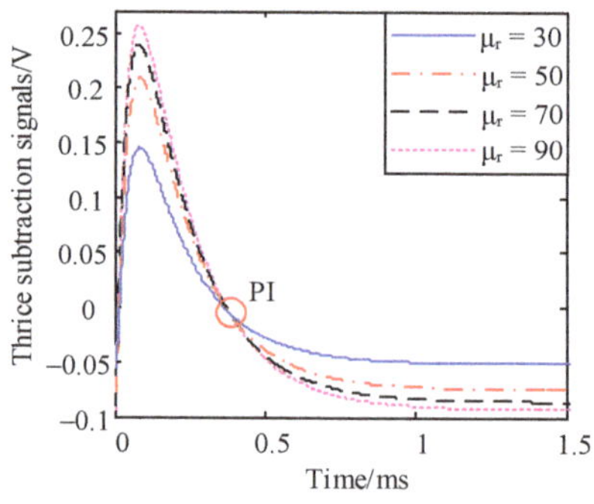

Figure 9. Thrice subtraction signals. Permeability invariance (PI) is shown by the red circle at the intersection of the signals.

4.2. Behavior of PI Feature

When conductivity and thickness of samples changed, the PI time and amplitude were changed, as shown in Figures 10 and 11. It was found that the PI amplitude and time changed when the thickness changed from about 0 to 2 mm and the conductivity changed from about 0 to 10 MS/m, which signified that the thickness and conductivity of sample could be measured by testing the PI amplitude and time. When the thickness was larger than 2 mm and conductivity changed from 0 to 3 MS/m, the PI time and amplitude changed with increasing conductivity. When the thickness was larger than 2 mm and the conductivity changed from 3 MS/m to 10 MS/m, the PI time did not change. The results demonstrated that the detection range of conductivity was from 0 to 10 MS/m when the thickness was smaller than 2 mm and the detection range of conductivity was from 0 to 3 MS/m when the thickness was larger than 2 mm. It also indicated that the conductivity measurement was only suitable for a certain range when sample thickness was determined.

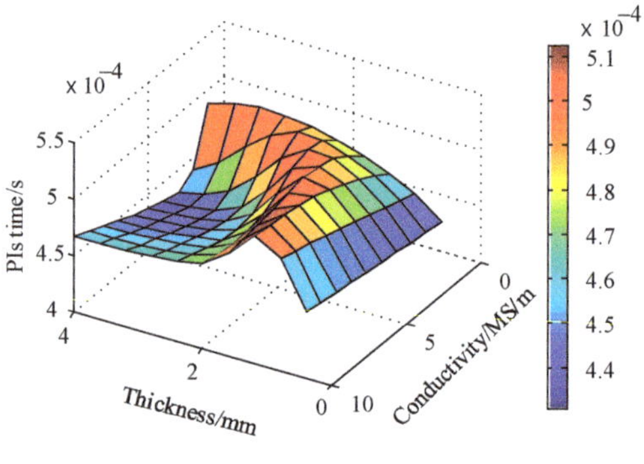

Figure 10. PIs time change with thickness and conductivity.

When the conductivity of sample was 3.2 MS/m, the PI time and amplitude curves are depicted with thickness in Figure 12. It was observed that the PI time and amplitude were changed when thickness was smaller than 3.6 mm. When sample thickness was larger than 4 mm, the PI time and amplitude were almost unchanged, which signified that the thickness of sample did not affect the PI feature with increasing thickness. The PI time and amplitude curves are shown with conductivity in Figure 13 for a thickness samples of 0.8 mm. It was found that the PI time and amplitude increased with increasing conductivity. The PIs changes along a curve with increasing conductivity. The results indicated that the PI time and amplitude could serve as an index of conductivity when sample thickness was constant, which signified that the conductivity of ferromagnetic samples could be determined by testing PI time or amplitude. In other words, the thrice subtraction method was a feasible program to obtain a PI feature for reducing the effect of permeability in conductivity measurements of ferromagnetic samples. The results indicated that the behavior of PI feature was affected by the conductivity and thickness of the samples. Thus, in practical measurement of sample conductivity, the thickness of sample should keep constant and the other variables should also remain unchanged in testing. When the thickness of sample was bigger than skin depth of eddy current in conductivity measurement, the thickness did not affect the change of PI feature, as shown in Figures 10 and 11. However, the measurement range of conductivity was changed. In order to measure the conductivity of the sample, the thickness of sample should be known and the size of sample should also be the same.

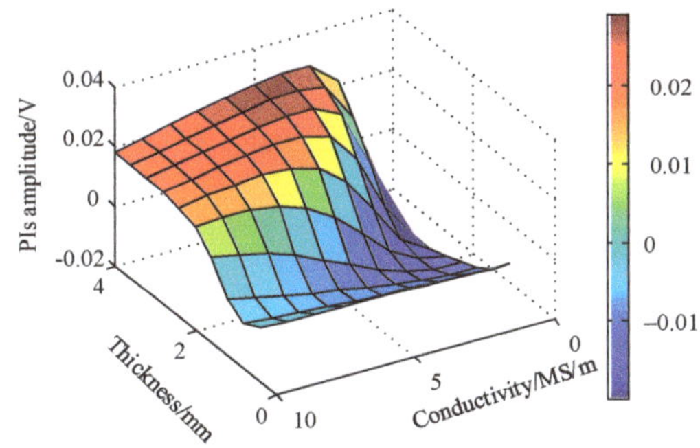

Figure 11. PIs amplitude change with thickness and conductivity.

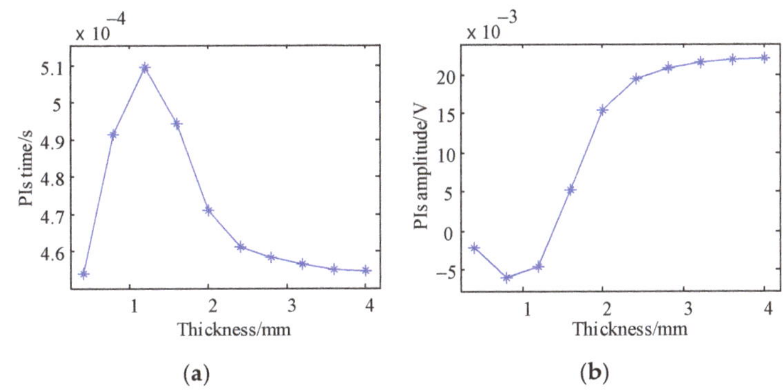

Figure 12. PI time and amplitude curves with thickness. (**a**) PI time curve; (**b**) PI amplitude curve.

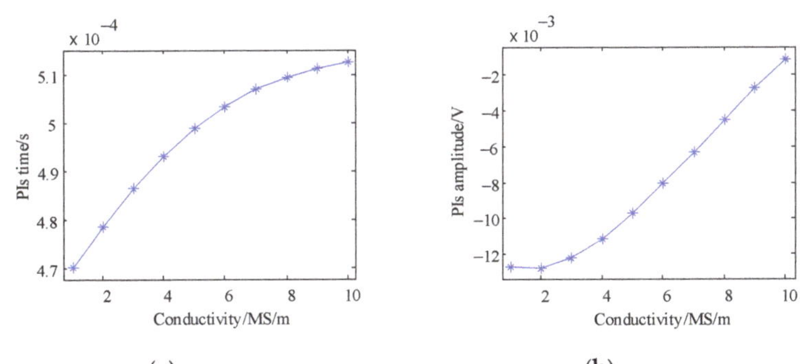

Figure 13. PI time and amplitude curves with conductivity. (**a**) PI time curve; (**b**) PI amplitude curve.

5. Experimental Setup and Results

5.1. Experimental Setup

As shown in Figure 14, the experimental setup included a computer, signal generator (DG1022Z), data acquisition card (NI USB-6356) and probe. The pulse excitation signal was obtained from the signal generator and applied to the probe. The data acquisition card communicated with computer by USB interface. The data acquisition card was set to trigger on the rising edge, which means that the PEC signal was recorded when the signal rising edge began. The probe signal was processed by low pass filter and was enlarged in LabVIEW software (LabVIEW 2015, NI, United States, 2015). The ferromagnetic material can be seen as many small magnetic domain structures. When the environmental magnetic field changes, the magnetic domain organization is ordered in the direction of the magnetic field. Therefore, a magnet can be used to change the environmental magnetic field so that the permeability of the ferromagnetic material is considered to have changed. In order to change the permeability of ferromagnetic samples, a magnet piece was used to change the magnetic field caused by material permeability. The magnet piece was only used to change the magnetic field in the testing environment, which means that any magnet piece could be used in testing. The distance between magnet and sample was set to change the size of the magnetic field, as shown in Figure 15.

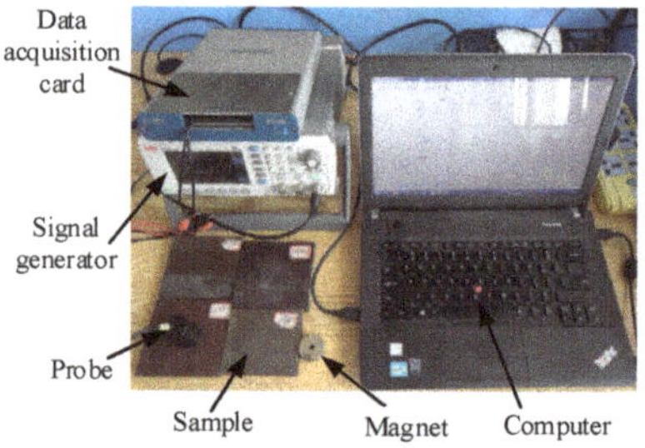

Figure 14. Experimental setup.

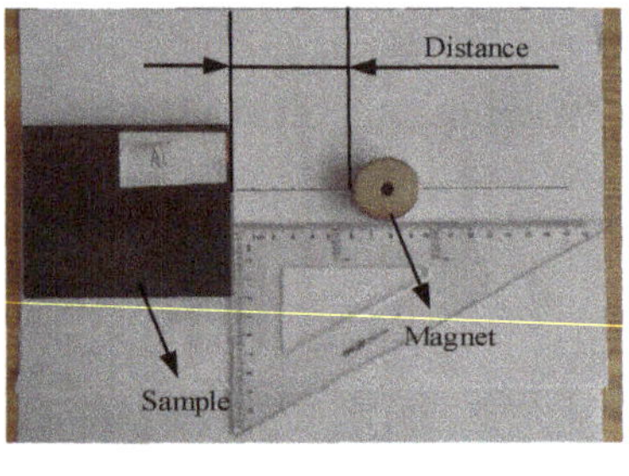

Figure 15. Sample and magnet.

5.2. Results and Discussion

In experiments, the pulse excitation had a 10 V amplitude, a 200 Hz frequency and a 50% duty cycle. The sampling frequency was 1 MHz and the sampling point was 2500. When the rising edge was triggered, the data acquisition card began to record the probe signals. The samples included 45# steel, 65Mn steel, A3 steel and nodular cast iron plates. The sample had a length of 100 mm, width of 100 mm and height of 10 mm. The conductivities of the samples are listed in Table 1, which were measured using the four-wire Kelvin resistivity measurement method [20,21] in a laboratory environment on samples with a polished top surface.

Table 1. Conductivity of experiment samples.

	65Mn Steel	45# Steel	A3 Steel	Nodular Cast Iron
Conductivity (MS/m)	4.47	3.76	3.41	1.56

Figure 16 shows the difference PEC signals when the distance between the magnet and sample changed. It was observed that the difference PEC signals integrally increased or decreased when the magnet distance changed. In terms of 45# steel, 65Mn steel, A3 steel and nodular cast iron samples, the difference PEC signals did not exhibit intersection phenomenon and the amplitude of difference PEC signals was obviously affected. The results were consistent with the simulation results, which indicated that the difference PEC signals were affected when permeability changes and the PI feature did not appear.

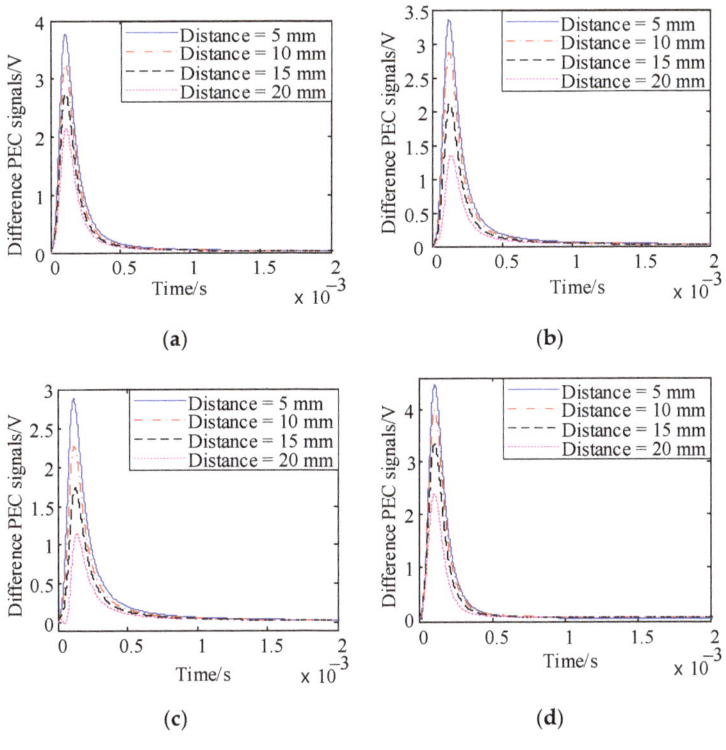

Figure 16. The difference PEC signals. (**a**) A3 steel; (**b**) 45# steel; (**c**) 65Mn steel; (**d**) nodular cast iron.

Based on the above differences PEC signals, the thrice subtraction method was used to obtain thrice subtraction signals and the PI feature. Figure 17 shows that a PI feature appeared in thrice subtraction signals in terms of 45# steel, 65Mn steel, A3 steel and nodular cast iron samples. At the PI feature, the thrice subtraction signals were almost independent of the permeability effect. Furthermore, the PI features were extracted from thrice subtraction signals, as shown in Figure 18. It was observed that the PI time and amplitude increased along a curve with increasing conductivity. The change rule of conductivity was consistent with the actual measurement. The result was in agreement with the simulation result under a certain condition, which demonstrated that by testing PI amplitude and time, the conductivity of ferromagnetic samples could be determined without the permeability effect. The results pointed that the thrice subtraction method was a feasible program to obtain a PI feature for reducing the permeability effect in the conductivity measurement of ferromagnetic samples.

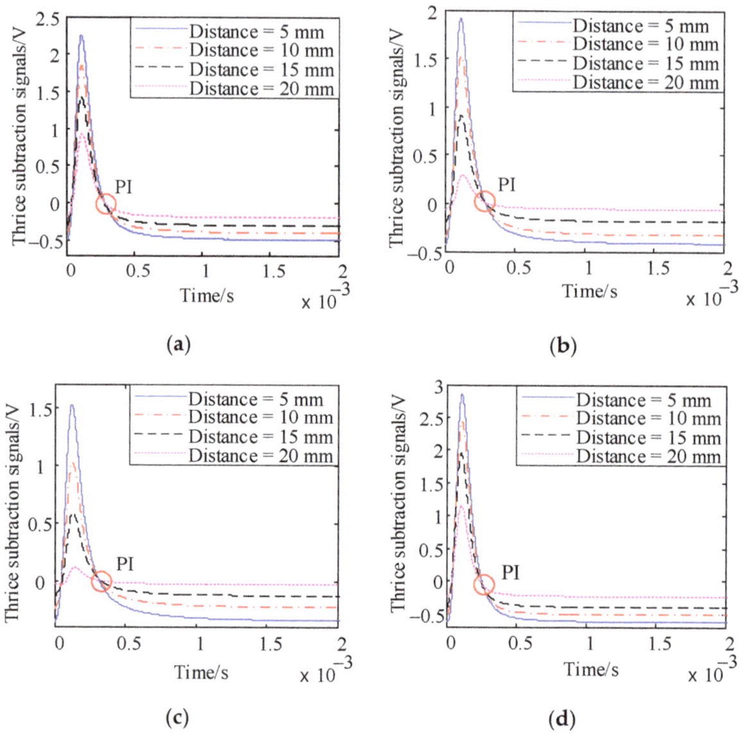

Figure 17. Thrice subtraction signals. (**a**) A3 steel; (**b**) 45# steel; (**c**) 65Mn steel; (**d**) nodular cast iron.

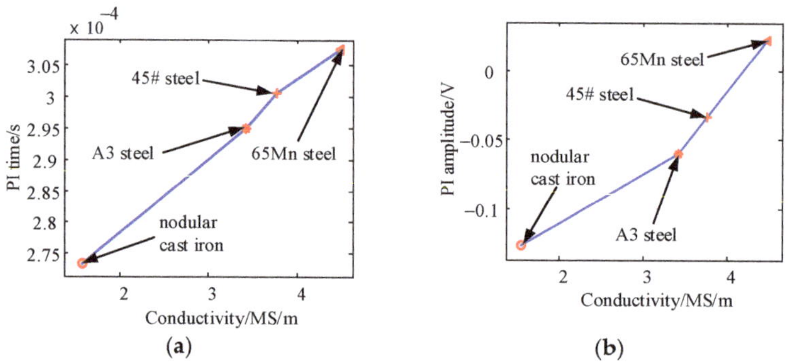

Figure 18. PIs with samples of different materials. (**a**) PI time curve; (**b**) PI amplitude curve.

In PEC testing, the formulation of the PEC model was established in a simplified condition. Thus, the simulation results fit one specific case in terms of a certain condition. The simulation results were available to verify the change law of the PEC response with different thickness, conductivity and permeability. The simulations may have led to the inconsistency with experimental results under the same conditions. On the other hand, due to the measurement error, there may have been deviation in the conductivity measured by the four-wire Kelvin resistivity measurement method. According to experiment and simulation results, it was feasible to verify the variation and trend of the conductivity in terms of real value. Although there were some shortcomings in the simulations and experiments, the change law of conductivity and variation of conductivity was consistent with the actual measurement, which demonstrated that the measurement results were feasible and the proposed method was effective.

6. Conclusions

Permeability, as an influence factor, will lead to measurement error in conductivity measurement of ferromagnetic samples when it changes. In this paper, the effect of permeability was considered in the PEC measurement of conductivity of ferromagnetic samples. In order to reduce the permeability effect, a thrice subtraction method was proposed to obtain a PI feature in conductivity measurement of ferromagnetic samples. The PI phenomenon was verified and the behavior of the PI feature was analyzed with changes in conductivity and thickness. Following that, the experiments were implemented to verify the availability of the thrice subtraction method and the practicality of the obtained PI feature in the conductivity measurement. The results demonstrated that the thrice subtraction method was a feasible program to obtain a PI feature for reducing the permeability effect. The PI feature has the potential to enable the measurement of conductivity of ferromagnetic samples without the permeability effect. The next step for further research is to measure ferromagnetic materials with unknown conductivity and evaluate the thickness of ferromagnetic samples.

Author Contributions: Conceptualization, D.W., M.F., B.C. and Z.X.; methodology, D.W.; software, D.W.; validation, M.F., B.C., P.W. and Z.X.; formal analysis, Z.X.; investigation, D.W. and Z.X.; resources, M.F. and B.C.; data curation, D.W.; writing—original draft preparation, D.W. and Z.X.; writing—review and editing, M.F., B.C. and P.W.; visualization, D.W.; supervision, M.F. and P.W.; project administration, B.C. and P.W.; funding acquisition, M.F.

Funding: This work was supported in part by the National Natural Science Foundation of China under grant 51677187, and in part by the Priority Academic Program Development of Jiangsu Higher Education Institutions.

Acknowledgments: The authors would like to thank the editors and reviewers for giving significant and insightful comments.

Conflicts of Interest: The authors declare no conflict of interest.

References

1. Lu, M.; Xu, H.; Zhu, W.; Yin, L.; Zhao, Q.; Peyton, A.; Yin, W. Conductivity Lift-off Invariance and measurement of permeability for ferrite metallic plates. *NDT E Int.* **2018**, *95*, 36–44. [CrossRef]
2. Wang, C.; Fan, M.; Cao, B.; Ye, B.; Li, W. Novel Noncontact Eddy Current Measurement of Electrical Conductivity. *IEEE Sens. J.* **2018**, *18*, 9352–9359. [CrossRef]
3. Rodrigues, N.M.; Rosado, L.S.; Ramos, P.M. A portable embedded contactless system for the measurement of metallic material conductivity and lift-off. *Measurement* **2017**, *111*, 441–450. [CrossRef]
4. Chen, X.; Lei, Y. Electrical conductivity measurement of ferromagnetic metallic materials using pulsed eddy current method. *NDT E Int.* **2015**, *75*, 33–38. [CrossRef]
5. Lefebvre, J.H.V.; Dubois, J.M.S. Lift-off point of intercept (LOI) behavior. *Rev. Prog. Quant. Nondestruct. Eval. Vols 24a 24b* **2005**, *760*, 523–530.
6. Lu, M.Y.; Zhu, W.Q.; Yin, L.Y.; Peyton, A.J.; Yin, W.L.; Qu, Z.G. Reducing the Lift-Off Effect on Permeability Measurement for Magnetic Plates from Multifrequency Induction Data. *IEEE Trans. Instrum. Meas.* **2018**, *67*, 167–174. [CrossRef]
7. Tian, G.Y.; He, Y.; Adewale, I.; Simm, A. Research on spectral response of pulsed eddy current and NDE applications. *Sens. Actuators A Phys.* **2013**, *189*, 313–320. [CrossRef]
8. Adewale, I.D.; Tian, G.Y. Decoupling the Influence of Permeability and Conductivity in Pulsed Eddy-Current Measurements. *IEEE Trans. Magn.* **2013**, *49*, 1119–1127. [CrossRef]
9. Dziczkowski, L. Elimination of Coil Liftoff from Eddy Current Measurements of Conductivity. *IEEE Trans. Instrum. Meas.* **2013**, *62*, 3301–3307. [CrossRef]
10. Lu, M.; Xie, Y.; Zhu, W.; Peyton, A.J.; Yin, W. Determination of the magnetic permeability, electrical conductivity, and thickness of ferrite metallic plates using a multi-frequency electromagnetic sensing system. *IEEE Trans. Ind. Inform.* **2018**. [CrossRef]

11. Yu, Y.T.; Zou, Y.; Al Hosani, M.; Tian, G.Y. Conductivity Invariance Phenomenon of Eddy Current NDT: Investigation, Verification, and Application. *IEEE Trans. Magn.* **2017**, *53*. [CrossRef]
12. Ye, C.; Su, Z.; Rosell, A.; Udpa, L.; Udpa, S.; Capobianco, T.; Tamburrino, A. A decay time approach for linear measurement of electrical conductivity. *NDT E Int.* **2019**, *102*, 169–174. [CrossRef]
13. Tian, G.Y.; Li, Y.; Mandache, C. Study of Lift-Off Invariance for Pulsed Eddy-Current Signals. *IEEE Trans. Magn.* **2009**, *45*, 184–191. [CrossRef]
14. Ulapane, N.; Alempijevic, A.; Miro, J.V.; Vidal-Calleja, T. Non-destructive evaluation of ferromagnetic material thickness using Pulsed Eddy Current sensor detector coil voltage decay rate. *NDT E Int.* **2018**, *100*, 108–114. [CrossRef]
15. Fan, M.; Huang, P.; Ye, B.; Hou, D.; Zhang, G.; Zhou, Z. Analytical modeling for transient probe response in pulsed eddy current testing. *NDT E Int.* **2009**, *42*, 376–383. [CrossRef]
16. Chen, X.; Lei, Y. Excitation current waveform for eddy current testing on the thickness of ferromagnetic plates. *NDT E Int.* **2014**, *66*, 28–33. [CrossRef]
17. Fan, M.B.; Cao, B.H.; Yang, X.F. Analytical time-domain model of transient eddy current field in pulsed eddy current testing. *Acta Phys. Sin.* **2010**, *59*, 7570–7574.
18. Mandache, C.; Lefebvre, J.H.V. Transient and harmonic eddy currents: Lift-off point of intersection. *NDT E Int.* **2006**, *39*, 57–60. [CrossRef]
19. Wood, J. Invariant pattern recognition: A review. *Pattern Recognit.* **1996**, *29*, 1–17. [CrossRef]
20. Bowler, N.; Huang, Y. Electrical conductivity measurement of metal plates using broadband eddy-current and four-point methods. *Meas. Sci. Technol.* **2005**, *16*, 2193–2200. [CrossRef]
21. Miccoli, I.; Edler, F.; Pfnur, H.; Tegenkamp, C. The 100th anniversary of the four-point probe technique: The role of probe geometries in isotropic and anisotropic systems. *J. Phys. Condens. Matter* **2015**, *27*, 223201. [CrossRef] [PubMed]

MDPI
St. Alban-Anlage 66
4052 Basel
Switzerland
Tel. +41 61 683 77 34
Fax +41 61 302 89 18
www.mdpi.com

Applied Sciences Editorial Office
E-mail: applsci@mdpi.com
www.mdpi.com/journal/applsci

www.ingramcontent.com/pod-product-compliance
Lightning Source LLC
Chambersburg PA
CBHW061842210926
51850CB00006B/951

9783036522920